Vaibhav A. N.

Introduction to Computational Biology

Introduction to Computational Biology

Maps, sequences and genomes

Michael S. Waterman
Professor of Mathematics and of Biological Sciences
University of Southern California
USA

CHAPMAN & HALL
London · Glasgow · Weinheim · New York · Tokyo · Melbourne · Madras

Published by Chapman & Hall, 2-6 Boundary Row, London SE1 8HN, UK

Chapman & Hall, 2-6 Boundary Row, London SE1 8HN, UK

Blackie Academic & Professional, Wester Cleddens Road, Bishopbriggs, Glasgow G64 2NZ, UK

Chapman & Hall GmbH, Pappelallee 3, 69469 Weinheim, Germany

Chapman & Hall USA, 115 Fifth Avenue, New York, NY 10003, USA

Chapman & Hall Japan, ITP-Japan, Kyowa Building, 3F, 2-2-1 Hirakawacho, Chiyoda-ku, Tokyo 102, Japan

Chapman & Hall Australia, 102 Dodds Street, South Melbourne, Victoria 3205, Australia

Chapman & Hall India, R. Seshadri, 32 Second Main Road, CIT East, Madras 600 035, India

First edition 1995

© 1995 Michael S. Waterman

Printed in Great Britain at the University Press, Cambridge

ISBN 0 412 99391 0

Apart from any fair dealing for the purposes of research or private study, or criticism or review, as permitted under the UK Copyright Designs and Patents Act, 1988, this publication may not be reproduced, stored, or transmitted, in any form or by any means, without the prior permission in writing of the publishers, or in the case of reprographic reproduction only in accordance with the terms of the licences issued by the Copyright Licensing Agency in the UK, or in accordance with the terms of licences issued by the appropriate Reproduction Rights Organization outside the UK. Enquiries concerning reproduction outside the terms stated here should be sent to the publishers at the London address printed on this page.

The publisher makes no representation, express or implied, with regard to the accuracy of the information contained in this book and cannot accept any legal responsibility or liability for any errors or omissions that may be made.

A catalogue record for this book is available from the British Library

∞ Printed on permanent acid-free text paper, manufactured in accordance with ANSI/NISO Z39.48-1992 and ANSI/NISO Z39.48-1984 (Permanence of Paper).

TO THE MEMORY OF MY GRANDFATHERS

AMOS PAYNE AND CHARLIE WATERMAN

Contents

Preface   **xiii**

0 Introduction   **1**
    0.1 Molecular Biology 3
    0.2 Mathematics, Statistics, and Computer Science 3

1 Some Molecular Biology   **5**
    1.1 DNA and Proteins 6
       1.1.1 The Double Helix 6
    1.2 The Central Dogma 7
    1.3 The Genetic Code 8
    1.4 Transfer RNA and Protein Sequences 12
    1.5 Genes Are Not Simple 16
       1.5.1 Starting and Stopping 16
       1.5.2 Control of Gene Expression 16
       1.5.3 Split Genes 17
       1.5.4 Jumping Genes 18
    1.6 Biological Chemistry 18

2 Restriction Maps   **29**
    2.1 Introduction 29
    2.2 Graphs 31
    2.3 Interval Graphs 33
    2.4 Measuring Fragment Sizes 38

3 Multiple Maps   **41**
    3.1 Double Digest Problem 42
       3.1.1 Multiple Solutions in the Double Digest Problem 43
    3.2 Classifying Multiple Solutions 48
       3.2.1 Reflections 48
       3.2.2 Overlap Equivalence 48
       3.2.3 Overlap Size Equivalence 51
       3.2.4 More Graph Theory 52

	3.2.5	From One Path to Another	53
	3.2.6	Restriction Maps and the Border Block Graph	56
	3.2.7	Cassette Transformations of Restriction Maps	58
	3.2.8	An Example	61

4 Algorithms for DDP 65
4.1 Algorithms and Complexity 65
4.2 DDP is NP-Complete 67
4.3 Approaches to DDP 68
    4.3.1 Integer Programming 68
    4.3.2 Partition Problems 69
    4.3.3 TSP 70
4.4 Simulated Annealing: TSP and DDP 70
    4.4.1 Simulated Annealing 70
    4.4.2 Traveling Salesman Problem 75
    4.4.3 DDP 76
    4.4.4 Circular Maps 78
4.5 Mapping with Real Data 79
    4.5.1 Fitting Data to a Map 80
    4.5.2 Map Algorithms 81

5 Cloning and Clone Libraries 83
5.1 A Finite Number of Random Clones 85
5.2 Libraries by Complete Digestion 85
5.3 Libraries by Partial Digestion 87
    5.3.1 The Fraction of Clonable Bases 88
    5.3.2 Sampling, Approach 1 91
    5.3.3 Designing Partial Digest Libraries 92
5.4 Genomes per Microgram 98

6 Physical Genome Maps: Oceans, Islands and Anchors 101
6.1 Mapping by Fingerprinting 102
    6.1.1 Oceans and Islands 102
    6.1.2 Divide and Conquer 110
    6.1.3 Two Pioneering Experiments 111
    6.1.4 Evaluating a Fingerprinting Scheme 114
6.2 Mapping by Anchoring 119
    6.2.1 Oceans, Islands and Anchors 119
    6.2.2 Duality Between Clones and Anchors 126
6.3 An Overview of Clone Overlap 127
6.4 Putting It Together 129

Contents

7 Sequence Assembly   **135**
- 7.1 Shotgun Sequencing 135
  - 7.1.1 SSP is NP-complete 137
  - 7.1.2 Greedy is at most Four Times Optimal 138
  - 7.1.3 Assembly in Practice 143
  - 7.1.4 Sequence Accuracy 145
  - 7.1.5 Expected Progress 147
- 7.2 Sequencing by Hybridization 148
  - 7.2.1 Other SBH Designs 154
- 7.3 Shotgun Sequencing Revisited 156

8 Databases and Rapid Sequence Analysis   **161**
- 8.1 DNA and Protein Sequence Databases 162
  - 8.1.1 Description of the Entries in a Sequence Data File 163
  - 8.1.2 Sample Sequence Data File 164
  - 8.1.3 Statistical Summary 166
- 8.2 A Tree Representation of a Sequence 167
- 8.3 Hashing a Sequence 168
  - 8.3.1 A Hash Table 169
  - 8.3.2 Hashing in Linear Time 170
  - 8.3.3 Hashing and Chaining 170
- 8.4 Repeats in a Sequence 171
- 8.5 Sequence Comparison by Hashing 172
- 8.6 Sequence Comparison with at most l Mismatches 176
- 8.7 Sequence Comparison by Statistical Content 180

9 Dynamic Programming Alignment of Two Sequences   **183**
- 9.1 The Number of Alignments 186
- 9.2 Shortest and Longest Paths in a Network 190
- 9.3 Global Distance Alignment 192
  - 9.3.1 Indel Functions 194
  - 9.3.2 Position-Dependent Weights 197
- 9.4 Global Similarity Alignment 198
- 9.5 Fitting One Sequence into Another 201
- 9.6 Local Alignment and Clumps 202
  - 9.6.1 Self-Comparison 206
  - 9.6.2 Tandem Repeats 207
- 9.7 Linear Space Algorithms 209
- 9.8 Tracebacks 212
- 9.9 Inversions 215
- 9.10 Map Alignment 219
- 9.11 Parametric Sequence Comparisons 223
  - 9.11.1 One-Dimension Parameter Sets 225
  - 9.11.2 Into Two-Dimensions 228

10	**Multiple Sequence Alignment**	**233**
	10.1 The Cystic Fibrosis Gene	233
	10.2 Dynamic Programming in r-Dimensions	236
	10.2.1 Reducing the Volume	237
	10.3 Weighted-Average Sequences	238
	10.3.1 Aligning Alignments	242
	10.3.2 Center of Gravity Sequences	242
	10.4 Profile Analysis	242
	10.4.1 Statistical Significance	244
	10.5 Alignment by Hidden Markov Models	245
	10.6 Consensus Word Analysis	248
	10.6.1 Analysis by Words	249
	10.6.2 Consensus Alignment	250
	10.6.3 More Complex Scoring	251
11	**Probability and Statistics for Sequence Alignment**	**253**
	11.1 Global Alignment	254
	11.1.1 Alignment Given	254
	11.1.2 Alignment Unknown	255
	11.1.3 Linear Growth of Alignment Score	256
	11.1.4 The Azuma-Hoeffding Lemma	257
	11.1.5 Large Deviations from the Mean	259
	11.1.6 Large Deviations for Binomials	261
	11.2 Local Alignment	263
	11.2.1 Laws of Large Numbers	263
	11.3 Extreme Value Distributions	275
	11.4 The Chein-Stein Method	278
	11.5 Poisson Approximation and Long Matches	280
	11.5.1 Headruns	280
	11.5.2 Exact Matching Between Sequences	282
	11.5.3 Approximate Matching	288
	11.6 Sequence Alignment with Scores	294
	11.6.1 A Phase Transition	294
	11.6.2 Practical p-Values	299
12	**Probability and Statistics for Sequence Patterns**	**305**
	12.1 A Central Limit Theorem	307
	12.1.1 Generalized Words	313
	12.1.2 Estimating Probabilities	313
	12.2 Nonoverlapping Pattern Counts	314
	12.2.1 Renewal Theory for One Pattern	314
	12.2.2 Li's Method and Multiple Patterns	318
	12.3 Poisson Approximation	321
	12.4 Site Distributions	323

Contents

    12.4.1 Intersite Distances 324

13 RNA Secondary Structure   **327**
  13.1 Combinatorics 327
    13.1.1 Counting More Shapes 332
  13.2 Minimum Free-energy Structures 334
    13.2.1 Reduction of Computation Time for Hairpins 336
    13.2.2 Linear Destabilization Functions 338
    13.2.3 Multibranch Loops 339
  13.3 Consensus folding 340

14 Trees and Sequences   **345**
  14.1 Trees 345
    14.1.1 Splits 347
    14.1.2 Metrics on Trees 351
  14.2 Distance 353
    14.2.1 Additive Trees 353
    14.2.2 Ultrametric Trees 357
    14.2.3 Nonadditive Distances 359
  14.3 Parsimony 361
  14.4 Maximum Likelihood Trees 367
    14.4.1 Continuous Time Markov Chains 367
    14.4.2 Estimating the Rate of Change 369
    14.4.3 Likelihood and Trees 372

15 Sources and Perspectives   **377**
  15.1 Molecular Biology 377
  15.2 Physical Maps and Clone Libraries 377
  15.3 Sequence Assembly 379
  15.4 Sequence Comparisons 379
    15.4.1 Databases and Rapid Sequence Analysis 379
    15.4.2 Dynamic Programming for Two Sequences 380
    15.4.3 Multiple Sequence Alignment 382
  15.5 Probability and Statistics 382
    15.5.1 Sequence Alignment 382
    15.5.2 Sequence Patterns 383
  15.6 RNA Secondary Structure 384
  15.7 Trees and Sequences 385

References   **387**

I Problem Solutions and Hints   **401**

II Mathematical Notation   **421**

Algorithm Index 423

Author Index 425

Subject Index 428

Preface

It is difficult to recall that only in 1953 was the famous double helical structure of DNA determined. Since then a stupendous series of discoveries have been made. The unraveling of the genetic code was only the beginning. Learning the details of genes and their discontinuous nature in eukaryotic genomes like ours has led to the ability to study and manipulate the material of that abstract concept of Mendel's, the gene itself. Learning to read the genetic material more and more rapidly has enabled us to attempt to decode entire genomes. As we approach the next century we also approach an incredible era of biology.

The rate of innovation in molecular biology is breathtaking. The experimental techniques that must be painstakingly mastered for the Ph.D. theses of one generation are usually routine by the time those students have students. The accumulation of data has necessitated international databases for nucleic acids, for proteins, and for individual organisms and even chromosomes. The crudest measure of progress, the size of nucleic acid databases, has an exponential growth rate. Consequently, a new subject or, if that is too grand, a new area of expertise is being created, combining the biological and information sciences. Finding relevant facts and hypotheses in huge databases is becoming essential to biology. This book is about the mathematical structure of biological data, especially those from sequences and chromosomes.

Titles in mathematics tend to be very brief, to the point of being cryptic. Titles in biology are often much longer and more informative, corresponding to the brief abstract that a mathematician might give. Correspondingly, a biologist's abstract might have the length and detail of a mathematician's introduction. Engaged in an effort to bridge the gulf between these until recently almost isolated cultures, my title reflects these conflicting traditions. "Introduction to Computational Biology" is the short title that might preface many disparate volumes. The remainder of the title "Maps, Sequences and Genomes" is to let the reader know that this is a book about applications in molecular biology. Even that is too short: "Introduction to computational biology..." should be "Introduction to computational, statistical, and mathematical molecular biology...".

As detailed in the first chapter, the intended reader should have had a first course in probability and statistics and should be competent at calculus. The ideas of algorithm and complexity from computer science would be helpful. As for biology, a beginning undergraduate course would be very useful. Every educated person should know that material in any case. The intent is to introduce someone who has mathematical skills to the fascinating structure of biological data and problems. It is not intended for those who like their subjects neatly fenced in and contained.

In such a rapidly developing subject there is a significant risk of instant obsolescence. I have tried to strike a balance between what I see as fundamentals unlikely to change and those data structures and problems whose relevance could be eliminated tomorrow by a clever piece of technology. For example, the basic nature of physical maps (such as restriction maps) will remain relevant. It is possible that the double digest problem might become dated although it has been of interest for 20 years now. Sequence assembly is also vulnerable to technology and will undergo many changes. Sequence comparison will always be of interest and dynamic programming algorithms are a good simple framework in which to imbed those problems. And so it goes; I have tried to present interesting mathematics that is motivated by the biology. It is not the last word and significant subjects are missing. Constructing evolutionary trees deserves a book that has not yet been written. Protein structure is a vast and often unmathematical subject that is not covered here. What I have tried to do is give some interesting mathematics relevant to the study of genomes.

A good deal of concern has been given to the topic of properly defining the area of study associated with this book. Even the name has not been settled on. Mathematical biology has not seemed satisfactory, due in part to the misadventures of earlier times, and the choice may have narrowed to *computational biology* and *informatics*. (If the latter designation wins out I hope it receives the French pronunciation.) More importantly, of what does the subject consist? There are three major positions: one, that it is a subset of biology proper and that any required mathematics and computer science can be made up on demand; two, that it is a subset of the mathematical sciences and that biology remains a remote but motivating presence; three, that there are genuine interdisciplinary components, with the original motivation from biology suggesting the mathematical problems, the solution of which suggest biological experiments and so on. My own view is that although the last is most rewarding activity, all three are not only worthwhile but inevitable and appropriate activities. Written to describe mathematical formulation and developments, I hope this book helps set the stage for more real interdisciplinary work in biology.

There are more people to thank than I have space for here, and I apologize in advance for significant omissions – they are inadvertent. Stan Ulam and Bill Beyer at Los Alamos were basic to getting me going in this area. Stan influenced several people with his belief that there was mathematics in the new biology although in typical Ulam style he did not give any details. From the beginning when I knew absolutely no biology Temple Smith put me onto good problems that had mathematical and statistical content and worked with me to solve them. Gian-Carlo Rota encouraged me in this work when it was not clear to anyone

Preface

else that there was real substance to the area. He, and later, Charlie Smith (then of System Development Foundation) have given me significant support in this work. My colleagues at the University of Southern California, without whom this book would be much shorter and less interesting, include Richard Arratia, Norman Arnheim, Caleb Finch, David Galas, Larry Goldstein, Louis Gordon, and Simon Tavaré. Gary Benson, Gary Churchill, Ramana Idury, Rob Jones, Pavel Pevzner, Betty Tang, Martin Vingron, Tandy Warnow, and Momiao Xiong, postdocs over the years, have been kind enough to teach me some of what they know and to make this a richer subject. My students Daniela Martin, Ethan Port, and Fengzhu Sun have read drafts, worked problems, and generally improved and corrected the text. Three talented and hard working people translated my successive drafts into LaTeX: Jana Joyce, Nicolas Rouquette, and Kengee Lewis. My work has been supported by the System Development Foundation, the National Institutes of Health and the National Science Foundation. Finally, I want to express my gratitude to Walter Fitch, Hugo Martinez, and especially to David Sankoff, pioneers in this subject who were there at the beginning and who stayed with it.

I close with an invitation to the reader to communicate errors to me. Donald Knuth in his wonderful volumes on *The Art of Computer Programming* offered at first $1 and later $2 for each error found by his readers. I would like to make a similar offer, but, in spite of my best efforts to eliminate mistakes, I doubt that I can afford to pay a sum proportional to the number of remaining errors and instead can only offer my sincere gratitude. I will make software, a list of errors, and other information about this book available via ftp or mosaic at http://hto-e.usc.edu.

Chapter 0

Introduction

Some of the impressive progress made by molecular biology was briefly mentioned in the Preface. Molecular biology is an experimental subject and although the material of living organisms obeys the familiar laws of chemistry and physics, there are few real universals in biology. Even the so-called universal genetic code that describes the mapping from triplets of nucleotides (the "letters of DNA") to amino acids (the "letters of protein") is not identical in all living systems. I once had a mathematical colleague mutter, "why don't they call it the almost universal genetic code?" The point is that evolution has found different solutions to various problems or it has modified structures differently in different but related species. Biologists often look for universals but expect that there will be variations of whatever they discover. To achieve rigor they carefully describe the organism and experimental conditions. In a similar way, mathematicians carefully state the assumptions for which they have been able to prove their theorems. In spite of this common desire to do work that is valid, mathematicians and biologists do not mix too often, as do, for example, mathematicians and physicists.

In the early part of this century, mathematical models elaborated by Fisher, Haldane, Wright, and others were at the forefront of biology. Today, the discoveries of molecular biology have left the mathematical sciences far behind. However, there are increasing connections between these fields, mostly due to the databases of biological sequences and the pressing need to analyze that data. Biology is at the beginning of a new era, promising significant discoveries, that will be characterized by the information-packed databases.

After a brief survey of molecular biology in Chapter 1, Chapters 2 to 4 will study restriction maps of DNA, rough landmark maps of the more detailed underlying sequences. Then clones and clone maps will be studied in Chapters 5 and 6. Making biological libraries of genomes and constructing "tilings" or maps of genomes is very important. Chapter 7 gives some problems associated with reading the DNA sequences themselves. Chapters 8 to 11 present aspects of comparing sequences for finding common patterns. Comparing two or more sequences

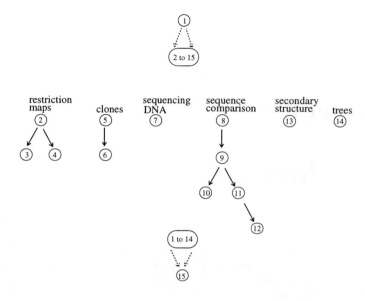

Figure 0.1: *Chapter dependencies*

is one of the most important mathematical applications in molecular biology. New developments in algorithms and in probability and statistics have resulted from these biological problems. The statistics of pattern counts in sequences is surprisingly subtle and is covered in Chapter 12. Molecular structure in biology is a central problem. Protein structure is a huge and largely unsolved problem that does not appear in this text. Instead, the more mathematically developed topic of RNA secondary structure is treated in Chapter 13. Finally, given a set of related sequences, we can try to infer their evolutionary history. This appears in Chapter 14. Classical genetics, genetic mapping, and coalescents deserve full treatments and are not covered in the present book.

Historical perspective is essential in research papers but clogs the flow of material in an introduction. So as not to entirely omit references to original material, some key references to original sources are given in Chapter 15. In addition a few recent papers at the forefront of today's research are also given. This subject develops quite rapidly and the reader should not assume that I have presented the last word on any given topic. Instead, these chapters should provide a beginning. Then go to the literature, the databases, and the laboratory.

There are many independent modules in this book as can be seen in Figure 0.1. It is not necessary to read from start to finish. A number of one-semester courses can be taken from these chapters. I recommend Chapters 1, 2, 6.1, 8, 9, 11.4-6, 14, and whatever of 12 can be fit into the remainder of the time.

Introduction

0.1 Molecular Biology

The next chapter gives a brief sketch of molecular biology. Ideally, the reader will have taken a beginning course in molecular biology and can skip to Chapter 2. In any case, I recommend the purchase of one of the excellent introductory texts. There is a tradition of superbly well-written books in molecular biology, perhaps tracing back to James D. Watson's first edition of *Molecular Biology of the Gene*. Here are some biology references.

Shorter Introductions to Molecular Biology

Berg, P., and M. Singer. *Dealing with Genes: The Language of Heredity*, University Science Books, Mill Valley, CA, 1992

Watson, J.D., M. Gilman, J. Witkowski, and M. Zoller. *Recombinant DNA*, 2nd ed., Scientific American Books, New York, 1992.

Introductions to Molecular Biology

Watson, J.D., N. Hopkins, J. Roberts, J.A. Steitz, and A. Weiner. *Molecular Biology of the Gene*, 4th ed. Benjamin-Cummings, Menlo Park, CA, 1987.

Lewin, B. *Genes IV*, Oxford University Press, Oxford, 1990.

Cell Biology

Alberts, B., D. Bray, J. Lewis, M. Raff, K. Roberts, and J.D. Watson. *Molecular Biology of the Cell*, 2nd ed., Garland, New York, 1989.

Darnell, J., H. Lodish, and D. Baltimore. *Molecular Cell Biology*, 2nd ed., Freeman, New York, 1990.

Biochemistry

Stryer, L. *Biochemistry*, 3rd ed. Freeman, New York, 1988.

Zubay, G. *Biochemistry*, 2nd ed. Macmillan, New York, 1988.

Lehninger, A. *Principles of Biochemistry*, Worth, New York, 1982.

0.2 Mathematics, Statistics, and Computer Science

It is assumed that the reader knows some mathematics. Although except for Chapter 11 the material is not at all deep, reading this book will require a certain level of familiarity with mathematical language and some ability to do mathematics. Much of our work will deal with discrete structures and involves some combinatorics. Most of the contents of the book could be divided into two categories: (1)

algorithms and (2) probability and statistics. The algorithms are usually intuitive and not hard to verify. The applications of probability vary in depth, and without some solid background at the upper undergraduate level, the reader should proceed with caution. Occasionally some more advanced ideas are used. Mathematics is largely not a spectator sport. Get involved and you will take away much more understanding and ability to apply these concepts to new situations.

Here are some good mathematical references.

Analysis

Rudin, W. *Principles of Mathematical Analysis*, 2nd ed., McGraw-Hill, New York, 1964.

Apostol, T. *Mathematical Analysis: A Modern Approach to Advanced Calculus*, Addison-Wesley, Reading, MA, 1957.

Beginning Probability

Feller, W. *An Introduction to Probability Theory and Its Applications*, 3rd ed., Vol. I, John Wiley and Sons, New York, 1968.

Chung, K.L. *Elementary Probability Theory with Stochastic Processes*, Springer-Verlag, New York, 1974.

Stochastic Processes

Karlin, S. and H.M Taylor. *A First Course in Stochastic Process*, 2nd ed., Academic Press, New York, 1975.

Ross, S. *Introduction to Probability Models*, 5th ed., Academic Press, San Diego, CA, 1993.

Advanced Probability

Chung, K.L. *A Course in Probability Theory*, Harcourt, Brace & World, New York, 1968.

Durrett, R. *Probability: Theory and Examples*, Wadsworth, Inc., Belmont, CA, 1991.

Computer Science

Aho, A.V., J.E. Hopcroft, and J.D. Ullman. *Data Structures and Algorithms*, Addison-Wesley, Reading, MA, 1983.

Baase, S. *Computer Algorithms: Introduction to Design and Analysis*, 2nd ed., Addison-Wesley, Reading, MA, 1988.

Crochemore, M. and W. Rytter. *Text Algorithms*, Oxford University Press, Oxford, 1994.

Chapter 1

Some Molecular Biology

The purpose of this chapter is to provide a brief introduction to molecular biology, especially to DNA and protein sequences. Ideally, the reader has taken a beginning course in molecular biology or biochemistry and can go directly to Chapter 2. Introductory textbooks often exceed 1000 pages; here we just give a few basics. In later chapters we introduce more biological details for motivation.

One of the basic problems of biology is to understand inheritance. In 1865 Mendel gave an abstract, essentially mathematical model of inheritance in which the basic unit of inheritance was a *gene*. Although Mendel's work was forgotten until 1900, early in this century it was taken up again and underwent intense mathematical development. Still the nature of the gene was unknown. Only in 1944 was the gene known to be made of DNA; and, it was not until 1953 that James Watson and Francis Crick proposed the now famous double helical structure for DNA. The double helix gives a physical model for how one DNA molecule can divide and become two identical molecules. In their paper appears one of the most famous sentences of science: "It has not escaped our notice that the specific pairing we have postulated immediately suggests a possible copying mechanism for the genetic material." That copying mechanism is the basis of modern molecular genetics. In the model of Mendel the gene was abstract. The model of Watson and Crick describes the gene itself, providing the basis for a deeper understanding of inheritance.

The molecules of the cell are of two classes: large and small. The large molecules, known as macromolecules, are of three types: DNA, RNA, and proteins. These are the molecules of most interest to us and they are made by joining certain small molecules together in polymers. We next discuss some of the general properties of macromolecules, including how DNA is used to make RNA and proteins. Then we give some more details of the biological chemistry that are the basis of these properties.

1.1 DNA and Proteins

DNA is the basis of heredity and it is a polymer, made up of small molecules called nucleotides. These nucleotides are four in number and can be distinguished by the four bases: adenine (A), ctyosine (C), guanine (G), and thymine (T). For our purposes a DNA molecule is a word over this four letter alphabet, $\mathcal{A} = \{A,C,G,T\}$. DNA is a nucleic acid and there is one other nucleic acid in the cell, RNA. RNA is a word over another four letter alphabet of ribonucleotides, $\mathcal{A} = \{A,C,G,U\}$ where thymine is replaced by uracil. These molecules have a distinguishable direction, and for reasons detailed later, one end (usually the left) is labeled $5'$ and the other $3'$.

Proteins are also polymers and here the word is over an alphabet of 20 amino acids. See Table 1.1 for a list of the amino acids and their one and three letter abbreviations. Proteins also have directionality.

How much DNA does an organism need to function? We can only answer a simpler question: How much DNA does an organism have? The intestinal bacterium *Escherichia coli* (*E. coli*) is an organism with one cell and has about 5×10^6 letters per cell. The DNA contained in the cell is known as the *genome*. In contrast to the simpler *E. coli*, the genome of a human is about 3×10^9 letters. Each human cell contains the same DNA.

Both RNA and proteins are made from instructions in the DNA, and new DNA molecules are made from copying existing DNA molecules. These processes are discussed next.

1.1.1 The Double Helix

The key feature of DNA that suggested the copying mechanism is the complementary basepairs; that is, the bases pair with A pairing T and G pairing C. This so-called pairing is by hydrogen bonds; more on that later. The idea is that a single word (or strand) of DNA (written in the $5'$ to $3'$ direction)

$$5' ACCTGAC 3'$$

is paired to a complementary strand running in the opposite direction:

$$5' ACCTGAC 3'$$
$$| \, | \, | \, | \, | \, | \, |$$
$$3' TGGACTG 5'$$

There are seven basepairs in this illustration. The A-T and G-C pairs are formed by hydrogen bonds, here indicated by a heavy bar. DNA usually occurs double stranded and its length is often measured by number of basepairs.

The three-dimensional structure is helical. In the next figure we show the letters or bases as attached to a string or backbone; note that the bars indicating the hydrogen bonds have been deleted. To properly view this figure, imagine a ribbon with edges corresponding to the backbones twisted into a helix.

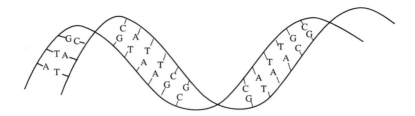

1.2 The Central Dogma

DNA carries the genetic material–the information required by an organism to function. (There are exceptions in the case of certain viruses where the genetic material is RNA.) DNA is also the means by which organisms transfer genetic information to their descendants. In organisms with a nucleus (eukaryotes), DNA remains in the nucleus; whereas proteins are made in the cytoplasm outside of the nucleus. The intermediate molecule carrying the information out of the nucleus is RNA. The information flow in biology is summarized by the "central dogma," put forward by Francis Crick in 1958:

> The central dogma states that once 'information' has passed into protein it cannot get out again. The transfer of information from nucleic acid to nucleic acid, or from nucleic acid to protein, may be possible, but transfer from protein to protein, or from protein to nucleic acid, is impossible. Information means here the precise determination of sequence, either of bases in the nucleic acid or of amino acid residues in the protein.

A schematic for the central dogma is:

$$\circlearrowleft \text{DNA} \longrightarrow \text{RNA} \longrightarrow \text{PROTEIN}$$

The loop from DNA to DNA means that the molecule can be copied. This is called replication. The next arrow is called transcription and the last translation. This chapter explores the arrows of the schematic in more detail.

Each of the arrows indicates making another macromolecule guided by the sequence of an existing macromolecule. The general idea is that one macromolecule can be used as a template to construct another. The fascinating details of these processes are basic to life. Understanding of templating will give us insight into the reasons for some interesting analytical studies. Today the central dogma has been extended. There are examples of genetic systems in which RNA templates RNA. Also, retroviruses can copy their RNA genomes into DNA by a mechanism called reverse transcription.

Making new molecules is called *synthesis*. When we look in detail we will see that certain proteins are required for the synthesis of both RNA and DNA. In other words, we are about to sketch a highly complex system. For now it is easy to see how DNA can be a template to make new DNA.

RNA is made single stranded. First the strands of the double helix are separated in a region by breaking the hydrogen bonds forming the basepairs. One strand of the DNA is used to template a single strand of RNA that is made by moving along the DNA. At the conclusion, the double stranded DNA remains as before and a single strand of RNA has been made. In the next illustration, the RNA is made from 5' to 3'. Note that where T's existed in the complementary DNA, U's exist in the complementary RNA.

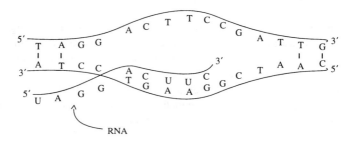

Making a new DNA from one already existing is called DNA *replication*. We begin with a double helix that has been separated into two single strands.

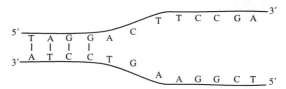

Then the single strands are used to template new double strands.

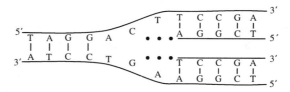

In this way two identical DNA molecules are made, each having one strand of the original molecule. Replication in this picture proceeds from right to left.

1.3 The Genetic Code

As soon as Watson and Crick proposed the double helix model of DNA in 1953, scientists began to study the problem of how a linear or helical DNA molecule

Some Molecular Biology

could encode a linear protein molecule. Cracking the genetic code became a hot topic and even attracted George Gamow (of the Big Bang Theory), a physicist. The sequence of insulin was the only protein sequence available and it was scrutinized very carefully. At that time it was not known that all amino acid sequences could be encoded in genes. Gamow, concentrating on the insulin sequence and the fact that 20 amino acids are used in protein sequences, discovered a very compelling code.

By example, consider the helix

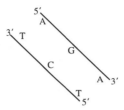

Gamow extracted the diamond

$$\begin{array}{ccc} & A & \\ C & & G \\ & T & \end{array}$$

and, reasoning that the code should be the same in either direction, he decided that

$$\begin{array}{ccc} & T & \\ G & & C \\ & A & \end{array}$$

should encode the same amino acid. The second diamond is obtained by rotating the first one by 180°. Let us count the diamonds which encode amino acids in this scheme. There are two basepairs $A \cdot T$ and $G \cdot C$. The number of diamonds with top and bottom base identical is, therefore, $\binom{4}{1} \times 2 = 8$. Otherwise, the top and bottom bases are unequal and the number of diamonds is $\binom{4}{2} \times 2 = 12$. In this scheme, the orientation of the basepair is not counted. When Gamow realized this he concluded that $20 = 8 + 12$, and he had found a candidate for the genetic code. The restrictions imposed on the possible sequences of amino acids by Gamow's scheme were severe, and his idea was rejected even before the genetic code was solved.

Crick's approach was to assume the code reads blocks of letters. These blocks cannot be less than 3 letters long: 4 and 4^2 are both less than 20, whereas $4^3 = 64$ exceeds 20. He probably came to this approach by reasoning that each strand was a template for the double helix, so should be sufficient to code proteins too. Crick decided that the genetic code should be "comma-free" – that the reading frame is determined by the blocks. Thus, if amino acids are encoded by triplets of nucleotides of DNA (*codons*) and if the code is comma-free, the reading frame of three consecutive nucleotides is

$$\underbrace{x_1 x_2 x_3}_{R_1} \underbrace{x_4 x_5 x_6}_{R_2} \underbrace{x_7 x_8 x_9}_{R_3} \cdots,$$

and not

$$x_1 \underbrace{x_2 x_3 x_4} \underbrace{x_5 x_6 x_7} \underbrace{x_8 x_9 x_{10}}$$

or

$$x_1 x_2 \underbrace{x_3 x_4 x_5} \underbrace{x_6 x_7 x_8} \underbrace{x_9 x_{10} x_{11}},$$

so that there is only one reading frame encoding $R_1 R_2 R_3 \ldots$.

Once again the magic number 20 comes out of counting. The assumption or requirement is that all possible amino acid sequences are possible. Clearly AAA, TTT, GGG, and CCC are all impossible because in AAAAAA there is no obvious reading frame. (There are four places to begin reading AAA.) Therefore if $4^3 = 64$ possible codons are being considered, we are left with $4^3 - 4 = 60$ to study.

Of those remaining, let XYZ be the codon. Clearly, to have a comma-free code, XYZXYZ must be read unambiguously, and whenever XYZ is a codon, YZX and ZXY are not. The number of remaining codons equals $1/3 \times 60 = 20$. Alas, it turns out that biology has found a different and less mathematically elegant solution.

The genetic code can be read from a single strand of RNA, and it is read 5' to 3'. The code is a triplet code: nonoverlapping successive blocks of three letters are translated into amino acids. There is a defined start or reading frame. Table 1.2 gives the genetic code in a compact form. There are three triplets–codons–that cause protein transcription to cease: UAA, UAG, and UGA. Viewed abstractly, the genetic code is a language in which 64 possible combinations of the 4 bases– uracil (U), cytosine (C), adenine (A), and guanine (G)–taken 3 at a time specify either a single amino acid or termination of the protein sequence. With 64 possible "words" and 21 possible "meanings," there is clearly the potential for different codons coding for identical amino acids. These 21 meanings are the 20 amino acids plus termination or stop. This is, in fact, the case: many pairs of codons that differ only in the third position base code for the same amino acid. On the other hand, a pair of codons differing only in the first or second position usually code for different amino acids.

RNA that is translated into protein is known as messenger RNA, mRNA. For example

mRNA UUUACUGCGGCC$\cdots$
protein Phe Tyr Cys Gly $\cdots$

A shift of one letter in reading the same nucleic acid sequence results in a very different amino acid sequence:

mRNA U UUUACUGCGGCC$\cdots$
protein $\cdots$ Phe Thr Ala Ala $\cdots$

Some Molecular Biology

amino acid	3 letter code	1 letter code
alanine	Ala	A
arginine	Arg	R
aspartic acid	Asp	D
asparginine	Asn	N
cysteine	Cys	C
glutamic acid	Glu	E
glutamine	Gln	Q
glycine	Gly	G
histine	His	H
isoleucine	Ile	I
leucine	Leu	L
lysine	Lys	K
methionine	Met	M
phenylalanine	Phe	F
proline	Pro	P
serine	Ser	S
threonine	Thr	T
tryptophan	Trp	W
tyrosine	Tyr	Y
valine	Val	V

Table 1.1: *Amino acid abbreviations*

The phase of codon reading is called the reading frame. There are three reading frames going 5' to 3'. Reading the complementary DNA strand, there are three reading frames in the opposite direction. Therefore, there are a total of six possible reading frames possible for double stranded DNA.

The genetic code was solved in a very interesting way. Extracts from bacteria were prepared except for the RNA template, mRNA. The extracts would then make a protein sequence when the experimenter added synthetic mRNA. They began by adding UUUUU··· and the proteins synthesized were composed of phenylalanine Phe · Phe ·,.... Because the reading frame is irrelevant, this suggests UUU codes for Phe. Next, try a mRNA such as UGUGUG..., where the results are polypeptides of Cys or Val. This still does not allow us to assign a unique codon so we try UUGUUG···, where the resulting polypeptides are made of Leu, Cys, and Val. This does not yield a codon assignment even though {UGU, GUG} ∩ {UUG, UGU, GUU}={UGU}. Next try UGGUGG··· where the peptides contain Trp, Gly, and Val. The common codon is {UGU, GUG} ∩ {UGG,GGU,GUG} = {GUG} so we assign GUG to Val, the common amino acid. Most of the genetic code was cracked in this manner.

Let $\mathbf{N} = \{A, C, G, U\}$ be the set of nucleic acids, $\mathbf{C} = \{(x_1x_2x_3) : x_i \in \mathbf{N}\}$,

Table 1.2: *The genetic code (64 triplets and corresponding amino acids) shown in its most common representation. The three codons marked* TC *are termination signals of the polypeptide chain.*

		2nd			
		U	C	A	G	
1st						3rd
U	Phe	Ser	Tyr	Cys	U
	Phe	Ser	Tyr	Cys	C
	Leu	Ser	TC	TC	A
	Leu	Ser	TC	Trp	G
C	Leu	Pro	His	Arg	U
	Leu	Pro	His	Arg	C
	Leu	Pro	Gln	Arg	A
	Leu	Pro	Gln	Arg	G
A	Ile	Thr	Asn	Ser	U
	Ile	Thr	Asn	Ser	C
	Ile	Thr	Lys	Arg	A
	Met	Thr	Lys	Arg	G
G	Val	Ala	Asp	Gly	U
	Val	Ala	Asp	Gly	C
	Val	Ala	Glu	Gly	A
	Val	Ala	Glu	Gly	G

and $\mathbf{A}$ be the set of amino acids and termination codon. The genetic code is simply a map $g : \mathbf{C} \to \mathbf{A}$.

1.4 Transfer RNA and Protein Sequences

As we have mentioned above, mRNA is read to make proteins. The amino acids are made available in the cell, and some are synthesized by the cell itself. With the amino acids and mRNA in the cell, there is an obvious question: How does a protein get made?

Part of the answer lies with the so-called adapter molecule, another RNA molecule known as transfer RNA (tRNA). Amino acids are linked to these smaller tRNA molecules of about 80 bases, and the tRNA then interacts with the codon of the mRNA. In this way, tRNA carries the appropriate amino acids to the mRNA. Obviously these reactions must be very specific. To understand this process it is

Some Molecular Biology

necessary to closely examine tRNA.

As RNA is single stranded, without the complimentary strand that DNA has, the molecule tends to fold back on itself to form helical regions. For example, 5' GGGGAAAAACCCC 3' can form the structure

```
G G G G A A
| | | |    A
C C C C A A
```

with a helix of 4 GC basepairs. This structure is known as a hairpin with a four basepair stem and a five base loop. Later in Chapter 13 we will study prediction of

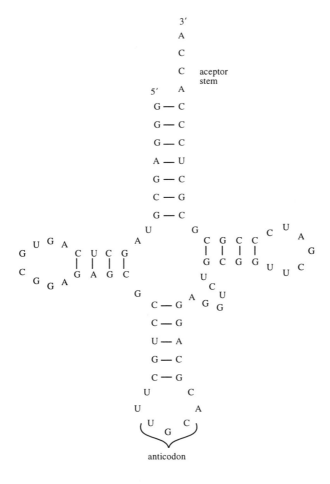

Figure 1.1: E. coli *Ala tRNA*

RNA structure. The longer sequence of a tRNA forms a more complex structure known as a cloverleaf. See Figure 1.1 for an *E. coli* tRNA associated with Ala. Next, the cloverleaf structure is given schematically, where only the backbone is shown.

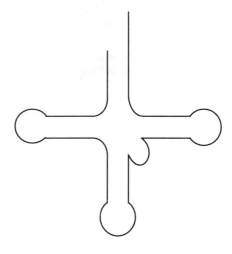

Figure 1.2: *tRNA schematic*

Actually this schematic only shows the simplest components of tRNA structure. There are some additional bonds formed and the entire structure becomes L shaped, with the 3' ACCA sequence at one end and the anticodon at the other. As the name indicates, an *anticodon* is complementary to the codon, and three basepairs can form between the codon in the mRNA and the anticodon of the tRNA. For example, for the Ala codon GCA, we have the following

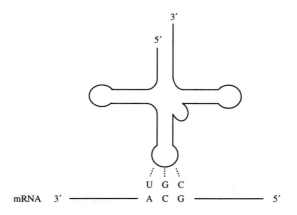

Some Molecular Biology

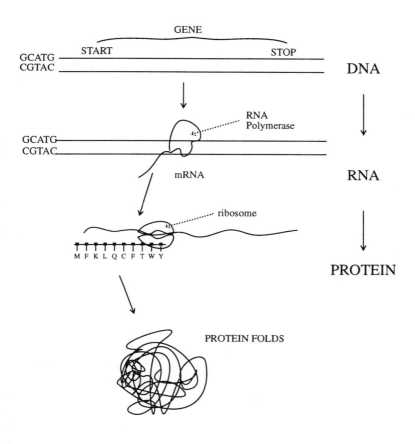

Figure 1.3: *DNA to RNA to protein*

Given that the codon GCA encodes Ala, the interaction between the tRNA and the mRNA seems almost inevitably to involve basepairing. The triplet UGC is the anticodon.

At the ribosome, mRNA is read and tRNA is utilized to make the protein sequence. From DNA to RNA to linear protein sequence to folded protein is shown in Figure 1.3.

1.5 Genes Are Not Simple

In this section we will look at some complexities and variations of how genes work in living systems. It is not possible to do more than briefly mention some of these fascinating topics, but the reference books contain much fuller treatments.

1.5.1 Starting and Stopping

In the genetic code (Table 1.2) there are three codons for "stop," signaling the end of the gene. Not mentioned there is the fact that genes begin with the so-called start codon, AUG, which codes for Met. As is often the case in biology, the story is not so simple, and the details are highly dependent on the organism. In this section we will describe a well-studied system, the bacterium *Escherichia coli* or *E. coli*.

A molecular complex of several proteins called RNA polymerase is required to transcribe mRNA from DNA. For reasons of efficiency and control, there are signals in DNA to start and stop RNA transcription. The canonical start pattern has specific small sequences in the DNA as follows:

The idea is that the polymerase binds to these two patterns and then is in position to proceed down the DNA, transcribing it into RNA. The sequences the polymerase binds to are called *promoter sequences*. The Met initiator codon is 10 or so bases beyond the mRNA start at +1. (There is no 0 in the numbering schemes of biological sequences.) These patterns are not precise in content or in location. Later in Chapter 10 we will study ways to discover these patterns in bacterial promoter sequences.

1.5.2 Control of Gene Expression

Proteins from different genes exist in widely varying amounts–sometimes in ratios of 1/1000. Gene expression could be controlled at two points: DNA $\to$ RNA or RNA $\to$ protein. One common way of regulating a gene is by a repressor, which affects the step DNA $\to$ RNA. Suppose that the gene exists to process a molecule such as the sugar lactose. When lactose is absent, a repressor molecule (another protein) binds the DNA, stopping the DNA $\to$ RNA step. When lactose is present, it binds to the repressor and, in turn, prevents it from binding DNA. Of course, when the expressed gene (the protein) has processed all the lactose molecules, the repressor is no longer inhibited by lactose and the repressor again binds DNA, shutting down the transcription of the gene.

This clever scheme allows the organism to only make protein to process lactose when needed, thereby saving much unneeded RNA and protein. This simple device is just one of a long and complex series of control mechanisms the cell has invented to deal with various environmental situations and various developmental stages.

1.5.3 Split Genes

Initially sequencing was done for *E. coli*, a member of the *prokaryotes*, organisms without a nucleus. When more rapid DNA sequencing began in 1976-1977, reading the genes of *eukaryotes*, organisms with a nucleus, was an obvious goal. There was soon a great surprise: The DNA encoding proteins was interrupted by noncoding DNA that somehow disappeared in the mRNA. Biologists, for example, expected $E_1 E_2 E_3$ to appear as one continuous coding region. Instead I_1 and I_2 split the gene into two pieces.

The so-called *exons*, E_1, E_2, and E_3, become an uninterrupted sequence, whereas the so-called *introns* I_1 and I_2 are spliced out and discarded. See Figure 1.4. A gene of 600 bases might be spread out over 10,000 bases of DNA. In yeast there is a tRNA gene that has 76 bps interrupted by a 14-bp intron. In a human gene, thyroglobin, 8500 bps are interrupted by over 40 introns of 100,000 bps.

Much remains to be learned about introns and exons. Why did they evolve? How can we recognize genes in uninterpreted DNA? What are the signals for splicing out the introns? These interesting questions do not yet have simple answers although much has been learned.

Originally it was supposed that most of the DNA encoded genes. It turns out to be true for viruses where it is important to be compact. In higher organisms, this is far from the case. Humans have around 5% of the genome used in protein coding. The function of much of the remaining DNA is unknown. Many people feel that much of it is "junk DNA," just sitting around not used for anything; others think that this DNA has important biological functions that are not yet understood.

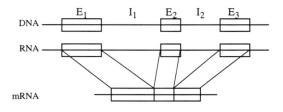

Figure 1.4: *Exons and introns*

1.5.4 Jumping Genes

One idea of molecular evolution is that it proceeds in small local steps. Our concept of a genome is that it is a blueprint for an organism. This concept is significantly altered by the discovery in both prokaryotic and eukaryotic genomes of segments of sequence that move from place to place in the genome. These sequences are known as transposable elements. They carry genes required for their movement or transposition, hence the name "jumping genes".

Much speculation has been made regarding the role of transposable elements. They of course can carry genetic material into new locations in the genome. In addition, as they often propagate themselves, they create identical or similar segments of DNA in various places in the genome. This can set the stage for duplication or deletion of the DNA between the transposable elements.

The role of transposable elements is not very clear. Some have suggested that transposable elements are "selfish DNA" and exist only for their own well being. That is, these elements could be viewed as mini-organisms themselves, living in the larger environment of genomic DNA.

In case this story sounds like an oddity, isolated to some obscure organisms, we point out that all organisms examined for transposable elements have been found to have them. Bacteria to humans, we all have jumping genes.

1.6 Biological Chemistry

It is now commonly understood that the molecules of biological organisms obey the standard and familiar laws of chemistry and physics. Until very recently it was thought otherwise, that there was a special set of laws of nature that apply to living organisms–perhaps a vital force. In fact, the chemistry of living organisms is special due to the requirements of organization and replication. In this chapter we will briefly touch on some of the basics of this chemistry. As pleasant as it is for mathematical scientists to view a DNA molecule as a long word over a four letter alphabet, it is very helpful to understand a little more of the basics.

The Basic Atoms

We will refer to the molecules of living organisms as biomolecules. Most biomolecules are made of only six different atoms: carbon, hydrogen, nitrogen, oxygen, phosphorus, and sulfur. Table 1.3 shows some properties of the atoms. These atoms combine into the fantastic variety of organisms by means of chemical bonds.

The most abundant elements in living organisms are carbon, hydrogen, nitrogen, and oxygen. The are found in all organisms. In addition, calcium (Ca),

Some Molecular Biology

Atom	# electrons in outer shell	usual # covalent bonds
carbon (C)	4	4
hydrogen (H)	1	1
nitrogen (N)	5	3,5
oxygen (O)	6	2
phosphorus (P)	5	3,5
sulfur (S)	6	2 (up to 6)

Table 1.3: *Covalent bonds*

chlorine (Cl), magnesium (Mg), phosphorus (P), sodium (Na), and sulphur are present in all organisms, but in much smaller amounts. Finally, cobalt (Co), copper (Cu), iron (Fe), manganese (Mn), and zinc (Zn) are present in small amounts and are essential for life. Other elements are utilized by some organisms in trace amounts.

Covalent Bonds

Recall that a covalent bond is formed when two atoms are held together because electrons in their outer shells are shared by both atoms. Covalent bonds are the strongest of the variety of bonds between biomolecules and contribute to great stability. The outer, unpaired electrons are the only ones that participate in covalent bonds. There cannot be more covalent bonds than electrons in the outer shell, but all outer shell electrons, by no means, need be used in covalent bonds.

The arrangement of the atoms in space is only hinted at in the chemical structure. The ball and stick model of methane has the hydrogen atoms at the points of a tetrahedron. Water approximates that structure due to two groups of two electrons that occupy the points with the two hydrogen atoms. Another complication of the water molecule comes from unequal sharing of electrons in the covalent bonds; such bonds are called dipolar.

The stability of the covalent bond can be quantized by the potential energy of the bond. The energy is given in the number of kilocalories in the bonds of a mole (6.02×10^{23} molecules), the units are kcal/mol. In these units an O–H has 110 kcal/mol, C–O has 84 kcal/mol, S–S has 51 kcal/mol, and C=O has 170 kcal/mol. The range is approximately 50 to 200. Most chemical bonds found in biological molecules are 100 kcal/mol.

Weak Bonds

We will discover that the structure or three-dimensional shape of a molecule is extremely important in biology. Often these structures as well as molecular interactions are stabilized by bonds much weaker than those discussed above. The energies of these weaker bonds are in the range of 1 to 5 kcal/mol. They

Name	water	methane	ethylene
Shorthand	H$_2$O	CH$_4$	H$_4$C$_2$
Chemical Structure	H—O—H	H—C(H)(H)—H	H$_2$C=CH$_2$
Ball and Stick Model	104.5	109.5	
Space Filling Model			

Table 1.4: *Molecular models*

exist in a range that allows them to be formed or broken easily. This is because the kinetic energy of molecules at physiological temperatures ($\approx 25°C$) is about 0.5 kcal/mol. Just a few of these weak bonds can stabilize a structure, but the structure can then be altered as necessary.

There are several types of weak bonds.

• **The hydrogen bond** is a weak electrostatic bond forming between an negatively charged atom (frequently oxygen) and a hydrogen atom that is already covalently bound. The hydrogen atom is positively charged. Two such bonds are

$$C = O \cdots H - N,$$

$$C = O \cdots H - O$$

The strengh of a hydrogen bond is from 3 to 6 kcal/mol.

• **The ionic bond** is formed between oppositely charged components of molecules. These bonds would often be very strong if they were not in water, which reacts with the components decreasing the bonding energy. In solids the bond could be 80 kcal/mol, whereas in solution it might be 1 kcal/mol.

• **Van der Waals interactions** occur when two atoms are close together. Random

Some Molecular Biology

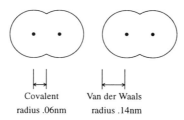

Figure 1.5: *Covalent bonds and Van der Waals interactions.*

movements of the electrons create dipole moments in one atom which generate an attractive dipole in the second. This causes a weak interaction that occurs between all molecules, polar or not. These interactions are nonspecific; that is, they do not depend on the specific identity of the molecules.

Figure 1.5 is a sketch of two O_2 molecules in Van der Waals contact. Van der Waals bonds have between 0.5 and 1.0 kcal/mol.

• **Hydrophobic interactions** occur among nonpolar molecules that cannot interact with water. The attraction is due to their aggregation in water, due to the noninteraction with water. Hydrophobic bond strength is between 0.5 and 3.0 kcal/mol.

Classes of Biomolecules

Many small molecules are present in organisms. They are needed for various reactions or are the product of these reactions. The general classes of small molecules are sugars, fatty acids, amino acids, and nucleotides.

Large molecules are built from the small molecules. The large molecule/small molecules relationships are polysaccharides/sugars, lipids/fatty acids, proteins/a--mino acids, nucleic acids/nucleotides. The last two large molecules, proteins and nucleic acids, are so big that they are known as macromolecules. This book is largely the mathematical study of these molecules and their biological properties.

Life is complex and the complexity requires these macromolecules. The amount of DNA required for life varies with the organism. Humans have 46 chromosomes, each if untangled and extended, about 4cm in length. Therefore the entire DNA in a nucleus of a single cell is about 2 meters in total length. The reason for this large size is the need to include all the information required to encode a human. While the human genetic material or genome is about 1000 times that of a bacterium, even bacterial genomes are large. Generally the size of the genome is an indicator of the size of the organism, but this is not a fixed rule.

(a)

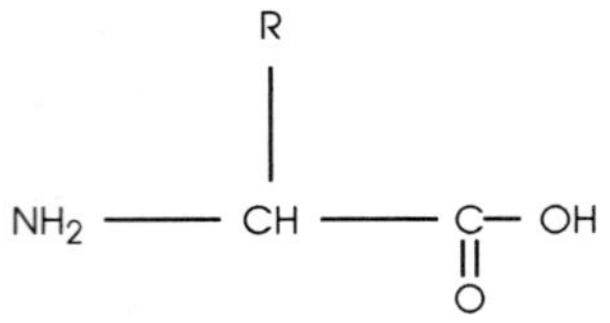

$$NH_2 - CH(R) - C(=O) - OH$$

(b) [detailed structure with N-H bonds, central C with R, H, and C(=O)-O-H]

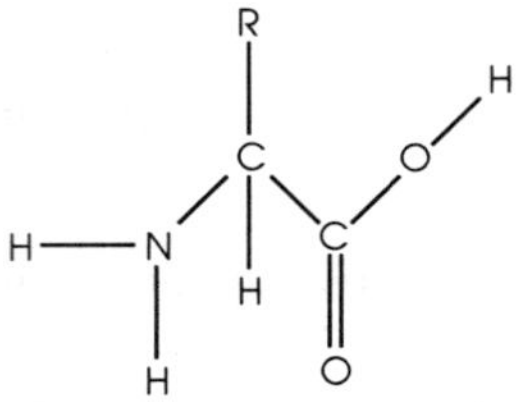

Figure 1.6: *General chemical structure of an amino acid (a) with more detail shown in (b)*

For example, the genomes of certain lilies and lungfish are about 100 times larger than the human.

Proteins

Proteins are the structural elements and the enzymatic elements of an organism–they are the working parts as well as the building material. These very important macromolecules are made of a sequence of molecules called amino acids.

There are 20 amino acids with the general chemical structures shown in Figure 1.6. The "R" in Figure 1.6 stands for the variable element or group which is known as a side chain, R group, or residue R. R gives the amino acid its identity; there are 20 R's and, consequently, 20 amino acids. COOH is known as the carboxyl group and NH_2 as the amino group. The central carbon is known as the α carbon. This atom is often used to locate an amino acid in a protein.

How do amino acids become proteins? There are many levels at which to approach this question. Here we focus on the most elementary chemical view and see in Figure 1.7 how three amino acids with residues R_1, R_2, and R_3 can be joined to form a three residue protein $R_1R_2R_3$ plus two water molecules.

Note that there is a chemically defined direction or orientation to this molecule. It "begins" with the amino end (N) and proceeds to the carboxyl end (C). The number of unique proteins is enormous, as there are 20^n proteins of length n.

The 20 amino acids structures are shown in Figure 1.8 with some of their properties.

Some Molecular Biology

Figure 1.7: *(a) Three amino acids with residues R_1, R_2, and R_3. (b) A protein or polypeptide with residues $R_1 R_2 R_3$, plus the two water molecules.*

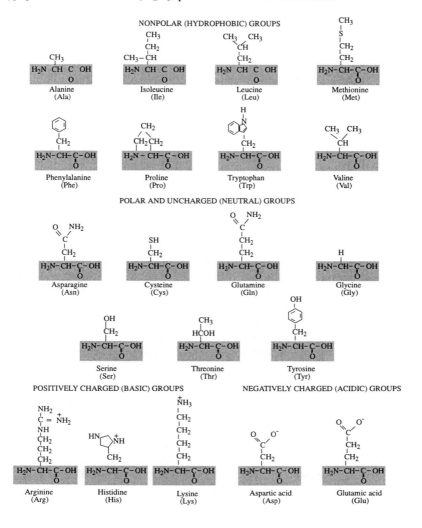

Figure 1.8: *Chemical structure of the 20 amino acids*

DNA

Deoxyribonucleic acid (DNA) is the carrier of genetic information for all organisms, except for some viruses. DNA consists of four different units called nucleotides. The components of a nucleotide are a phosphate group, a pentose (a 5-carbon sugar) and an organic base. The four different bases determine the identity of a nucleotide.

The general picture of a nucleotide is

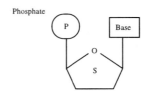

In more detail, the structure of the sugar S (a pentose) is

2-Deoxyribose

The carbon atoms are often not shown in the figure; only one appears in the above sugar structure. The numbers $1'$ to $5'$ refer to the carbon atom locations in the sugar. The carbon atoms $5'$ and $3'$ are used to define the orientation of the molecule.

In Figure 1.9 the structure of the bases in DNA is shown with the two types purines and pyrimidines illustrated. Purines have two rings, whereas pyrimidines have one ring. Bonds are represented by straight lines. In the bases, a carbon is present (and not shown) where two lines intersect. Bonds with no atom at the end have a hydrogen atom at the end and are hydrogen atoms.

A single strand of the DNA molecule is formed by the sequence phosphate-sugar-phosphate-····-sugar, with the $5'$ carbon of the sugar linked to the phosphate and the $1'$ carbon linked to the base. See Figure 1.10. Note that there is a definite direction to the chain, conventionally noted as $5'$ to $3'$.

Two chains joined by hydrogen bonds between so-called complementary bases form the DNA molecule. The complementary bases form the basepairs A-T and C-G in Figure 1.11. The hydrogen bonds are shown by dotted lines.

Finally we are ready to join the two strands into a complete DNA molecule in Figure 1.12. The strands must be of opposite orientation and, for a perfect fit, must be of complementary sequence. Given a sequence of one strand, it is obvious how to predict the other strand.

Some Molecular Biology

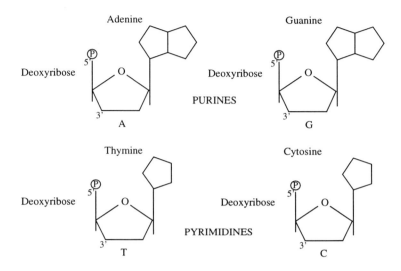

Figure 1.9: *Purines and pyrimidines*

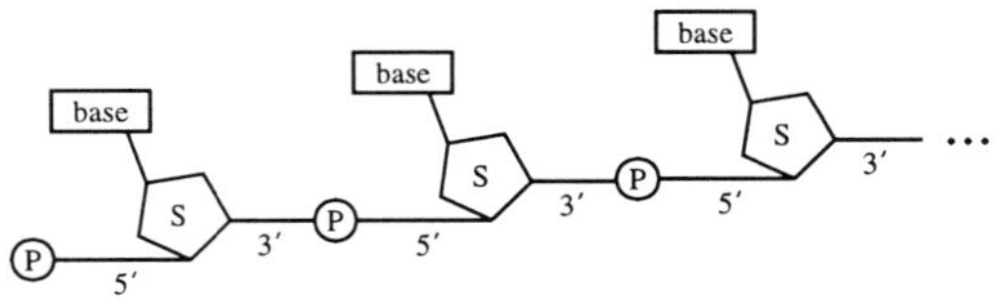

Figure 1.10: *DNA molecule*

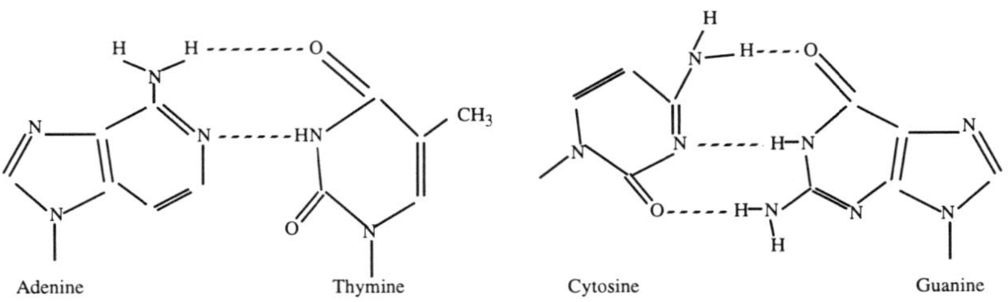

Figure 1.11: *Basepairs*

Figure 1.12: *Complete DNA molecule*

RNA

To describe ribonucleic acid (RNA), there is not much to do formally. The sugar in RNA is ribose instead of 2-deoxyribose.

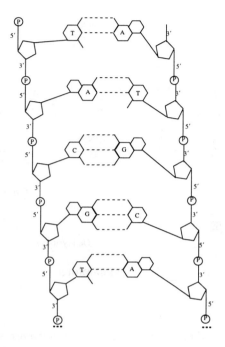

Ribose

In place of thymine, we have the pyrimidine base uracil.

Problems

Problem 1.1 Update the schematic for the central dogma according to the information given in the text.

Problem 1.2 We introduced the technique of transcribing synthetic mRNA in order to solve some of the genetic code. The synthetic mRNA was periodic in nature: XXXX..., XXYXXY..., XYYXYY..., etc. Derive all the information you can about the genetic code using the two letters A and C. Clearly define the synthetic mRNAs and their protein products. Recall that only the presence of amino acids could be detected, not the sequence.

Problem 1.3 There are 20^n different possible proteins of n residues. If the N- and C-terminus were indistinguishable and $R_1 R_2 \cdots R_n$ did not have a given orientation, $R_1 R_2 \cdots R_n$ would be indistinguishable from $R_n \cdots R_2 R_1$. How many different proteins would exist then?

Problem 1.4 Suppose noncoding DNA is randomly generated by independent identically distributed letters of probabilities p_A, p_G, p_C, and p_T. In a sequence of length n, what is the expected number of start codons reading 5' to 3'? Each of these start codons marks the beginning of a sequence of codons of amino acids ending with a stop codon. What is the probability distribution of the length X of one of these potential coding regions, counting the start and stop codons?

Problem 1.5 Splicing removes introns from transcribed RNA. Suppose the set up is $E_1 I_1 E_2 \cdots I_{k-1} E_k$. It is possible to make other protein sequences by so-called alternate splicing; that is, instead of removing I_1 we might remove $I_1 E_2 I_2$, resulting in the spliced sequence $E_1 E_3 E_4 \cdots E_k$. (We assume that each exon $E_1, \ldots, E_k$ can be translated individually.) (i) By allowing all possibilities of alternate splicing, how many distinct protein sequences can be made? (ii) If only one removal of adjacent exons is allowed, such as $I_1 E_2 I_2 E_3 I_3$, how many distinct sequences can be made?

Chapter 2

Restriction Maps

2.1 Introduction

When foreign DNA is introduced into a bacterium, it is usually unable to perform any genetic functions. One reason for this is that bacteria have evolved efficient means to protect themselves from invading DNA. A group of enzymes known as restriction endonucleases or restriction enzymes perform this function by cleaving the DNA, thus "restricting" the activity of the invading DNA. The bacteria's own DNA is protected from its arsenal of restriction enzymes by another class of enzymes that modifies or methylates the host DNA. Restriction enzymes cleave unmethylated DNA. One way to view restriction enzymes is as the bacterial immune system. This class of restriction enzymes is invaluable to the practice of molecular biology because they always cut DNA at short specific patterns in the DNA. These patterns are called restriction sites.

Table 2.1 contains some examples of these sites with details of their cleavage patterns. About 300 restriction enzymes have been found and they cut at about 100 distinct restriction sites. Note that these sites are palindromes, as are most known restriction sites, although a palindrome in molecular biology means the $5' \to 3'$ sequence on the top strand is identical with the $5' \to 3'$ sequence on the bottom strand; that is, a palindrome in molecular biology is a word that is equal to its reverse complement.

Details of the cutting are indicated in Table 2.1. Sites such as *Hae*III are cut leaving blunt ends

```
       G G C C      G G  C C
       | | | | ─→  | |   | |
       C C G G      C C   G G
```

whereas others such as *Eco*RI leave overhangs, also known as sticky ends:

```
     G A A T T C              G          A A T T C
     | | | | | |     ─────→   |                  |
     C T T A A G              C T T A A          G
```

Microorganism	Restriction Enzyme Name	Restriction Site
Bacillus amyloliquefaciens H	*Bam*HI	G|G A T C C C C T A G|G
Brevibacterium albidum	*Bal*I	T G G|C C A A C C|G G T
Escherichia coli RY13	*Eco*RI	G|A A T T C C T T A A|G
Haemophilus aegyptius	*Hae*II	Pu G C G C|Py Py|C G G C Pu
Haemophilus aegyptius	*Hae*III	G G|C C C C|G G
Haemophilus influenzae R_d	*Hind*II	G T Py|Pu A C C A Pu|Py T G
Haemophilus influenzae R_d	*Hind*III	A|A G C T T T T C G A|A
Haemophilus parainfluenzae	*Hpa*I	G T T|A A C G A A|T T G
Haemophilus parinfluenzae	*Hpa*II	C|C G G G G C|C
Providencia stuartii 164	*Pst*I	C T G C A|G G|A C G T C
Streptomyces albus G	*Sal*I	G|T C G A C C A G C T|G

Table 2.1: *Restriction enzymes*

The first three letters of a restriction enzyme refers to the organism, the fourth letter (if any) refers to the strain, and the Roman numerals index the restriction enzymes from the same organism.

These enzymes are essential to the practice of molecular biology, as they make

Restriction Maps

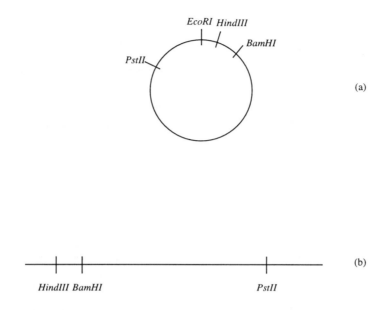

Figure 2.1: *Circular (a) and linear (b) restriction maps of pBR322*

possible critical manipulations. Our topic in Chapters 2 to 4 is to describe some mathematical problems that arise in connection with making maps of the locations of restriction sites. Restriction maps show the location or approximate location of a selection of restriction sites along linear or circular DNA. Later in the course we will show a restriction map of *E. coli* which has approximately 7000 sites. For now we show in Figure 2.1(a) a circular restriction map of a famous plasmid in (pBR322, 4363 bps) in *E. coli*, whereas in (b) the map is made linear by cutting at the *Eco*RI site. In Figure 2.2 is a linear restriction map of bacteriophage λ of 48,502 bps. Note that we have started measuring length in bps or basepairs.

To study the mathematics of restriction maps it is essential to introduce some concepts from graph theory.

2.2 Graphs

Graphs, a topic from discrete mathematics and computer science, provide very natural mathematical models for several data structures and relationships in biology, including restriction maps. We take this opportunity to formalize some notation.

A *graph G* is a set of *vertices V* and *edges E*, where $e \in E$ implies $e = \{u, v\}$, where $u \in V$ and $v \in V$. See Figure 2.3(a). In the case $e = \{u, v\} \in E$, u and

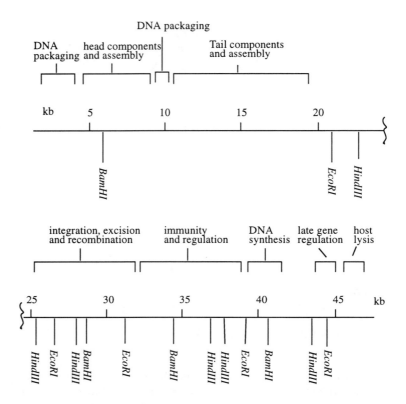

Figure 2.2: *Restriction map of bacteriophage* λ. *Genetic functions are indicated at the top of the figure by brackets.*

v are said to be adjacent. The *degree* of a vertex v is the number of distinct edges e such that $v \in e$. Directed graphs are graphs where the edges have a direction $e = (u, v)$ that is ordered.

A *bipartite graph* is a graph such that $V = V_1 \bigcup V_2$, $V_1 \bigcap V_2 = \emptyset$, and $e = \{u, v\} \in E$ implies $u \in V_1$ and $v \in V_2$ or $v \in V_1$ and $u \in V_2$. See Figure 2.3(b). Two graphs G and H are *isomorphic*, $G \cong H$, if there is a one-to-one correspondence between the vertices that preserves adjacency.

Many graphs arise from linear structures. Let S be a set and $\mathcal{F} = \{S_1, S_2, \ldots, S_n\}$ be a family of distinct nonempty subsets of S. The *intersection graph of* $\mathcal{F}$, $\mathcal{I}(\mathcal{F})$, is defined by $V(\mathcal{I}(\mathcal{F})) = \mathcal{F}$ with S_i and S_j adjacent when $S_i \bigcap S_j \neq \emptyset$. G is an *intersection graph* on S if there exists $\mathcal{F}$ such that $G \cong \mathcal{I}(\mathcal{F})$. An *interval graph* is a graph that is isomorphic to some $\mathcal{I}(\mathcal{F})$, where $\mathcal{F}$ is a family of intervals on $\mathbb{R}$, the real line.

Restriction Maps

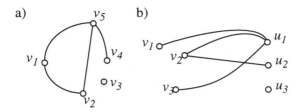

Figure 2.3: *(a) The graph G with $V = \{v_1, v_2, \ldots, v_5\}$ and $E = \{\{v_1, v_2\}, \{v_1, v_5\}, \{v_2, v_5\}, \{v_5, v_4\}\}$. (b) A bipartite graph with $V_1 = \{v_1, v_2, v_3\}$ and $V_2 = \{u_1, u_2, u_3\}$.*

2.3 Interval Graphs

The study of interval graphs has its origin in a paper of Benzer in 1959 who was studying the structure of bacterial genes. At that time it was not known whether or not the collection of DNA composing a bacterial gene was linear. It is now well known that such genes are linear along the chromosome, and Benzer's work was basic in establishing this fact. Essentially, he obtained data on the overlap of fragments of the gene and showed the data consistent with linearity. Of course, there is no longer active interest in Benzer's problem, but we discuss here a special class of interval graphs central to the modern practice of molecular biology. These graphs arise in connection with restriction maps which show the location of certain sites (short specific sequences) on a specific DNA.

To make these ideas specific, we graphically present a restriction map, which we will refer to as an $A \wedge B$ map, with three occurrences of restriction site A and four occurrences of restriction site B:

Next we show the maps for A and B separately and refer to them as the A map and B map, respectively:

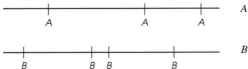

Biologists, when constructing restriction maps, can identify individual intervals between sites, as restriction enzymes cleave the DNA into these intervals, but they cannot directly observe the order of these intervals. Instead, they try to establish whether or not an A interval overlaps a B interval and, from this overlap data, construct the map. Formally we say two intervals overlap if their interiors have nonempty intersection. Frequently this overlap data comes from determining which A and B intervals contain the various $A \wedge B$ intervals. In fact, frequently the most difficult aspect of restriction map construction is determining the overlap data. This difficult problem is not pursued until later chapters.

Next the intervals are arbitrarily labeled:

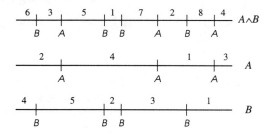

Label the components of the A map by $A_1, A_2, \ldots$ and of the B map by $B_1, B_2, \ldots$. Define the incidence matrix $I(A, B)$ whose (i, j)-th entry is 1 if $A_i \cap B_j \neq \emptyset$ and is 0 otherwise. For the above example and labeling

$$I(A,B) = \begin{pmatrix} 1 & 0 & 1 & 0 & 0 \\ 0 & 0 & 0 & 1 & 1 \\ 1 & 0 & 0 & 0 & 0 \\ 0 & 1 & 1 & 0 & 1 \end{pmatrix};$$

$$I(A, A \wedge B) = \begin{pmatrix} 0 & 1 & 0 & 0 & 0 & 0 & 0 & 1 \\ 0 & 0 & 1 & 0 & 0 & 1 & 0 & 0 \\ 0 & 0 & 0 & 1 & 0 & 0 & 0 & 0 \\ 1 & 0 & 0 & 0 & 1 & 0 & 1 & 0 \end{pmatrix};$$

$$I(B, A \wedge B) = \begin{pmatrix} 0 & 0 & 0 & 1 & 0 & 0 & 0 & 1 \\ 1 & 0 & 0 & 0 & 0 & 0 & 0 & 0 \\ 0 & 1 & 0 & 0 & 0 & 0 & 1 & 0 \\ 0 & 0 & 0 & 0 & 0 & 1 & 0 & 0 \\ 0 & 0 & 1 & 0 & 1 & 0 & 0 & 0 \end{pmatrix}.$$

As mentioned above, $I(A, A \wedge B) = (x_{ik})$ and $I(B, A \wedge B) = (y_{jk})$ are frequently known, whereas $I(A, B)$ is desired. The next proposition relates these matrices.

Proposition 2.1 *For the incidence matrices defined above*

$$I(A, B) = I(A, A \wedge B) I^T(B, A \wedge B), \tag{2.1}$$

where $I^T(B, A \wedge B)$ is the transpose of $I(B, A \wedge B)$.

Proof. Note that the (i, j) element of this matrix product equals the number of $A \wedge B$ intervals in both the i-th A interval and the j-th B interval; that is,

$z_{ij} = \sum_k x_{ik} y_{jk}$. But the $A \wedge B$ intervals are formed by intersection of A intervals with B intervals so that x_{ik} and y_{jk} are both 1 if and only if $A_i \cap B_j = (A \wedge B)_k$. This happens for at most one k. ∎

Having shown that $I(A, B)$ is easily obtained from $I(A, A \wedge B)$ and $I(B, A \wedge B)$, we now turn to characterizing $I(A, B)$ and then present an algorithm for constructing restriction maps from $I(A, B)$. Two equivalent characterizations are discussed and then collected in Theorem 2.1.

The matrix $I(A, B)$ tells us when an A interval and a B interval have an $A \wedge B$ interval in common, or, equivalently, when the interiors of the A interval and B interval intersect. Thus, constructing a restriction map from $I(A, B)$ is equivalent to finding an interval representation for a certain graph $G(A, B)$ which is obtained in the following natural way: The vertex set $V(A, B)$ of $G(A, B)$ consists of the union of the set of A intervals and the set of B intervals, and the edge set $E(A, B)$ consists of the unordered pairs $\{A_i, B_j\}$ for each A intervals, A_i, and B intervals, B_j, which overlap. The graph $G(A, B)$ is completely defined by $I(A, B)$.

If $G(A, B)$ arises from a restriction map, we need only delete the endpoints of the pieces in the A map and the B map to obtain an (open) interval representation of the graph $G(A, B)$. Thus $G(A, B)$ is an interval graph. As the interiors of the A intervals are disjoint, the interiors of the B intervals are disjoint, and the interior of every A interval (respectively, B interval) overlaps the interior of some B interval (respectively, A interval), it follows that $G(A, B)$ is bipartite with no isolated vertices.

Conversely, one may construct a restriction map for any bipartite interval graph G without isolated vertices by drawing together the intervals representing vertices in each of the two parts until the A and B intervals correspond to A and B maps. Figure 2.4 is the graph for our above example.

Figure 2.4: *Interval graph $G(A, B)$*

We next observe that $G(A, B)$ is a connected graph unless the A and B restriction sites coincide. In general, if the A and B restriction sites coincide k times (not counting the ends), then $G(A, B)$ will have $k+1$ connected components.

There is also a $0 - 1$ matrix formulation characterizing interval graphs. In our special case it can be shown that for the graphs $G(A, B)$ that can arise, the associated $0 - 1$ matrix $I(A, B)$ can be put into a particularly nice form. Specifically, if the rows and columns of $I(A, B)$ are permuted in accordance with an ordering of the edges of $G(A, B)$ 1's in $I(A, B)$ will belong to one of a collection of $k + 1$ staircase shapes going from top left to lower right, where k is the number of components of $G(A, B)$. Each row or column has its 1's consecutive and meets precisely one of the staircases. There can be no 2×2 all ones submatrix. Here is such a permutation for our example:

$$A \begin{array}{c} \\ 2 \\ 4 \\ 1 \\ 3 \end{array} \begin{pmatrix} 4 & 5 & \overset{B}{2} & 3 & 1 \\ 1 & 1 & 0 & 0 & 0 \\ 0 & 1 & 1 & 1 & 0 \\ 0 & 0 & 0 & 1 & 1 \\ 0 & 0 & 0 & 0 & 1 \end{pmatrix}$$

Note that the matrix is now in "staircase form" with the 1's as the staircase. These characterizations of restriction maps are now collected in the following theorem.

Theorem 2.1 *The following statements are equivalent:*

(i) The bipartite graph $G(A, B)$ is the graph constructed from some restriction map.

(ii) $G(A, B)$ is a bipartite interval graph with no isolated vertices.

(iii) $I(A, B)$ can be transformed by row and column permutations into staircase form with each row or column having 1's in precisely one of these staircases.

For more information on interval graphs in general see the book by Golumbic (1980).

Interval graphs in general can be recognized and their representations can be found in time which is linear in the number of vertices plus edges. For the class of graphs considered here, we can provide a recognition and representation algorithm which is quite simple. Like the algorithm for the general problem, it requires only linear time and storage. Later in the book, we will become a little more precise about algorithms. Right now, just think of an algorithm as a way to accomplish a task, such as constructing a map from a graph.

Because we have a bipartite interval graph of nonintersecting intervals in each vertex set, each vertex has at most two adjacent vertices v with $\deg(v) \geq 2$. This is key for step 3 below. Define

$$L = \{v : \deg(v) \geq 2\}.$$

The reason we are not separating A vertices from B vertices shows up when we try to find an "end" of the map:

Note that of the two leftmost pieces, only the bottom has $\deg = 2$. In fact ends may have $\deg \geq 2$ but have only one adjacent vertex of $\deg \geq 2$.

Restriction Maps

Algorithm 2.1 (Restriction Map)

```
   set L₀ = L.
1. find an end.
   find v ∈ L₀ such that only one adjacent u ∈ L₀.  if no
   such u exists, go to step 3.
2. find connected component.
   set v₁ = v and v₂ = u.  extend the sequence of edges
   {v₁,v₂}{v₂,v₃}···{v_{r-1},v_r} to maximum r, where all v_i ∈ L₀.
   remove {v₁,...,v_r} from L₀.  go to step 1.
3. fill in the component.
   for v ∈ L ∩ L₀ᶜ and u ∈ Lᶜ with {u,v} ∈ E put {v,u} between
   {v_i,v} and in the component edge list.
4. single intervals.
   for v ∈ L₀ for all u_i ∈ Lᶜ with {u_i,v} ∈ E add {u₁,v}{u₂,v}···
   to list of components.
5. isolated pairs.
   for {v,u} ∈ E with v,u ∈ Lᶜ add the edge {v,u} to the list
   of components.
6. done.
```

The algorithm can be modified to provide recognition of these graphs.

Consider the graph from our example. Suppose the vertices are listed $A_1, A_2, A_3, A_4, B_1, B_2, B_3, B_4, B_5$, and edges are listed $\{A_1, B_1\}, \{A_1, B_3\}, \{A_2, B_4\}, \{A_2, B_5\}, \{A_3, B_1\}, \{A_4, B_2\}, \{A_4, B_3\}, \{A_4, B_5\}$. The algorithm steps 1 and 2 give

vertex v	A_1	A_2	A_3	A_4	B_1	B_2	B_3	B_4	B_5
degree deg(v)	2	2	1	3	2	1	2	1	2
$L(v)$	B_1, B_3	B_4, B_5	B_1	B_2, B_3, B_5	A_1, A_3	A_4	A_1, A_4	A_2	A_2, A_4.

Here $L(v) = \{u : \{u,v\}$ is an edge$\}$. In steps 1 and 2, we encounter the end A_2 which leads to vertices B_5, A_4, B_3, A_1, and B_1. No other ends are to be found. So far this gives the following edge ordering:

$$\{A_2, B_5\}, \{B_5, A_4\}, \{A_4, B_3\}, \{B_3, A_1\}, \{A_1, B_1\}.$$

Now we go through the edge list and insert each edge which contains a vertex of deg = 1 and vertices in $L \cap L_0^c$ previously involved in the ordering. The new edge must be adjacent to a previous edge involving the vertex in $L \cap L_0^c$. We insert it between previous edges if there is a choice, or else on the left or right if the vertex in $L \cap L_0^c$ was a left or right end, respectively. It happens that in our example we now obtain the component

$$\{A_2, B_4\}, \{A_2, B_5\}, \{A_4, B_5\}\{A_4, B_2\}, \{A_4, B_3\}, \{A_1, B_3\}, \{A_1, B_1\}, \{A_3, B_1\}.$$

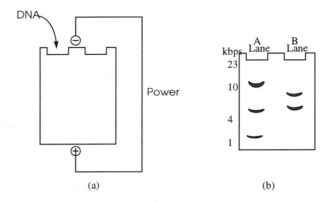

Figure 2.5: *(a) Electrophoresis setup; (b) resulting gel*

2.4 Measuring Fragment Sizes

In order to perform most experiments in biology it is necessary to have many ($\cong 10^8$) identical copies of the same DNA molecule. Usually this feature is not critical to mathematical understanding, but it is important for an understanding of the experimental data. To further explore restriction mapping, we must discuss how restriction fragment length measurements are obtained. To begin, assume that the DNA is present in many identical copies.

The size or length of the DNA is measured by a process known as gel electrophoresis. The gel refers to a solid matrix, usually agarose or polyacrylamide, which is permeated with a liquid buffer. Recall that DNA is a negatively charged molecule. When the gel is placed under an electric field, the DNA migrates toward the positive pole. See Figure 2.5(a) for this general setup.

It turns out that DNAs migration distance is a function of size. By using DNA of known lengths, we can calibrate gel electrophoresis and estimate DNA of unknown lengths. The DNA is left under the field a fixed time and the migration distance is measured. The DNA can be located by staining the gel with ethidium bromide, which causes the DNA to fluoresce and be visible under ultraviolet light. Another technique is to tag the DNA with a radioactive label and then to expose the x-ray film to the gel. The resulting gels appear in Figure 2.5(b).

The relationship between migration distance and DNA size or length is not precisely understood. The simplest useful model is that migration distance D is linear with the logarithm size or length L; that is, $D \approx a + b \log(L)$, where $b < 0$. The negative slope is easy to rationalize: Long DNAs get tangled up in the gel matrix and do not migrate far, whereas smaller DNAs thread through the matrix much more easily.

Because a large number of identical DNAs are moving through the gel it is not

Restriction Maps

surprising that the migration distance D is a smear, not a point; that is, the scientist cannot measure D too precisely. Therefore, let us assume that D can be measured but has a normal distribution $N(\mu_D, \sigma_D^2)$. When $a + b\log(L)$ has a normal distribution, statisticians say L has a lognormal distribution. An interval that D will fall into 95% of the time is $\mu_D \pm 2\sigma_D$. Therefore, $a + b\log(L) \in \mu_D \pm 2\sigma_D$ or, because $b < 0$,

$$L \in e^{\frac{\mu_D - a}{b} \pm \frac{2\sigma_D}{b}} = \left(e^{-\frac{2\sigma_D}{b}} \times e^{\frac{\mu_D - a}{b}}, e^{+\frac{2\sigma_D}{b}} \times e^{\frac{\mu_D - a}{b}}\right).$$

Now

$$L \approx \left(e^{\frac{\mu_D - a}{b}}\left(1 - \frac{2\sigma_D}{|b|}\right), e^{\frac{\mu_D - a}{b}}\left(1 + \frac{2\sigma_D}{|b|}\right)\right)$$

when $\frac{2\sigma_D}{|b|}$ is small. In this case we will read length "within $\frac{2\sigma_D}{|b|} \times 100\%$" of the true length. Interestingly enough, biologists often report their measurements in this "within $x \times 100\%$" form.

The implications of this bit of statistics is that we cannot expect measurement errors for fragment length to be independent of length. Small fragments can be measured very accurately, whereas larger ones can have very large measurement errors. An example of gel electrophoresis is shown below. The migration distances in 1000 basepairs (1 kbp) for known DNAs are indicated on the leftmost gel.

Problems

Problem 2.1 From the following map decomposition and labeling, find $I(A, B)$, $I(A, A \wedge B)$, $I(B, A \wedge B)$, verify Proposition 2.1, and find a staircase representation of $I(A, B)$.

```
    2    4  5   1     3      A
  ──┼────┼──┼───┼─────┼──

    2     3       4      1   B
  ──┼─────┼───────┼──────┼──

  1 2   3  4 5    6    7 8   A∧B
  ─┼─┼──┼──┼─┼────┼────┼─┼──
```

Problem 2.2 From

$$I(A,B) = \begin{array}{c} \\ 1 \\ 2 \\ 3 \\ 4 \\ 5 \end{array} \begin{array}{c} 1\ 2\ 3\ 4\ 5\ 6\ 7\ 8 \\ \left(\begin{array}{cccccccc} 0 & 1 & 0 & 0 & 0 & 0 & 0 & 0 \\ 0 & 0 & 0 & 1 & 0 & 0 & 0 & 0 \\ 1 & 0 & 0 & 0 & 0 & 0 & 1 & 1 \\ 0 & 1 & 1 & 0 & 0 & 1 & 0 & 0 \\ 0 & 1 & 0 & 0 & 1 & 0 & 0 & 0 \end{array}\right) \end{array}$$

use Algorithm 2.1 to find the restriction map.

Problem 2.3 Restriction maps can have more than two enzymes. Let $A, B,$ and C be three restriction enzymes, and define $I(A, B, C)$ with (i, j, k) entry $= 1$ if $A_i \cap B_j \cap C_k \neq \emptyset$ and 0 otherwise. (i) State and prove Proposition 2.1 for $I(A, B, C)$. (ii) State the generalization for e enzymes $A^{(1)}, A^{(2)}, \ldots, A^{(e)}$.

Chapter 3

Multiple Maps

The locations of restriction sites can be determined in several ways. The essential tool is the capability of digesting DNA with one or more restriction enzymes along with the capability of measuring the resulting DNA fragments by gel electrophoresis. In this chapter we will focus on two generic restriction enzymes, A and B.

Our earlier discussion of interval graphs via overlap data has a basis in experimental methods. The overlap $A_i \cap B_j$ can be determined, in some cases, in the following way. First, DNA is digested with enzyme A and then each A fragment, A_i, is digested with enzyme B. If all $A \wedge B$ sizes are unique, then each $A \wedge B$ fragment is uniquely assigned to an A_i. The uniqueness of $A \wedge B$ fragments is usually given by uniqueness of the fragment lengths. If this experiment is repeated with the enzymes applied in the opposite order, then each $A \wedge B$ fragment is uniquely assigned to a B_j. These experiments give data for the interval graph methods of Chapter 2.

Maps are not usually determined by direct experimental data on overlaps $A_i \cap B_j$, due to the difficulty of performing the experiments. Instead, the two single and one double digests are done, and the three batches of DNA are run in three lanes of a gel. In Section 3.1, we describe the problem and prove a theorem regarding the nonuniqueness of solutions to the problem of determining restriction maps from these data. Later sections give more detailed classifications of these multiple solutions.

A simple proposition will prove useful. Define $A \vee B$ to be the set of fragments created by those cut sites common to both the A and the B maps. Often $A \wedge B$ is referred to as the *meet* of A and B, and $A \vee B$ as the *join* of A and B. We further abuse notation by setting $A = \{A_1, \ldots, A_n\}$ and $B = \{B_1, \ldots, B_m\}$ as well as $[A, B]$ to denote a specific ordering of fragments in a restriction map.

Proposition 3.1 $|A \wedge B| = |A| + |B| - |A \vee B|$.

Proof. To enumerate the number of elements in a set of fragments, create the set S of all fragment endpoints, including the beginning and end of the DNA which need not be enzyme cut sites. If S_A is created from $\{A_1, \ldots, A_n\}$, then $|S_A| = n + 1$.

By inclusion-exclusion,

$$|S_A \cup S_B| = |S_A| + |S_B| - |S_A \cap S_B|.$$

The cut sites in $S_A \cup S_B$ create $A \wedge B$, and the sites in $S_A \cap S_B$ create $A \vee B$. By counting fragments by their left endpoints,

$$|S_A \cup S_B| = |A \wedge B| + 1,$$
$$|S_A| = |A| + 1,$$
$$|S_B| = |B| + 1,$$

and

$$|S_A \cap S_B| = |A \vee B| + 1$$

The inclusion-exclusion formula implies the result. ∎

3.1 Double Digest Problem

The problem we consider, the multiple digest problem, is as follows. We discuss the simplest case involving linear DNA, two digests, and no measurement error. We will refer to this problem as the *double digest problem*, or DDP. A restriction enzyme cuts a DNA molecule of length L bps at all occurrences of a short specific pattern, and the lengths of the resulting fragments are recorded. In the double digest problem, we have as data the list of fragment lengths when each enzyme is used alone, say

$$\mathbf{a} = \|A\| = \{a_i : 1 \leq i \leq n\} \quad \text{from the first digest,}$$
$$\mathbf{b} = \|B\| = \{b_i : 1 \leq i \leq m\} \quad \text{from the second digest,}$$

as well as a list of double digest fragment lengths when the restriction enzymes are used in combination and the DNA is cut at all occurrences of both restriction patterns, say

$$\mathbf{c} = \|A \wedge B\| = \|C\| = \{c_i : 1 \leq i \leq l\}.$$

Only fragment length information is retained. We will write $\mathrm{DDP}(\mathbf{a}, \mathbf{b}, \mathbf{c})$ to denote the problem of finding maps $[A, B]$ such that $\|A\| = \mathbf{a}$, $\|B\| = \mathbf{b}$, and $\|C\| = \|A \wedge B\| = \mathbf{c}$. In general $\|A\|, \|B\|$, and $\|C\|$ will be multisets; that is, there may be values of fragment lengths that occur more than once. We adopt the convention that the sets $\|A\|, \|B\|$, and $\|C\|$ are ordered, that is, $a_i \leq a_j$ for $i \leq j$, and likewise for the sets $\|B\|$ and $\|C\|$. Of course,

$$\sum_{1 \leq i \leq n} a_i = \sum_{1 \leq i \leq m} b_i = \sum_{1 \leq i \leq l} c_i = L,$$

Multiple Maps

as we are assuming that fragment lengths are measured in number of bases with no errors. Given the above data, the problem is to find orderings for the sets A and B such that the double digest implied by these orderings is, in a sense made precise below, C. This is a mathematical statement of a problem that is sometimes solved by exhaustive search. In Chapter 4, we will study algorithms for constructing maps from length data.

We may express the double digest problem more precisely as follows. For σ a permutation of $(1, 2, \ldots, n)$ and μ a permutation of $(1, 2, \ldots, m)$, call (σ, μ) a *configuration*. By ordering A and B according to σ and μ, respectively, we obtain the set of locations of cut sites

$$S = \left\{ s : s = \sum_{1 \leq j \leq r} a_{\sigma(j)} \text{ or } s = \sum_{1 \leq j \leq t} b_{\mu(j)}; 0 \leq r \leq n, 0 \leq t \leq m \right\}.$$

Because we want to record only the location of cut sites, the set S is not allowed repetitions, that is, S is not a multiset. Now label the elements of S such that

$$S = \{s_j : 0 \leq j \leq l\} \text{ with } s_i \leq s_j \text{ for } i \leq j.$$

The double digest implied by the configuration (σ, μ) can now be defined by

$$C(\sigma, \mu) = \{c_j(\sigma, \mu) : c_j(\sigma, \mu) = s_j - s_{j-1} \text{ for some } 1 \leq j \leq l\},$$

where we assume, as usual, that the set is ordered by size in the index j. The problem then is to find a configuration (σ, μ) such that $C = C(\sigma, \mu)$, where $C = A \wedge B$ is determined by experiment.

3.1.1 Multiple Solutions in the Double Digest Problem

In many instances, the solution to the double digest problem is not unique. For example, with

$$\mathbf{a} = \|A\| = \{1, 3, 3, 12\},$$
$$\mathbf{b} = \|B\| = \{1, 2, 3, 3, 4, 6\},$$

and

$$\mathbf{c} = \|C\| = \|A \wedge B\| = \{1, 1, 1, 1, 2, 2, 2, 3, 6\},$$

two distinct solutions are shown in Figure 3.1.

Because the A and B fragment orders determine the $C = A \wedge B$ fragments and the order of those fragments, the simplest combinatorics gives $n!m!$ map configurations. However, in our example, 3 is repeated twice in the A digest and 3 is also repeated twice in the B digest. Under the assumption that we cannot distinguish fragments of equal lengths, in our example of $n = 4$ and $m = 6$ there

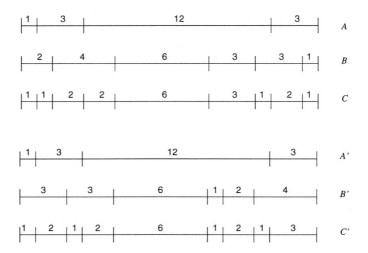

Figure 3.1: *Multiple solutions*

are $4!6!/2!2! = 4320$ map configurations. As we will discuss later, there are only 208 distinct solutions for these data. We now demonstrate that this phenomenon of multiple solutions is far from isolated; rather it is to be expected.

Below, we use a powerful result from probability theory, the Kingman subadditive ergodic theorem, to prove that the number of solutions to the double digest problem as formulated in Section 3.1 increases exponentially as a function of length under the probability model stated below.

For reference, we state a version of the subadditive ergodic theorem here.

Theorem 3.1 (Kingman) *For s, t non-negative integers with $0 \leq s \leq t$, let $X_{s,t}$ be a collection of random variables which satisfy the following:*

(i) *Whenever $s < t < u$, $X_{s,u} \leq X_{s,t} + X_{t,u}$,*

(ii) *The joint distribution of $\{X_{s,t}\}$ is the same as that of $\{X_{s+1,t+1}\}$,*

(iii) *The expectation $g_t = E[X_{0,t}]$ exists and satisfies $g_t \geq Kt$, for some constant K and all $t > 1$.*

Then the finite $\lim_{t \to \infty} X_{0,t}/t = \lambda$ exists with probability one and in the mean.

To motivate this theorem, recall the usual strong law of large numbers (SLLN) which treats independent, identically distributed (iid) random variables $W_1, W_2, \ldots$ where $\mu = E(W_i)$ and $|\mu| < \infty$. The SLLN asserts that

$$\frac{W_1 + W_2 + \cdots + W_n}{n} \to \mu$$

Multiple Maps 45

with probability 1.

Set
$$U_{s,t} = \sum_{s+1 \leq i \leq t} W_i.$$

It is easy to see that (i) is satisfied:

$$U_{s,u} = \sum_{s+1 \leq i \leq t} W_i + \sum_{t+1 \leq i \leq u} W_i$$

$$= U_{s,t} + U_{t,u}.$$

As W_i are iid, (ii) is evidently true. Finally, $g(t) = E(U_{0,t}) = t\mu$, so that (iii) holds with $\mu = K$. Therefore the limit

$$\lim_{t \to \infty} \sum_{1 \leq i \leq t} W_i/t$$

exists and is constant with probability 1. Note that this setup does not allow us to conclude that the limit is μ. This is a price of relaxing the assumption of additivity. Now we turn to the problem of the multiplicity of solutions to DDP.

It is necessary to impose a probability model. Labeled sites $1, 2, 3, \ldots$ are cut by two restriction enzymes independently with probability p_A and p_B, respectively, with $p_i \in (0, 1)$. These sites are the sugar-phosphate backbone *between* the successive bases. Let a *coincidence* be defined to be the event that a site is cut by both restriction enzymes; such an event occurs at each site independently with probability $p_A p_B > 0$, and at site 0 by definition. On the sites $1, 2, 3, \ldots$ there are an infinite number of such events with probability 1.

Theorem 3.2 *Assume the sites for two restriction enzymes are independently distributed with probability of a cut p_A and p_B, respectively, and $p_i \in (0, 1)$. Let $Y_{s,t}$ be the number of solutions between the s-th and the t-th coincident cut sites. Then there is a finite constant $\lambda > 0$ such that*

$$\lim_{t \to \infty} \frac{\log(Y_{0,t})}{t} = \lambda.$$

Proof. For $s, u = 0, 1, 2, \ldots$ with $0 \leq s \leq u$, we consider the double digest problem for only that segment located between the s-th and u-th coincidence. Let $Y_{s,u}$ denote the number of solutions to the double digest problem for this segment; that is, with $A_{s,u}$ and $B_{s,u}$ the sets of fragment lengths given by the first and second single digests, respectively, for only that part of the segment between the s-th and u-th coincidence, and $C_{s,u}$ the set of fragment lengths produced when both enzymes are used in combination for this same subsegment, $Y_{s,u}$ is the number of orderings of the sets $A_{s,u}$ and $B_{s,u}$ that produce $C_{s,u}$.

It is clear that whenever $s < t < u$, given a solution for the segment between the s-th and t-th coincidence and a solution for the segment between the t-th and

u-th coincidence, one has a solution for the segment between the s-th and u-th coincidence. Hence
$$Y_{s,u} \geq Y_{s,t} Y_{t,u}.$$
We note that the inequality may be strict as $Y_{s,u}$ counts solutions given by orderings where fragments initially between, say, the s-th and t-th coincidence now appear in the solution between the t-th and u-th coincidence. Letting
$$X_{s,t} = -\log Y_{s,t}$$
we have $s \leq t \leq u$ implies $X_{s,u} \leq X_{s,t} + X_{t,u}$.

The assumption that the cuts occur independently and with equal probability in each digest imply condition (ii) in the hypotheses of the theorem.

Finally, to show that condition (iii) of Kingman's theorem is satisfied, let $n_i, i = 1, 2, \ldots$, be the length of the segment between the $(i-1)$-st and i-th coincidence; note the n_i are independent and identically distributed random variables with $E[n_i] = 1/p_A p_B$. The length of the segment from the start until the t-th coincidence is given by $m(t) = n_1 + n_2 + \cdots + n_t$. There are $2^{(m(t)-1)}$ ways for either the first or second restriction enzyme to cut the remaining $m(t) - 1$ sites between 0 and $m(t)$, and so the total number of pairs of orderings of $A_{0,t}$ and $B_{0,t}$ is bounded above by $4^{m(t)}$. Note that not all of these orderings are solutions. Therefore,
$$Y_{0,t} \leq 4^{m(t)}$$
or
$$X_{0,t} \geq -(\log 4) m(t)$$
so
$$E[X_{0,t}] \geq Kt, \quad \text{where } K = -\log(4)/p_A p_B.$$

We may now conclude $X_{0,t}/t \to \lambda$ with probability 1. Observing that the existence of $\lim_{t \to \infty} \log(Y_{0,t})/t$ is a tail event, independent of any finite number of events, we see that λ is constant.

In addition, we may show that $\lambda > 0$ by the following argument. Iterating
$$Y_{s,u} \geq Y_{s,t} Y_{t,u},$$
we obtain
$$Y_{0,t} \geq \prod_{i=1}^{t} (Y_{i-1,i})$$
and so
$$E[\log(Y_{0,t})]/t \geq E[\log(Y_{0,1})].$$

Because the example with multiple solutions depicted in Figure 3.1 has positive probability of occurring under the probability model considered,
$$P(Y_{0,1} \geq 2) > 0.$$

This fact, together with the observation that by construction $Y_{0,1} \geq 1$, yields $E[\log(Y_{0,1})] = \mu > 0$. Taking limits, we obtain $\lambda \geq \mu > 0$.

∎

Obviously $\lambda \geq 0$, but it is important to prove $\lambda > 0$. Otherwise $Y_{0,t}$ could have constant or polynomial growth and $\lim_{t \to \infty} \frac{\log Y_{0,t}}{t} = 0$.

Theorem 3.3 *Assume the sites for two restriction enzymes A and B are independently distributed with probability of cut p_A and p_B respectively, and $p_i \in (0, 1)$. Let Z_l be the number of solutions for a segment of length l beginning at 0. Then*

$$\lim_{l \to \infty} \log \frac{Z_l}{l} = \lambda p_A p_B.$$

Proof.

For l, define t_l by

$$m(t_l) \leq l < m(t_l + 1).$$

By definition of t_l,

$$Y_{0,t_l} \leq Z_l \leq Y_{0,t_l+1}$$

and

$$\lim_{l \to \infty} \left\{ \log \frac{Y_{0,t_l}}{t_l} \cdot \frac{t_l}{l} \right\} \leq \lim_{l \to \infty} \log \frac{Z_l}{l} \leq \lim_{l \to \infty} \left\{ \log \frac{Y_{0,t_l+1}}{t_l + 1} \cdot \frac{t_l + 1}{l} \right\}.$$

Of course, $\log(Y_{0,t}/t) \to \lambda$ as $t \to \infty$. It remains to observe that

$$m(t_l) = \sum_{i=1}^{t_l} n_i \leq l < m(t_l + 1) = \sum_{i=1}^{t_l+1} n_i$$

and

$$\frac{1}{t_l} \sum_{i=1}^{t_l} n_i \leq \frac{l}{t_l} < \frac{t_l + 1}{t_l} \frac{1}{t_l + 1} \sum_{i=1}^{t_l+1} n_i,$$

so that

$$\lim_{l \to \infty} \frac{l}{t_l} = \frac{1}{p_A p_B}.$$

This proves that

$$\lim_{l \to \infty} \frac{\log(Z_l)}{l} = \lambda p_A p_B.$$

The consequence of this result is that for a segment of length m we have the approximation

$$Z_m \approx \exp(\gamma m), \text{ where } \gamma = p_A p_B \lambda;$$

that is, the number of solutions to the double digest problem increases exponentially fast as a function of the length of the segment. This is not good news for biology, where it is desirable to accurately map long stretches of DNA.

3.2 Classifying Multiple Solutions

The last section gave a proof that one should expect many solutions to DDP for long DNAs. The difficulty with such results is that nothing explicit has been given. The exponential growth rate is not known, and constants from Kingman's theorem are notoriously hard to determine. Because we saw a small example with a multiple solution, it might be expected that these phenomena are not restricted to extremely long DNAs. Even more problematic is the fact that nothing is given about the classification of the multiple solutions and about the combinatorics. The proof depends on coincidences of cuts; if all multiplicities of solutions were that simple, it would be desirable to attempt mapping large DNAs. Of course, the example had no coincident cut sites, so they cannot comprise an essential feature of all multiple solutions. As might be expected, the structure of multiple solutions is very complex. We proceed now to give some types of equivalent solutions with examples. As usual, we are guided by the nature of the data to make the definitions.

3.2.1 Reflections

Whenever $\sigma = (\sigma_1 \cdots \sigma_n)$ and $\mu = (\mu_1 \cdots \mu_m)$ is a solution to the DDP, then $\sigma' = (\sigma_n \sigma_{n-1} \cdots \sigma_1)$ and $\mu' = (\mu_m \mu_{m-1} \cdots \mu_1)$ also solve DDP. We call (σ', μ') the *reflection* of (σ, μ). The pairs of maps $[A, B]$ and $[A', B']$ in Figure 3.2 are reflections of each other, and they are both solutions of the same double digest problem. In a very real sense, they represent the same solution to the problem, as they differ only by an arbitrary choice of orientation. No fragment length data could possibly distinguish one from the other. It is quite reasonable to consider the set of solutions modulo the reflection relation.

3.2.2 Overlap Equivalence

We discussed in Chapter 2 and at the beginning of this chapter overlap data for DDP. Overlap data is knowledge of whether $A_i \cap B_j = \emptyset$ or not. Many distinct maps can have the same overlap data and it is a simple matter to describe them. These solutions with the same overlap data will be called *overlap equivalent*.

Multiple Maps

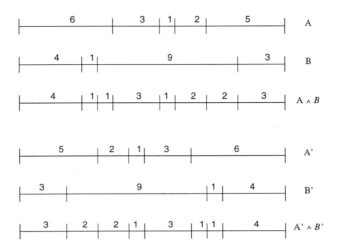

Figure 3.2: *Reflections*

If a map M has $t - 1$ coincident cut sites, then there are t (connected) components in the map. The components can be permuted in $t!$ ways and any subset of the components can be reflected, and clearly we obtain a solution that is overlap equivalent to the map. Therefore, M is one of $2^t t!$ overlap equivalent solutions. The number of components is $t = |A \vee B|$. If a component has one fragment in the A or B digest, then permutations and reflections are equivalent. If s equals the number of such components, 2^t should be 2^{t-s}.

Another way in which overlap equivalent solutions can occur is described as follows. For each B_j, let $\mathcal{A}_j = \{A_l : A_l \subset B_j\}$, and for each A_i, let $\mathcal{B}_i = \{B_k : B_k \subset A_i\}$. Note that permutation of members of $\mathcal{A}_j$ or $\mathcal{B}_i$ give maps overlap equivalent to the map M. We have the following theorem.

Theorem 3.4 *If a map has $t = |A \vee B|$ components, the number of overlap equivalent maps is*

$$2^{t-s} t! \prod_{j=1}^{m} |\mathcal{A}_j|! \prod_{i=1}^{n} |\mathcal{B}_i|!.$$

Proof. The interval graph representation of restriction maps implies that, within a component, the only duplicate maps that are overlap equivalent are reflections and permutations of fragments of one digest that are uncut by the second enzyme. ∎

In Figure 3.3, we show two of the $2^3 3! = 48$ distinct overlap equivalent solutions which result from all rearrangements of components. In Figure 3.4, we

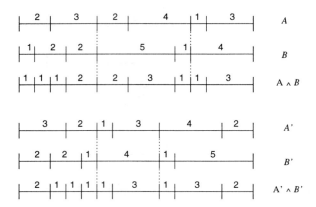

Figure 3.3: *Component rearrangements*

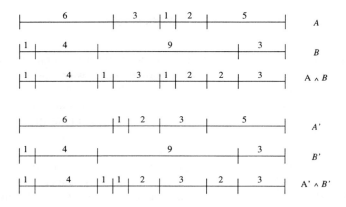

Figure 3.4: *Permutations in intervals*

show two of the $2 \times 2!3! = 24$ different solutions which result from permutations within intervals of uncut fragments and reflections.

The last theorem actually contains a characterization of overlap equivalence classes.

Theorem 3.5 *Overlap equivalence classes can be generated by permutations and reflections of components along with permutations of single digest fragments contained entirely within the single digest fragments of the other enzyme.*

Multiple Maps 51

3.2.3 Overlap Size Equivalence

In our discussion of overlap equivalence, we used the fact that $A_i \cap B_j$ was known to be empty or nonempty. Recall that, generally, sizes of fragments are all that is known. This motivates the definition of the *overlap size data* of a map to be

$$\{(|A_{i_s}|, |B_{j_s}|, |C_s|) : C_s = A_{i_s} \cap B_{j_s}\}.$$

Two solutions to DDP with data $\{a_1 \cdots a_n\}$, $\{b_1 \cdots b_m\}$, and $\{c_1 \cdots\}$ are said to be *overlap size equivalent* if they have the same set of overlap size data. The following propositions make formal evident facts.

Proposition 3.2 *When $\{a_1, a_2, \ldots\}$ and $\{b_1, b_2, \ldots\}$ each have all elements distinct, then overlap equivalence is identical with overlap size equivalence.*

Proposition 3.3 *Overlap equivalence implies overlap size equivalence.*

But when the A or B digest contains multiple fragments of the same length, then overlap size data give less information about the map than the overlap data. This loss of map information corresponds to our inability to experimentally separate and thus distinguish between different pieces of DNA having the same length in a given digest.

Given a solution $[A, B]$ to size data, the problem of describing the set of all solutions which are overlap size equivalent to $[A, B]$ is much more difficult than describing those solutions which are overlap equivalent to $[A, B]$. For example, in Figure 3.5 the overlap equivalence classes of the pairs $[A, B]$ and $[A', B']$ are disjoint from each other, each containing $3!(3!)^3 = 1296$ maps, whereas $[A, B]$ and $[A', B']$ are overlap size equivalent. The overlap size data for $[A, B]$ and $[A', B']$ are listed next:

$[A, B]$	$[A', B']$
(2, 15, 2)	(2, 15, 2)
(4, 15, 4)	(7, 15, 7)
(9, 15, 9)	(6, 15, 6)
(7, 15, 7)	(4, 15, 4)
(3, 15, 3)	(3, 15, 3)
(5, 15, 5)	(8, 15, 8)
(6, 15, 6)	(9, 15, 9)
(1, 15, 1)	(1, 15, 1)
(8, 15, 8)	(5, 15, 5)

This simple example indicates one of the essential difficulties in trying to describe the overlap size equivalence class of any arbitrary restriction map $[A, B]$: The uncut fragments no longer need be permuted only within intervals. Suppose that fragments B_i and B_j of B have the same length. Let $\mathcal{A}_i$ and $\mathcal{A}_j$ be the intervals of uncut fragments of A contained in B_i and B_j, respectively, and let L_i be the sum of the lengths of the fragments in $\mathcal{A}_i$. Then, in the process of finding

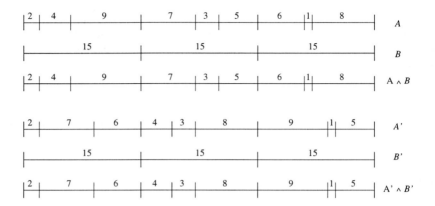

Figure 3.5: *Overlap size equivalence*

all solutions which are overlap size equivalent to $[A, B]$, one must determine all subsets S of $\mathcal{A}_i \cup \mathcal{A}_j$ such that the sum of the lengths of the elements of S is equal to L_i. But this is a version of the set partition problem which is known to be a very difficult computational problem. (See Chapter 4.)

As this approach to classifying solutions has severe computational limitations, we turn to another approach we call cassette equivalence classes. First, we introduce some graph theory and prove in Section 3.2.5 a relevant theorem. Then, in Sections 3.2.6 and 3.2.7, these ideas are applied to restriction maps.

3.2.4 More Graph Theory

We *edge color* the graph $G(V, E)$ in l colors by the function

$$f : E \to \{1, 2, \ldots, l\}.$$

A *path* $Q = x_1 x_2 \cdots x_m$ satisfies $\{x_i x_{i+1}\} \in E$ for $1 \leq i \leq m - 1$. We say that Q is a *cycle* if $x_1 = x_m$. Denote the reflection of Q by $Q^r = x_m x_{m-1} \cdots x_1$.

A path or cycle is called *alternating* if the colors of consecutive edges are distinct:

$$f(x_i, x_{i+1}) \neq f(x_{i+1}, x_{i+2}).$$

A path or cycle is called *Eulerian* if every $e \in E$ is traversed by Q exactly once. Set $d_c(v, E)$ to be the number of c *colored* edges of E incident to v. Clearly, the degree of v in E $d(v) = d(v, E)$ satisfies

$$d(v, E) = \sum_{c=1}^{l} d_c(v, E).$$

Multiple Maps

A vertex is not *balanced* if

$$\max_c d_c(v, E) \leq d(v, E)/2$$

and a *balanced graph* is a graph with every vertex balanced.

The following theorem will be useful for studying restriction maps.

Theorem 3.6 (Kotig) *Let G be an edge colored, connected graph with vertices of even degree. Then there exists an alternating Eulerian cycle in G if and only if G is balanced.*

Corollary 3.1 *If $G(V, E)$ is an edge bicolored ($l = 2$) connected graph, there is an alternating Eulerian cycle in G if and only if $d_1(v, E) = d_2(v, E)$ for all $v \in V$.*

3.2.5 From One Path to Another

Let $F = x_1 \cdots x_i \cdots x_j \cdots x_k \cdots x_n \cdots x_m$ be an alternating path in an edge bicolored graph G, with $x_k = x_i$ and $x_n = x_j$. Set $F = F_1 F_2 F_3 F_4 F_5$ with $F_1 = x_1 \cdots x_i$, $F_2 = x_i \cdots x_j$, $F_3 = x_j \cdots x_k$, $F_4 = x_k \cdots x_n$, and $F_5 = x_n \cdots x_m$. The transformation

$$\phi : F = F_1 F_2 F_3 F_4 F_5 \rightarrow F^* = F_1 F_4 F_3 F_2 F_5$$

is called an *order exchange* if $\phi(F) = F^*$ is an alternating path. See Figure 3.6. Also denote the *order reflection* $F = F_1 F_2 F_3 \rightarrow F^* = F_1 F_2^r F_3$ if $x_i = x_j$ F^* is an alternating path. See Figure 3.7. Let $X = x_1 \cdots x_m$ and $Y = y_1 \cdots y_m$ be arbitrary cycles in G. The interval $x_{i+1} \cdots x_{i+l}$ of X coincides with $y_{j+1} \cdots y_{j+l}$ of Y if $x_{i+k} = y_{j+k}$ for $1 \leq k \leq l$, where $i + k$ and $j + k$ are taken modulo m. We define the maximum l of vertices in coincident intervals between X and Y to be the index ind(X, Y).

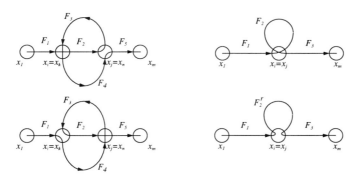

Figure 3.6: *Order exchange*   Figure 3.7: *Order reflection*

Theorem 3.7 *Two alternating Eulerian cycles X and Y in an edge bicolored graph G can be transformed into each other by a sequence of order transformations, that is, by order exchanges and order reflections.*

Proof. Let X and Y be the cycles of the theorem. Define

$$\mathcal{C} = \{X_i : X_i \text{ is obtained from } X \text{ by order transformations}\},$$

where $X = x_1 \cdots x_m$. Choose $X^* \in \mathcal{C}$ having the maximum length coincident interval with Y:

$$\text{ind}(X^*, Y) = \max\{\text{ind}(X_i, Y) : X_i \in \mathcal{C}\}.$$

The theorem states that $\text{ind}(X^*, Y) = m$.

Suppose the theorem is false, namely that $\text{ind}(X^*, Y) = l < m$. Label the graph so that the coincident intervals of $X^* = x_1 \cdots x_m$ and $Y = y_1 \cdots y_m$ begin with x_1 and y_1.

Set $v = x_l = y_l$, $w = x_{l+1}$, $u = y_{l+1}$, $e_1 = (v, w)$, and $e_2 = (v, u)$, $u \neq w$, so that e_1 and e_2 are the first distinct edges of X^* and Y. See Figure 3.8.

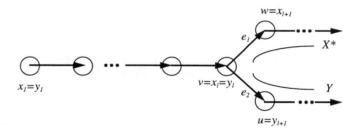

Figure 3.8: X^* and Y diverge at $x_l = y_l = v$

Because X^* and Y are alternating paths, $f(x_{l-1}, x_l) \neq f(x_l, x_{l+1}) = f(e_1)$ and $f(y_{l-1}, y_l) \neq f(y_l, y_{l+1}) = f(e_2)$. As $(x_{l-1}, x_l) = (y_{l-1}, y_l)$, we have $f(e_1) = f(e_2)$. The edge $\{v, u\}$ must be in X^* because it is Eulerian. There are two cases:

Case i. The edge e_2 in X^* has direction $x_i = u$, $x_{i+1} = v$. Thus $X^* = x_1 \cdots vw \cdots uv \cdots x_m$. Set $F_1 = x_1 \cdots v$, $F_2 = vw \cdots uv$, and $F_3 = v \cdots x_m$. Because (v, w) and (v, u) have the same colors,

$$\phi(F) = F_1 F_2^r F_3 = X^{**}$$

is an order reflection. Therefore, $X^{**} \in \mathcal{C}$ and $\text{ind}(X^{**}, Y) > \text{ind}(X^*, Y)$, which is a contradiction.

Case ii. Suppose, instead, that the edge e_2 in X^* has direction $x_n = v$, $x_{n+1} = u$.

$$X^* = x_1 \cdots x_l w \cdots x_n x_{n+1} \cdots x_m,$$

where $x_l = x_n = v$, $x_{n+1} = u$. Define

$$X^* = X_1 X_2 X_3,$$

where

$$X_1 = x_1 \cdots x_l,$$
$$X_2 = x_l \cdots x_n,$$

and

$$X_3 = x_n \cdots x_m.$$

We now introduce a Lemma.

Lemma 3.1 *There is a vertex x_j, $j > n$, in X_3 that is also in X_2.*

Proof. X^* and Y coincide until $x_l = y_l = v$. Because $x_{l+1} = w \neq y_{l+1} = u$, the path $x_{l+1} \cdots x_m$ must contain the edge $\{u, v\}$. In fact, $(x_n, x_{n+1}) = (v, u)$ and $(y_l, y_{l+1}) = (v, u)$. So there exists a minimum $i > l$ such that y_i is a vertex in $X_2 = x_l \cdots x_n$. This means that $(y_{i-1}, y_i) \notin X_2$ so that this edge $(y_{i-1}, y_i) \in X_3$ as it cannot belong to X_1. This means there is a y_i such that y_i is a vertex in X_2 and in X_3. ∎

Now write

$$X^* = F_1 F_2 F_3 F_4 F_5,$$

where

$$F_1 = x_1 \cdots x_l$$
$$F_2 = x_l \cdots x_k \qquad x_k = x_j$$
$$F_3 = x_k \cdots x_n \qquad x_l = x_n$$
$$F_4 = x_n \cdots x_j$$
$$F_5 = x_j \cdots x_m.$$

Suppose $f(x_{k-1}, x_k) = f(x_{j-1}, x_j)$. Then $f(x_k, x_{k+1}) \neq f(x_{j-1}, x_j)$ and the order exchange

$$X^{**} = F_1 F_4 F_3 F_2 F_5$$

is an alternating cycle if the following hold:

$$f(x_{l-1}, x_l) \neq f(x_n, x_{n+1})$$
$$f(x_{j-1}, x_j) \neq f(x_k, x_{k+1})$$
$$f(x_{n-1}, x_n) \neq f(x_l, x_{l+1})$$
$$f(x_{k-1}, x_k) \neq f(x_j, x_{j+1}).$$

Now $f(x_k, x_{k+1}) \neq f(x_{j-1}, x_j)$ by assumption, so the second condition holds. The fourth follows easily from this. For the first condition, $f(x_{l-1}, x_l) \neq f(x_n, x_{n+1}) = f(v, w) = f(v, u) = f(x_n, x_{n+1})$. The third condition easily follows. Because the initial $l + 1$ vertices of X^{**} and Y coincide, we have ind$(X^{**}) >$ ind(X^*, Y) which is a contradiction.

Suppose, instead, that $f(x_{k-1}, x_k) \neq f(x_{j-1}, x_j)$. Consider

$$F_1 F_2 F_3 F_4 F_5 \to F_1 F_2 (F_3 F_4)^r F_5 = F_1 F_2 F_4^r F_3^r F_5.$$

This is an order reflection, by use of the assumption $f(x_{k-1}, x_k) \neq f(x_j, x_{j-1})$ and $f(x_k, x_{k+1}) \neq f(x_j, x_{j+1})$. Finally,

$$F_1 (F_2 F_4^r)^r F_3^r F_5 \to F_1 F_4 F_2^r F_3^r F_5$$

is also an order reflection because, checking the edges,

$$f(x_l, x_{l-1}) \neq f(x_n, x_{n+1})$$

and

$$f(x_{l+1}, x_l) \neq f(x_n, x_{n+1}).$$

Again, $F_1 F_4 F_2^r F_3^r F_5 = X^{**}$ satisfies ind$(X^{**}, Y) >$ ind(X, Y) for the final contradiction. ∎

3.2.6 Restriction Maps and the Border Block Graph

Now we apply these general results to restriction maps.

The bases of the DNA being mapped can be considered an interval $[1, N]$, with fragments intervals $[i, j]$, $i \leq j$. Order can be defined by $[i, j] \leq [k, l]$ if $i \leq k$. $A = \{A_1, A_2, \ldots, A_n\}$ is now just a set of non empty disjoint intervals called the *blocks* of A, where $\cup A_i = [1, N]$ and $A_i < A_j$ if $i < j$. Note that we have required A to be ordered here. The double digest problem DDP(**a**, **b**, **c**) is still just the problem of finding maps $[A, B]$ satisfying $\mathbf{a} = \|A\|$, $\mathbf{b} = \|B\|$, and $\mathbf{c} = \|C\|$.

In this section, we define a new graph on restriction maps, the border block graph, after which we show that every restriction map is an alternating Eulerian path in the border block graph. We will assume for simplicity that there are no common cut sites, that is, $|A \wedge B| = |A| + |B| - 1$.

The inclusion of block A_i is the set of intervals contained in A_i:

Multiple Maps 57

$$\mathcal{I}(A_i) = \{C_j : C_j \subset A_i\}.$$

Our interest is when the inclusion contains at least two blocks. Obviously $|\mathcal{I}(A_i) \cap \mathcal{I}(B_j)| = 0$ or 1.

When $|\mathcal{I}(X)| > 1$, $C^* \in \mathcal{I}(X)$ is called a border block when

$$C^* = \min_{C_j \in \mathcal{I}(X)} C_j$$

or

$$C^* = \max_{C_j \in \mathcal{I}(X)} C_j.$$

The set of all border blocks of (A, B) is $\mathcal{B}$. Clearly, $C_1 \in \mathcal{B}$ and $C_l \in \mathcal{B}$ because we have assumed $|A \vee B| = 1$.

Lemma 3.2 *(i) Every border block belongs to exactly two $\mathcal{I}(X)$, $|\mathcal{I}(X)| > 1$, except C_1 and C_l which belong to one, and*

(ii) every $\mathcal{I}(X)$ with $|\mathcal{I}(X)| > 1$ contains exactly two border blocks.

Define $\mathcal{I}^*(X)$ to be the set of border blocks of $\mathcal{I}(X)$ with $|\mathcal{I}(X)| > 1$. Let the set of all border blocks be

$$\widehat{\mathcal{I}} = \{\mathcal{I}^*(X) : |\mathcal{I}(X)| > 1\}$$

and

$$V = \{|C_k| : C_k \in \mathcal{B}\}$$

be the set of border block lengths. Here, if $C_\alpha \neq C_\beta$ satisfy $|C_\alpha| = |C_\beta|$, they correspond to the same element of V. The graph H is an edge bicolored graph (with colors A and B), $H(V, E)$, with each edge in E corresponding to a pair of border blocks in $\mathcal{B}$.

$$E = \{(|C_i|, |C_j|) : (C_i, C_j) = \mathcal{I}^*(X), |\mathcal{I}(X)| > 1\}$$

Two vertices in V might be joined by more than one edge. The color is A if $X = A_i$, for some i, and the color is B if $X = B_j$, for some j. The graph H is called the border block graph of (A, B). See Figure 3.9 for an example of H.

Lemma 3.3 *All vertices of H except for $|C_1|$ and $|C_l|$ are balanced.*

If $|C_1|$ and $|C_l|$ are not balanced, then

$$|d_A(|C_1|, E) - d_B(|C_1|, E)| = 1$$

and

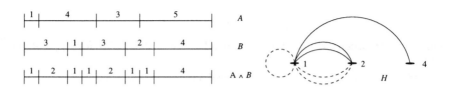

Figure 3.9: *Restriction map and H, the border block graph (A solid, B dashed).*

$$|d_A(|C_l|, E) - d_B(|C_l|, E)| = 1.$$

This lemma and these equalities imply that by adding one or two edges, H can be transformed into a balanced graph.

Order the border blocks in $\mathcal{B}$:

$$C_{i_1} \leq C_{i_2} \leq \cdots C_{i_m}.$$

From Lemmas 3.2 and 3.3 it follows that $m = |\widehat{\mathcal{I}}| + 1$.

Theorem 3.8 *The path $P = |C_{i_1}||C_{i_2}| \cdots |C_{i_m}|$ is an alternating Eulerian path in H.*

Proof. For each two consecutive blocks $C_{i_k} C_{i_{k+1}}$ in $\mathcal{B}$ there is $\mathcal{I}(X) \in \widehat{\mathcal{I}}$ containing these blocks. ∎

3.2.7 Cassette Transformations of Restriction Maps

In this section, we introduce cassette transformations that define an equivalence relation on restriction maps. Each member of an equivalence class solves the same DDP=DDP(**a**, **b**, **c**). The correspondence between restriction maps, the border block graph and alternating Eulerian paths characterizes equivalence classes defined by these cassette transformations.

As usual, A and B are single digests and $C = A \wedge B = \{C_1, \ldots, C_l\}$. For each pair i, j with $1 \leq i \leq j \leq l$ define

$$I_C = \{C_k : C_i \leq C_k \leq C_j\},$$

which is the set of intervals from C_i to C_j. The cassette defined by I_c is the pair of sets of intervals (I_A, I_B), the sets of all blocks of A and B, respectively, that contain a block of I_C. Define m_A and m_B to be the minimal elements of the leftmost blocks of I_A and I_B respectively. The *left overlap* is defined to be $m_A - m_B$. The *right overlap* is defined similarly, by substituting maximal for minimal, and rightmost for leftmost.

Multiple Maps

If two disjoint cassettes of solution $[A, B]$ to DDP$(\mathbf{a}, \mathbf{b}, \mathbf{c})$ have identical left and right overlaps and if, when the overlaps are non zero, the DNA comprising the overlap is a single double digest fragment, then they can be *exchanged* as in Figure 3.10 and we obtain a new solution $[A', B']$ to DDP$(\mathbf{a}, \mathbf{b}, \mathbf{c})$. Also if the left and right overlaps of a cassette have the same absolute value but different sign, then the cassette can be *reflected* as in Figure 3.11 and we obtain a new solution $[A'', B'']$ to DDP$(\mathbf{a}, \mathbf{b}, \mathbf{c})$.

Analogous to the inclusion sets $\mathcal{I}(X)$, define the inclusion sizes to be multisets

$$\mathcal{I}_s(X) = \{|C_i| : C_i \in \mathcal{I}(X)\},$$

and for $|\mathcal{I}(X)| > 1$, define the border block sizes as

$$\mathcal{I}_s^*(X) = \{|C_i| : C_i \in \mathcal{I}^*(X)\}.$$

For $|\mathcal{I}(X)| = 1$, set $\mathcal{I}_s^*(X) = \{0, 0\}$.

The inclusion bordersize data is correspondingly defined by

$$\mathcal{I}^* D = (\{(\mathcal{I}_s^*(A_i), \mathcal{I}_s(A_i)) : A_i \in A\}, \{(\mathcal{I}_s^*(B_j), \mathcal{I}_s(B_j)) : B_j \in B\}).$$

These data $\mathcal{I}^*$ distinguishes the border block sizes from those of the fragment.

It is clear that $\mathcal{I}^* D$ determines the border block graph uniquely. Next we remark that cassette transformations do not change $\mathcal{I}^* D$.

Lemma 3.4 *Suppose $[A', B']$ is obtained from $[A, B]$ by a sequence of cassette transformations. Then $\mathcal{I}^* D[A, B] = \mathcal{I}^* D[A', B']$.*

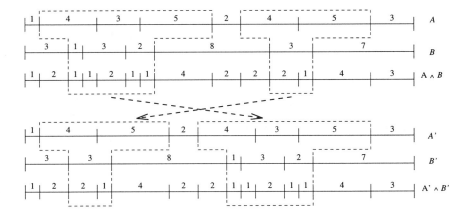

Figure 3.10: *Cassette exchange*

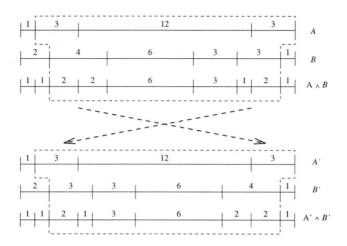

Figure 3.11: *Cassette reflection*

The next lemma states that order exchanges and reflections in the border block graph correspond to cassette exchanges and reflections in restriction maps.

Lemma 3.5 *Let H be the border block graph of $[A, B]$ and P be the alternating Eulerian path in H corresponding to $[A, B]$.*

  (i) *Let $[A', B']$ be obtained from $[A, B]$ by a cassette exchange (reflection) and P' be the alternating Eulerian path corresponding to $[A', B']$. Then there is an order exchange (reflection) taking P to P'.*

  (ii) *Let P' be obtained from P by means of an order exchange (reflection). Then there is a cassette exchange (reflection) taking $[A, B]$ to $[A', B']$, where P' corresponds to $[A', B']$.*

Finally, we can introduce an equivalence relation, cassette equivalence on the set of all solutions to DDP. We have $[A, B] \equiv [A', B']$ if and only if there is a sequence of cassette transformations and permutations of non border block uncut fragments transforming $[A, B]$ into $[A', B']$. This equivalence relation partitions the set of solutions into equivalence classes, where each equivalence class corresponds to the border block graph (Lemma 3.4). The next theorem characterizes equivalent restriction maps.

Theorem 3.9 $[A, B] \equiv [A', B']$ *if and only if $\mathcal{I}^* D[A, B] = \mathcal{I}^* D[A', B']$.*

Proof. Assume $\mathcal{I}^* D[A, B] = \mathcal{I}^* D[A', B']$. Then the border block graphs of $[A, B]$ and $[A', B']$ coincide, call it H. By Theorem 3.8, the maps $[A, B]$ and

Multiple Maps 61

$[A', B']$ correspond to alternating Eulerian paths P and P' in H. By Theorem 3.7 there is a sequence of order transformations taking P to P'. Therefore by Lemma 3.5 (ii) there is a sequence of cassette transformations taking $[A, B]$ to $[A', B']$. The other implication follows from Lemma 3.4. ∎

3.2.8 An Example

Recall the example from Section 3.1.1 which appears again as Figure 3.12. This problem has 208 different solutions which fall into 26 different overlap equivalence classes: 13 classes with 4 members each, and 13 classes of 12 members. The solution $[A, B]$ in Figure 3.12 has an overlap equivalence class containing four elements, which are generated by the reflection of the whole pair, and the several of the uncut fragments of length 3 and 6 in B. The overlap equivalence class of the solution $[A', B']$ contains 12 elements: $3! = 6$ permutations of uncut fragments in B multiplied by a factor of 2 for the reflection of the pair.

Somewhat surprisingly, the overlap size equivalence classes do not correspond precisely to the overlap classes even in this rather small problem. There are 25 overlap size equivalence classes of solutions to this DDP: 11 classes of 4 members, 13 classes of 12 members, and 1 class having 8 members. The solution $[A, B]$ in Figure 3.12 is a member of this unique class of eight solutions, which is the union of two different four-element overlap equivalence classes.

When we move to cassette equivalence classes, the situation changes. The 208 solutions fall into 18 cassette equivalence classes. The largest has 36 members, 2 have 24 members, 1 has 20 members, 4 have 12 members, 4 have 8 members, and 6 have 4 members. The cassette equivalence class corresponding to $[A, B]$ has

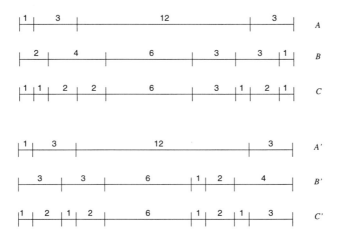

Figure 3.12: *Multiple solutions*

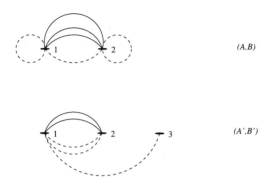

Figure 3.13: *Border block graph for Figure 3.12*

36 members, the largest, whereas that corresponding to $[A', B']$, has 24 members. See Figure 3.13 for the corresponding cassette equivalence classes.

We now consider the set of all pairs of restriction maps $[A, B]$ with $\mathbf{a} = \{1, 3, 3, 12\}$ and $\mathbf{b} = \{1, 2, 3, 3, 4, 6\}$. There are $4!6!/2!2! = 4320$ different pairs having this data, with 36 different vectors $\mathbf{c} = \{c_1 \ldots\}$, that is, there are 36 different double digest problems $\text{DDP}(\mathbf{a}, \mathbf{b}, \mathbf{c})$ having these values of $\mathbf{a}$ and $\mathbf{b}$.

Problems

Problem 3.1 DNA molecules also come in circular form. We digest such a molecule with enzymes A and B. The quantities $|A|, |B|, |A \wedge B|, and |A \vee B|$ are defined as in the linear case. (i) If A and B have no coincident cut sites, state and prove a formula for $|A \wedge B|$. (ii) For the general case with $|A \vee B| > 0$, state and prove a formula for $|A \wedge B|$.

Problem 3.2 If p_A and p_B belong to $(0, 1)$, show that there are an infinite number of coincidences with probability 1.

Problem 3.3 Suppose we are mapping three enzymes A, B, and C and that the sites are independently distributed with cut probabilities p_A, p_B, and p_C respectively. Although other definitions are possible, define a solution as a permutation of A, B, and C fragments to give the same set of triple digest fragments. Define $Y_{s,t}$ and $X_{s,t}$ as in Theorem 3.2.

(i) State the three enzyme generalization of Theorem 3.2.

(ii) Find K for the proof of the theorem in (i).

Problem 3.4 Present an example to show $P(Y_{0,1} \geq 2) > 0$ in the proof of the theorem of Problem 3.3.

Multiple Maps

Problem 3.5 Prove the inequality $Y_{0,t_l} \leq Z_l \leq Y_{0,t_l+1}$ that appears in the proof of Theorem 3.3.

Problem 3.6 In view of Theorem 3.4, check the number of overlap equivalent solutions in Figures 3.3 and 3.4.

Problem 3.7 Change the definition of cassette exchange so that overlap size equivalence is preserved.

Problem 3.8 Use Figure 3.10 to show cassette exchange does not always result in overlap size equivalence.

Problem 3.9 Given the following map and restricting yourself to cassettes that include the length 5 fragments, find a cassette interchange that results in cassette equivalent maps. (You might want to xerox the figures so that you can play with possible cassette interchanges.)

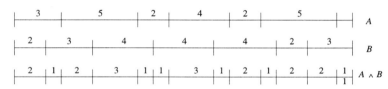

Problem 3.10 For a restriction map with $A \wedge B = \{C_1, C_2, \ldots C_l\}$ give an upper bound for the number of cassettes.

Problem 3.11 Prove that if the graph is bicolored ($l = 2$) and balanced, then $d_1(v, E) = d_2(v, E)$.

Problem 3.12 If cassette exchange is defined as exchange of cassettes with identical left and right overlaps, give an example to show the border block graphs need not be identical.

Chapter 4
Algorithms for DDP

In this chapter we study methods for making restriction maps. Initially, we discuss algorithms and measures of their difficulty. Then we formulate different approaches to the double digest problem (DDP) that was described in the last chapter.

4.1 Algorithms and Complexity

In many areas of modern science we are faced with the task of solving a problem P by computation. Assigning airplanes to passengers, for example, is a nontrivial problem now routinely solved by computers. Throughout this course we will be considering various problems that must be solved by computation. Problem DDP is one, of course. Before the specifics of DDP are considered, we will give a brief general discussion of algorithms.

The method of solving a computational problem is called an *algorithm*. Knuth in his classic volumes on *The Art of Computer Programming* lists five important features of an algorithm that distinguish it from commonly used words like recipe, procedure, and so on. (1) An algorithm must stop after a *finite* number of steps. This requirement shows the influence of computers; we are not interested in methods that take literally forever. (2) All steps of an algorithm must be precisely defined. This shows awareness that it must be possible for algorithms be coded by the reader, or else the algorithm is not well defined. (3 and 4) Input to the algorithm (zero or more) must be specified, as must output. There must be at least one output. Finally, Knuth lists (5) effectiveness by which he means that the operations of the algorithm are basic–they could be done by a person using pencil and paper in finite time. Be aware that finite time might be trillions of years.

As we will be dealing with combinatorial problems on finite sets, the problem can always be handled in a finite number of steps. In the precomputer era, all finite problems tended to be lumped together. Not today, however. The old saying "there are many roads to Rome" holds true for a computational problem. There are many roads, and some of them are a good deal shorter or more scenic than others.

There are various ways of measuring the effectiveness of an algorithm. One is the time required to execute the algorithm on a given computer model. Computer models began with Charles Babbage in the early and middle 1800s. Technology held Babbage back, as he had to design a mechanical and not an electronic computer. The idea of programming computers is due to his contemporary, Ada Augusta, Countess of Lovelace. The first computer model of the modern era is that of Turing in the 1930s who conceived of an automaton with an infinite paper tape of squares and instructions that could cause the tape to move one square to the left or right and to mark the square or to erase an already marked square. These simple four operations are very powerful and have a linear, sequential memory. Modern existing computers have random access memories, and models of these computers also exist.

Let us for the moment assume a set of basic operations. Given our problem P, we ask for the number of basic operations needed by an algorithm to solve P. Essential to this discussion is a measure of problem size. Usually we will loosely indicate problem size by the number of nodes in a graph or the number of elements in a sequence. One can be more precise, but this will suit our purposes. To measure the number of operations some nice notation has been developed. For example, the algorithm, for a problem of size n,

for $i = 1$ to n
$x_i = x_i + 1$

takes n additions, whereas the algorithm

for $i = 1$ to n
  for $j = 1$ to n
    $x_{ij} = 2x_{ij}$

takes n^2 multiplications. Here, the idea of a loop is introduced along with the notation $a \leftarrow b$ for assignment, which means replace a by b. $x_i \leftarrow x_i + 1$ means that 1 is added to the number in location x_i. Clearly, the usual = notation is inadequate: $2 = 2 + 1$ is both wrong and confusing. If both addition and multiplication are considered basic operations, we say that the first algorithm has time complexity $O(n)$ (read, of order n) and the second has time complexity $O(n^2)$. Note that time complexity is measured as a function of n, the so-called problem size. A problem P has polynomial complexity $O(n^k)$ if $0 < k < \infty$ satisfies

$$k = \min\{l : \text{ some algorithm solves P in time complexity } O(n^l)\}.$$

A problem which has no polynomial time algorithms is called an *intractable* problem. It is of interest to note that intractable problems are essentially independent of the computer model. Classes of problems in logic have been shown

Algorithms for DDP

intractable. Of great interest to us is another class of problems for which no polynomial time algorithms are known but of which there is no proof that they do not have polynomial time algorithms. This class is known as NP-complete problems and all problems in the class are of equivalent difficulty. If one of these problems could be solved in polynomial time, they all could be. If one could be shown to require exponential time, they all would require exponential time.

We present two problems in the class NP-complete. The first is the traveling salesman problem (TSP). We are given a finite graph whose vertices are cities and all edges between all vertices exist and have positive weights (distances). The TSP is to find the shortest route to visit all cities. Obviously we can solve the problem by considering all $n!$ routes, but this solution is not polynomial.

Our second NP-complete problem is presented in the next section, in which we also prove that DDP is NP-complete.

4.2 DDP is NP-Complete

Our second NP-complete problem is the partition problem.

In the partition problem, we are given a finite set F, say $|F| = n$, and a positive integer $f(a)$ for each $a \in F$ and wish to determine whether there exists a subset $F' \subset F$ such that

$$\sum_{a \in F'} f(a) = \sum_{a \in F - F'} f(a).$$

Of course, if $\sum_{a \in F} f(a) = J$ is not divisible by two, there can be no such subset F'.

To prove a problem P is NP-complete there are two requirements: (1) A possible solution must be checked in polynomial time. (2) An algorithm that solves P must solve one of the NP-complete problems.

Theorem 4.1 *DDP is NP-complete.*

Proof. Let the problem data be $\mathbf{a} = \|A\|$, $\mathbf{b} = \|B\|$, and $\mathbf{c} = \|C\|$, where $\mathbf{c}$ is the double digest data. The number of elements in $\mathbf{c}$ is $|A \wedge B|$.

To verify that (σ, μ) solves DDP, find the elements of $C(\sigma, \mu)$ by first finding the locations G of the double digest points $\sum_{1 \leq i' \leq i} a_{i'}$ and $\sum_{1 \leq j' \leq j} b_{j'}$ for $1 \leq i \leq |A|-1$ and $1 \leq j \leq |B|-1$ along with 0 and L. $G = \{g_0, g_1, \ldots, g_{|A \wedge B|}\}$ is a multiset. Then, if G is ordered,

$$C(\sigma, \mu) = \{c_j(\sigma, \mu) : c_j(\sigma, \mu) = g_j - g_{j-1} \text{ for some } 1 \leq j \leq |A \wedge B|\}.$$

A set can be ordered in time less than the square of the number of elements. (See Problem 4.7): Now check that $\mathbf{c} = C(\sigma, \mu)$ by first ordering $\mathbf{c}$ and $C(\sigma, \mu)$ and then checking that the i-th elements of the ordered sets are equal.

To complete our proof, we use DDP to solve the set partition problem. Using the notation introduced above, consider as input to DDP the data

$$A = \{f(a_k) : 1 \le k \le n\},$$
$$B = \{J/2, J/2\},$$
$$C = A \wedge B = A.$$

It is clear that any solution to DDP with these data yields a solution to the partition problem through the order of the implied digest C. ∎

4.3 Approaches to DDP

Due to the basic difficulty of DDP, we are not likely to find a polynomial time solution. Still, other *NP*-complete problems have been approached by useful heuristic methods. In some cases these methods have been of great practical value. It is wise to look at the structure of the problem to guide us. Three approaches suggest themselves.

4.3.1 Integer Programming

First, we introduce a bit of notation, assuming a solution. We "assign" double digest fragments $c_1, \ldots, c_{|A \wedge B|}$ by

$$(c_1, \ldots, c_{|A \wedge B|})E = (a_1, \ldots, a_n),$$

where E is a $|A \wedge B| \times n$ matrix of zeros and ones. We have $a_i = \sum_{k=1}^{|A \wedge B|} c_k e_{ki}$. Similarly,

$$(c_1, \ldots, c_{|A \wedge B|})F = (b_1, \ldots, b_m),$$

where F is an $|A \wedge B| \times m$ matrix of zeros and ones. Obviously we wish to assign each C fragment once and only once to each single digest. Therefore, we have the additional identities $\sum_{j=1}^{n} e_{ij} = 1$ and $\sum_{k=1}^{m} f_{ik} = 1$.

The above equations are wishful thinking, as they derive properties of a solution. However, DDP is solved by finding consistent solutions to the two systems:

$$\text{minimize } \{\alpha + \beta\}$$

where $\alpha, \beta \in I^+$,

$$-\alpha \le a_i - (cE)_i \le \alpha \quad \text{for all } i = 1, \ldots, n,$$
$$e_{ij} \in \{0, 1\} \quad \text{for all } i, j,$$
$$\sum_{k=1}^{n} e_{ik} = 1 \quad \text{for all } i = 1, \ldots, |A \wedge B|,$$
$$-\beta \le b_j - (cF)_j \le \beta \quad \text{for all } j = 1, \ldots, m,$$
$$f_{ij} \in \{0, 1\} \quad \text{for all } i, j,$$
$$\sum_{k=1}^{m} f_{ik} = 1 \quad \text{for all } i = 1, \ldots, |A \wedge B|.$$

This is a problem in integer linear programming and the packages available have not proved too useful for DDP.

If measuring error proportional to size is desired, then for the A digest, for example,
$$-\alpha \leq a_i - (cE)_i \leq \alpha$$
could become
$$-\alpha \leq \frac{a_i - (cE)_i}{\epsilon a_i} \leq \alpha$$
or
$$-\epsilon a_i \alpha \leq a_i - (cE)_i \leq -\epsilon a_i \alpha.$$

Because ϵ and a_i are both constants in this linear system involving the variables α, β, and $E = (e_{ij})$, the system remains a linear integer programming problem.

4.3.2 Partition Problems

Another approach to DDP is to consider it as a complex, interrelated partition problem. Each a_i is the sum of disjoint c_k:

$$a_1 = \sum_{k \in R_1} c_k$$
$$\vdots$$
$$a_n = \sum_{k \in R_n} c_k$$

where $\cup R_i = \{1, 2, \ldots, |A \wedge B|\}$ and $R_k \cap R_j = \emptyset, i \neq j$.

Likewise

$$b_1 = \sum_{k \in S_1} c_k$$
$$\vdots$$
$$b_m = \sum_{k \in S_m} c_k$$

when $\cup S_i = \{1, 2, \ldots, |A \wedge B|\}$ and $S_i \cap S_j = \emptyset, i \neq j$.

Thus it is quite natural to think of DDP as a partition problem. In fact this approach is just a restatement of the integer programming formulation. Several programs are based on such an approach. Our theorem on multiple solutions should sound a note of caution. Because many solutions to the two systems are likely to exist, each system alone will have many more solutions. Some scheme to simultaneously and consistently solve both at once is desired.

Overlap is easy to handle because the i-th A fragment overlaps the j-th B fragment if and only if $R_i \cap S_j \neq \phi$. Our earlier interval graph methods give a rapid means of obtaining a map from overlap data and for testing possible solutions for consistency with a restriction map, that is, interval graph.

4.3.3 TSP

The TSP minimizes the cost of permutation of $\{1, 2, \ldots, n\}$. Problem DDP has two permutations, one of $\{1, \ldots, n\}$ and one of $\{1, 2, \ldots m\}$, but any computational scheme to minimize the salesman's tour might be adapted to DDP. A salesman's tour is identified with a permutation of the A digest, say. Thus, the idea is to have two salesmen (named A and B) who work together to minimize some route cost. They both tour cities, n cities (for A) and m cities (for B), and their reward is the goodness of fit to the double digest data C. In this hybrid problem, the cities A visits and the cities B visits are disjoint–they could be in different countries. We consider one such approach in the next section.

4.4 Simulated Annealing: TSP and DDP

4.4.1 Simulated Annealing

Suppose we are to minimize the function $f : V \to R$, where $|V| < \infty$. The method of simulated annealing was proposed by Metropolis and others in 1953 and has the following statistical mechanical interpretation. It must be emphasized that statistical mechanics is providing motivation only. The algorithm is developed by analogy with statistical physics. The set V can be thought of as the set of all possible configurations of some physical system; the quantity $f(v)$ is the energy of the system when in configuration v with T playing the role of temperature. The Gibbs distribution from statistical physics then gives the probability of finding the system in a particular configuration at some given temperature. We will explicitly present the Gibbs distribution below. At high temperature, the system can be found in any of its states with approximately equal probabilities whereas at low temperature, it is more likely that the system will be in a low energy configuration.

Let us contrast two standard optimization techniques. The simplest gradient method for discrete optimization chooses a point v from the neighborhood of the current point w. If $f(v) < f(w)$, the algorithm moves to v. Otherwise, it stays at w. When no neighboring points yield smaller $f(v)$, then the value $f(w)$ is returned as the minimum. There are many options in constructing a gradient algorithm, but the general idea is to go downhill until at least a local minimum is achieved. A difficulty with the algorithm is that once a local minimum is located, the algorithm terminates. In contrast, take simple Monte Carlo methods for finding the minimum. Here, points from V are chosen at random with probability $1/|V|$. The minimum values of $f(v)$ are recorded as the random sampling proceeds.

Algorithms for DDP 71

This method has an advantage over deterministic searching in that the function evaluations can be spread over the configuration space in an unbiased manner. Local minima present no particular problems, but a large $|V|$ makes it unlikely to find a minima. Thus, the appealing features of the gradient method are lost. Simulated annealing combines both these approaches in a novel manner.

We present now a description of the simulated annealing algorithm which we will also call the Metropolis algorithm. Let V be a finite set of elements, and f be a function that assigns a real number to each element of V. Suppose we wish to find an element $v^* \in V$ that corresponds to the global minimum value of f; that is, find $v^* \in V$ such that $f(v^*) = \min_{v \in V} f(v)$. For any $T > 0$, let π_T be the Gibbs distribution over V, given by

$$\pi_T(v) = \exp\{-f(v)/T\}/Z,$$

where Z, the partition function, is chosen such that $\sum_{v \in V} \pi_T(v) = 1$; that is,

$$Z = \sum_{v \in V} \exp\{-f(v)/T\}.$$

Note that for large values of T the distribution tends to be uniform over V, whereas for small values of T the favorable elements of V, that is, those elements of V for which $f(v)$ is small, are weighted with a large probability.

In fact,

$$\pi_\infty(v) = \lim_{T \to \infty} \pi_T(v) = \frac{1}{|V|}$$

and $\pi_\infty(v)$ corresponds to simple Monte Carlo sampling. At the other extreme,

$$\pi_0(v) = \lim_{T \to 0+} \pi_T(v) = \begin{cases} 0 & \text{if } f(v) > \min\{f\} \\ |\{w : f(w) = \min\{f\}\}|^{-1} & \text{if } f(v) = \min\{f\} \end{cases}$$

and $\pi(v)$ is uniform on $\{w : f(w) = \min_{v \in V} f(v)\}$.

It is not obvious how to perform a simulation of the distribution π_T without computing all $f(v), v \in V$. This would defeat the purpose of the simulation which is to estimate the minimum without computing all $f(v)$. A method can be based on the theory of Markov chains. Essentially a *Markov chain* is a sequence of random variables $\{X_n\}_{n \geq 0}$ where the probabilities are specified in the following manner. Let $\mu = (\mu_1, \mu_2, \ldots, \mu_{|V|})$ satisfy

$$\mu(i) = \mathbb{P}(X_0 = i).$$

μ is called the *initial distribution* of the chain and $\Sigma_i \mu(i) = 1$. Then

$$\mathbb{P}(X_0 = i_0, X_1 = i_1, \ldots, X_n = i_n) = \mu(i_0) p(i_0, i_1) p(i_1, i_2), \ldots, p(i_{n-1}, i_n).$$

We have defined the $|V|^2$ *transition probabilities* by

$$p(i, j) = \mathbb{P}(X_{k+1} = j | X_k = i).$$

Observe that conditioning on the past up to and including k-th state X_k is equivalent to conditioning on the k-th state. Define $P = (p(i,j))$. The n-step transition probabilities

$$p^{(n)}(i,j) = \mathbb{P}(X_{n+m} = j | X_m = i)$$

are obtained from raising P to the n-th power: $P^n = (p^{(n)}(i,j))$.

The period of a state $d(i)$ is defined to be the greatest common divisor of all $k \geq 1$ such that $p^k(i,i) > 0$. All states i and j for which there is a path of positive probability from i to j and from j to i (i and j communicate) have the same period $d(i) = d(j)$. A Markov chain is *aperiodic* if $d(i) = 1$ for all states i. The following theorem is standard in the theory of Markov chains.

Theorem 4.2 *Let P be the transition matrix of an aperiodic finite Markov chain with all pairs of states communicating. Then there is a unique probability vector $\pi = (\pi_1, \pi_2, \ldots)$ called the equilibrium or stationary distribution of P satisfying*

(a) $\lim_{n \to \infty} \mu P^n = \pi$, *for all initial distributions μ,*

and

(b) $\pi P = \pi$.

The interpretation of (a) is that whatever the initial distribution on the states, wherever the chain starts at, as the process proceeds from state to state, it will have limiting distribution π. The equation in (b) is a method by which π can be verified to be the equilibrium distribution. Assuming (a), it is easy to prove (b): $\pi = \lim_{n \to \infty} \mu P^n = \lim_{n \to \infty} \mu P^{n-1} P = \pi P$.

We now turn to defining the Metropolis algorithm. The idea is to construct a Markov chain with state space V that has π_T as its stationary distribution. The first step is to define for all $v \in V$ a set of neighbors $N_v \subset V$ where transitions from v have the property that the resulting Markov chain has all pairs of states communicating. This means that for all $v, w \in V$ we must find, for some k, $v_1, v_2, \ldots v_k$ satisfying

$$v_1 \in N_w, v_2 \in N_{v_1}, \ldots, v_k \in N_{v_{k-1}}, v \in N_{v_k}.$$

Then, if the transition probabilities satisfy

$$\mathbb{P}_T(X_n = v | X_{n-1} = w) = p_T(w, v) > 0$$

for all $v \in N_w$, all pairs of states communicate. We also require

$$v \in N_w \text{ if and only if } w \in N_v$$

and

$$| N_v | = | N_w |.$$

Algorithms for DDP

Finally, we define $p_T(w,v)$ by

$$p_T(w,v) = 0 \text{ if } v \text{ is not in } N_w$$

and

$$p_T(w,v) = \frac{\alpha \exp\{-(f(v) - f(w))^+/T\}}{|N_w|} \text{ if } v \in N_w, v \neq w,$$

and $p_T(w,w)$ is fixed by the requirement

$$\sum_{v \in N_w} p_T(w,v) = 1/\alpha.$$

Theorem 4.3 (Metropolis) *The Markov chain $p_T(v,w)$ with state space V defined above has equilibrium distribution π_T,*

$$\pi_T(v) = \frac{\exp\{-f(v)/T\}}{Z}.$$

Proof. We have only to prove π_T satisfies (b) of Theorem 4.2. First, we show π_T satisfies the balance equation

$$p_T(w,v)\pi_T(w) = p_T(v,w)\pi_T(v).$$

For $v = w$, the equation is trivial. For $v \neq w$,

$$p_T(w,v)\pi_T(w) = \alpha \frac{\exp\{-(f(v) - f(w))^+/T\}}{|N_w|} \frac{\exp\{-f(w)/T\}}{Z}$$
$$= \alpha \frac{\exp\{-[(f(v) - f(w))^+ + f(w)]/T\}}{|N_w|Z}$$
$$= \alpha \frac{\exp\{-\max\{f(v), f(w)\}\}}{|N_w|Z}.$$

As $|N_v| = |N_w|$, the last expression is symmetric in v and w, and the balance equation holds. To finish the proof, sum the balance equation over $w \in V$:

$$\sum_{w \in V} p_T(w,v)\pi_T(w) = \sum_{w \in V} p_T(v,w)\pi_T(v)$$
$$= \pi_T(v) \sum_{w \in V} p_T(v,w) = \pi_T(v).$$

∎

Therefore, a probabilistic solution to the problem of locating an element $v \in V$ for which $f(v)$ is minimized is given by sampling from the distribution π_T for small $T > 0$. In practice, as the function f may be expensive computationally, the Markov chain is simulated in the following way: When at w, a neighbor of w is selected from N_w uniformly, say v, and $f(v)$ is computed. The move to v is then accepted with probability

$$p = \exp\{-(f(v) - f(w))^+/T\}$$

and the new state of the chain is v, otherwise the move is rejected and the state of the chain remains w. The transition probabilities are consistent with $p_T(w, v)$ defined above.

More recently, the idea of cooling the system has been introduced in the hope that in the limit the distribution $\pi_0 = \lim_{T \to 0+} \pi_T$ will be obtained. We showed above that π_0 is that distribution that distributes mass one uniformly over the states of minimum energy. In this way, the algorithm resembles the physical process of annealing, or cooling, a physical system. As in the physical analog, the system may be cooled too rapidly and become trapped in a state corresponding to a local energy minimum. Recently, a theorem was proved as to the rate at which cooling could take place and obtain π_0 as a limiting distribution. The cooling version of the algorithm is called the extended Metropolis algorithm.

Theorem 4.4 (Geman and Geman) *For the Metropolis algorithm defined as above at stage n use the transition probabilities for temperature T_n. If $\lim_{n \to \infty} T_n = 0$ and $T_n \geq c/\log(n)$ where c is a constant depending on f, then*

$$\left(\lim_{n \to \infty} \mathbb{P}(X_n = v)\right)_{v \in V} = \pi_0.$$

Heuristic. The proof is a bit too involved to present in its entirety here, but it is possible to give a very accurate idea about the required $1/\log(n)$ cooling rate. Recall that the Metropolis algorithm can "jump" out of local minima with positive probability. If $\Delta = f(v) - f(w) > 0$ is the required jump to escape these minima, the associated probability is

$$e^{-\Delta/T_n} = e^{-(f(v)-f(w))^+/T_n}.$$

The probability the local minima is never left is

$$\prod_{n \geq 1}(1 - e^{\Delta/T_n}) = \prod_{n \geq 1}(1 - p_n).$$

Calculus tells us that

$$1 - p_i \leq 1 - p_i + \frac{p_i^2}{2} - \frac{p_i^3}{3!} + \cdots = e^{-p_i}$$

and

$$\prod_{n=1}^{N}(1 - p_n) \leq \exp\left\{-\sum_{n=1}^{N} p_i\right\}.$$

Algorithms for DDP

If $\sum_{n=1}^{\infty} p_i = \infty$, then

$$\lim_{n \to \infty} \prod_{n \geq 1}^{N} (1 - p_n) = 0.$$

Because we wish to escape local minima with probability 1, we want

$$\sum_{n=1}^{\infty} p_n = \sum_{n=1}^{\infty} e^{-\Delta/T_n} = \infty.$$

Setting $T_n = \frac{c}{\log(n)}$ yields

$$\sum_{n=1}^{\infty} e^{-\Delta/T_n} = \sum_{n=1}^{\infty} \frac{1}{n^{\Delta/c}}.$$

Setting $\Delta/c \leq 1$ makes the series diverge. ∎

The Metropolis algorithm yields a general, that is, problem nonspecific, way to attack many difficult combinatorial optimization problems. It should be noted that in order to implement the simulated annealing algorithm, the user has control over the energy function and the neighborhood structure on V. The success or failure of the algorithm may depend on these choices.

4.4.2 Traveling Salesman Problem

The TSP minimizes the cost associated with permutations of $\{1, 2, \ldots, n\}$. Problem DDP has two permutations, one of $\{1, \ldots, n\}$ and one of $\{1, 2, \ldots m\}$, but any computational scheme to minimize a salesman's route might be adapted to DDP. We consider one such approach in the next section. A special version of the simulated annealing algorithm (the extended Metropolis algorithm) has been applied to the TSP, and for large problems, this method is competitive to with other leading approaches to the TSP.

In the traveling salesman problem, known to belong to the class of *NP*-complete problems conjectured to have no polynomial time solution, we wish to find a path, or tour, of minimal length to be taken by a salesman required to visit each of n cities, labeled $1, 2, \ldots, n$, and then return home. The set V in this case may be taken to be S_n, the set of all permutations of $\{1, 2, \ldots, n\}$, where for each permutation $\sigma \in S_n$, we identify the corresponding configuration given by the tour, taken in the order dictated by σ. The energy may be taken to be the total length of the tour, although any monotone transformation of this quantity will also serve.

We now choose a neighborhood structure for S_n motivated by a deterministic algorithm for the traveling salesman problem. There is a set of neighborhood structures known as k-opt for each $k \geq 2$. The permutation is broken into $k + 1$ pieces by choosing k "links." Then each piece except the initial and final ones can be reversed or exchanged. We say that the tour σ is k-opt, $1 \leq k \leq n$, for all tours in the k-opt neighborhood; the tour given by σ is the shortest. Thus, every tour is 1-opt and only the true best tours are n-opt. Note that the requirement to return to the starting city makes all tours $(i_1 + \Delta, i_2 + \Delta, \ldots, i_n + \Delta)$ equivalent, where addition is moldulo (n).

It is easily seen that a tour $\sigma = (i_1, i_2, \ldots, i_n)$ is 2-opt if and only if it yields the shortest tour of all tours which are elements of

$$N(\sigma) = \{\tau \in S_n : \tau = (i_1, i_2, \ldots, i_{j-1}, i_k, i_{k-1}, \ldots, i_{j+1}, i_j, i_{k+1}, \ldots, i_n)\}$$
for some $1 \leq j \leq k \leq n$.

It is not hard to see that given any initial tour σ_0 and any final tour $\sigma_n = (j_1, j_2, \ldots, j_n)$, we can obtain σ_n from σ_0 through a sequence of permutations $\sigma_1, \sigma_2, \ldots, \sigma_{n-1}$ such that $\sigma_k \in N(\sigma_{k+1})$ for $k = 0, 1, \ldots, n-1$ as follows. Given σ_k such that $\sigma_k = (j_1, j_2, \ldots, j_k, l_{k+1}, \ldots, l_m, l_{m+1}, \ldots, l_n)$, where $j_{k+1} = l_m$, say, invert l_{k+1} through l_m to obtain $\sigma_{k+1} = (j_1, j_2, \ldots, j_k, j_{k+1}, l_{m-1}, \ldots, l_{k+1}, l_{m+1}, \ldots, l_n)$. Thus we see that this notion of neighborhood yields an aperiodic Markov chain with communicating states in the algorithm described above. The other requirements desired of a neighborhood structure listed above are satisfied trivially: $|N_\sigma| = |N_\mu|$ for all $\sigma, \mu \in S_n$, and $\sigma \in N_\mu$ if and only if $\mu \in N_\sigma$.

4.4.3 DDP

In order to implement the simulated annealing algorithm as described above, an energy function and a neighborhood structure are required. Recall that we are given the single digest lengths $A = \{a_1, \ldots, a_n\}$ and $B = \{b_1, \ldots, b_m\}$ as well as double digest length $C = \{c_1, \ldots, c_{|A \wedge B|}\}$ as data. Any member of $V = \{(\sigma, \mu) : \sigma \in S_n \text{ and } \mu \in S_m\}$ corresponds to a permutation of $a_1, \ldots, a_n$ and $b_1, \ldots, b_m$ and, therefore, to a map. Earlier we defined $C(\sigma, \mu) = \{c_1(\sigma, \mu), c_2(\sigma, \mu), \ldots, c_{|A \wedge B|}(\sigma, \mu)\}$ to be an ordered $(c_1 \leq c_2 \leq \ldots)$ listing of the double digest implied by (σ, μ). Obviously, we have a solution when $C = C(\sigma, \mu)$.

To set up an extended Metropolis algorithm for this situation, we need to define f and N_V. We take as our energy function f the chi-squared-like criteria

$$f(v) = f(\sigma, \mu) = \sum_{1 \leq i \leq |A \wedge B|} (c_i(\sigma, \mu) - c_i)^2 / c_i;$$

note that if all measurements are error-free then f attains its global minimum value of zero for at least one choice of (σ, μ).

Following the extended Metropolis method for TSP as above, we define the set of neighbors of a configuration (σ, μ) by

$$N(\sigma, \mu) = \{(\tau, \mu) : \tau \in N(\sigma)\} \cup \{(\sigma, v) : v \in N(\mu)\}.$$

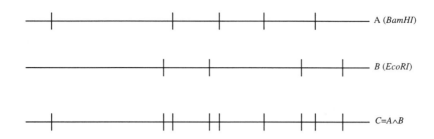

Figure 4.1: *Restriction map of* λ

Here $N(\rho)$ are the neighborhoods used in the discussion of the traveling salesman problem above.

With these ingredients, the algorithm was tested on exact, known data from the bacteriophage lambda with restriction enzymes *BamHI* and *EcoRI*. See Chapter 2 for a multiple enzyme map of lambda. The complete sequence and consequently the map information about lambda is known. *BamHI* cuts at G*GATCC whereas *EcoRI* cuts at G*AATTC where the *'s indicate where the enzymes cut bonds. Lambda is 48502 bps in length and each enzyme cuts at 3' sites.

BamHI	EcoRI
5509	21230
22350	26108
27976	31751
34503	39172
41736	44976

Consequently, we can derive the sets A, B, and C:

$A = \{5509, 5626, 6527, 6766, 7233, 16841\}$,
$B = \{3526, 4878, 5643, 5804, 7421, 21230\}$,
$C = \{1120, 1868, 2564, 2752, 3240, 3526, 3758, 3775, 4669, 5509, 15721\}$.

The reason for using this example is that the answer is known and that it is a reasonably sized region to restriction map. A computer-generated map of lambda appears in Figure 4.1.

Temperature was not lowered at the rate $c/\log(n)$ as suggested by Theorem, but, for reasons of practicality was instead lowered at the rate $1/n$. On three separate trials using various annealing schedules, the solution was located after 29702, 6895 and 3670 iterations from random initial configurations.

The algorithm was tested further on simulated data constructed by the random model introduced earlier. On a segment of length n with sites one unit apart labeled $1, 2, \ldots, n$, assume the restriction enzyme used in the first and second single digest makes a cut at site i independently with probability p_A and p_B respectively. This model can be justified on the grounds that a segment of DNA can be approximated as a string of independent, identically distributed random variables with values in a four-letter alphabet, although higher-order Markov chains frequently fit real data better. In addition, although in a real segment, sites cut by different restriction enzymes never exactly coincide, our model allows this to occur. This feature of our model is justified by the fact that DNA segments lengths can seldom be measured precisely and that two different enzymes can cut at sites very close together. On data generated by this model the algorithm was able to locate solutions to large problems in a small number of iterations. For example, on a problem of size $(16!16!)/(2!)^7(3!)^2(4!) = 3.96 \times 10^{21}$, a solution was located in only 1635 iterations. It must be emphasized, however, that any study of the algorithm's performance under the above probability model is confounded by the presence of multiple solutions to the exact problem in many instances. Recall, for example, the simulated problem of size 4320 that was found to have 208 distinct exact solutions. This problem was presented in Chapters 2 and 3. The same feature of many exact solutions must also be a property of the problem of size 3.96×10^{21} mentioned above.

The algorithm presented in this section is probably the most efficient known algorithm for exact data. Alas, it does not perform as well on realistic data. The reasons for this are discussed in the next section.

4.4.4 Circular Maps

DNA occurs in closed circular forms and mapping circular DNA is a natural problem. The object is to find circular arrangements of the single digest fragments that give the double digest fragments. Assume the measurements are without error, as before.

It is not enough to specify (σ, μ) for the single digest permutations, as the single digests can rotate relative to one another. Distinguish a fragment in each single digest and consider the points in a circular arrangement where the distinguished fragments are first encountered when moving counterclockwise. If p is the counterclockwise distance between the point in the A digest to that in the B digest, then a configuration is specified by (σ, μ, p), where, without loss of generality, $\sigma(1)$ and $\mu(1)$ index the distinguished fragments. The implied double digest can be obtained by an easy reduction to the linear case. Lay out the A fragments in the order $\sigma(1), \sigma(2), \ldots$. Now find the point in the μ order of the B fragments that is $L - p$ counterclockwise from the point that begins the $\mu(1)$ fragment. Introduce a cut at this point and use it as the left end of the B digest, corresponding to the left end of $\mu(1)$ and the A digest.

The cost function is chosen to be the same as the linear case. Define two configurations to be neighbors if one can be obtained from the other by reversing the order of any sequence of fragments in either digest. In Prob-

Algorithms for DDP

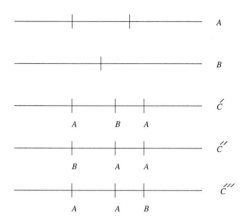

Figure 4.2: *Site orders*

lem 4.6, it is shown that the single digests can be rotated through multiples of $g = \gcd\{a_1, a_2, \ldots, a_n, b_1, \ldots, b_m\}$. It is unlikely that $g > 1$, but if this condition holds, we can introduce random rotations; that is, after reversing the order of any sequence of fragments, rotate either digest through any distance up to L.

4.5 Mapping with Real Data

It has been repeatedly emphasized that we have been studying an idealized situation where restriction fragments are measured exactly. Of course, measurement errors make a difficult problem even harder. Below, we will explore just why that is the case. Interestingly, it is necessary to say precisely just what it means to fit the data to a map. Although a graph-theoretic definition of restriction map was given in Chapter 2, our study of making maps using data must take into account the relationship of the data to the map.

Before we begin that task we should point out that one of the main problems in mapping is getting incorrect permutations of closely spaced sites.

In Figure 4.2, we have indicated three possible solutions, C', C'', and C'''. It might seem strange that the single digests can imply more than one double digest or be consistent with more than one. If, however, the short fragment of the A digest is in the measurement error of the longer A and the B fragments, then the order of the two A and the B sites cannot be determined. This problem motivates the next discussion.

4.5.1 Fitting Data to a Map

In this section we assume all fragments are of unique size. Essentially, we have been counting maps by permutations of single digest fragments:

$$|A|!|B|!.$$

As pointed out above, this does not allow us to determine a reasonable double digest $C(\sigma, \mu)$ implied by A and B.

The next logical step is to permute the double digest fragments in our map combinatorics,

$$|A|!|B|!|A \wedge B|!$$

now being the new number of maps. Still we have the same problem as in Figure 4.2; we cannot assign the order of sites along the linear DNA. Note that in the case of Figure 4.2 we have assigned one of the 4! orders of double digest fragments, but we have not determined which of C', C'', and C''' is the correct map.

Finally, we are forced to assign site labels to cut sites on the double digest. This further increases the number of map assignments to

$$|A|!|B|!|A \wedge B|! \binom{|A \wedge B| - 1}{|A| - 1}.$$

Here we assume $|A \wedge B| = |A| + |B| - 1$. For the case of Figure 4.2, we have

$$3!2!4! \binom{4-1}{3-1} = 3!2!4! \binom{3}{2}$$

map assignments. To see the factor $\binom{3}{2}$, see Figure 4.3 where C'' is designated.

The idea of this discussion is that we must assign data to a map. The purpose for doing so is to check the fit of the data to a map. The permutations and subset

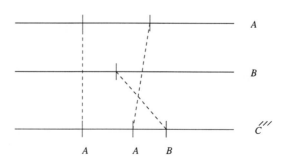

Figure 4.3: *Assignment of site order*

calculations above do allow us to find subsets R_i such that

$$a_i \approx \sum_{k \in R_i} c_k$$

and S_j such that

$$b_j \approx \sum_{l \in S_j} c_l.$$

The goodness-of-fit measure given by

$$\sum_{i=1}^{n} \left(a_i - \sum_{k \in R_i} c_k \right)^2 \Big/ a_i$$

and

$$\sum_{j=1}^{m} \left(b_j - \sum_{l \in S_j} c_l \right)^2 \Big/ b_j$$

is a reasonable measure of how well one of the configurations designates a map.

4.5.2 Map Algorithms

A popular algorithm for map construction uses the $|A|!|B|!$ method: All the single digest permutations are tried. In addition to being very slow, this does not work too well, as is to be expected from the presence of measurement errors.

Many other methods have been proposed. One method builds the single digest permutations fragment by fragment from left to right. Bounds on site locations are computed by adding the cumulative error bounds on single fragment lengths. A hypothesized order is rejected when no double digest fragment can be made to fit the map. After a few fragments, the error bounds are so large as to give no help in cutting down the $|A|!|B|!$ permutations. Besides that problem, it is not clear how to designate the full assignment of data to a map.

Simulated annealing, which was very successful for the exact length problem, has also been tried here with much less success. The reason for this is probably the huge configuration space, but there is no way to prove that a clever neighborhood structure and function will not turn out to be successful.

Problems

Problem 4.1 For the map given below, find the integer programming matrices E and F in Section 4.3.1.

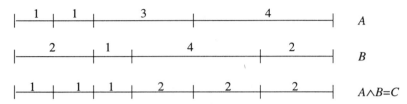

Problem 4.2 For the map given in Problem 4.1, find the partitions $R_1 R_2 R_3 R_4$ and $S_1 S_2 S_3 S_4$ of $\{1, 1, 1, 2, 2, 2\}$ as required in Section 4.3.2 on partition problems.

Problem 4.3 There are three states: 1, 2, 3. We move from state to state according to the following rules: When in state 1, move to state 2 with probability 1/2, otherwise remain in state 1. When in state 2, move to 1 or 3 each with probability 1/2. When in state 3, move to 1 or 2 each with probability 1/2. (i) Find the transition matrix P. (ii) Prove the Markov chain is communicating and aperiodic. (iii) Find the equilibrium distribution π.

Problem 4.4 Let $L = \{(i,j) : 0 \leq i,j \leq n\}$. Describe a neighborhood structure $N_{i,j}$ with all pairs of states communicating with (i) $|N_{i,j}| = 5$ and (ii) $|N_{i,j}| = 3$.

Problem 4.5 Let the function f be defined on $C = \{0,1\}^n$. Describe a Metropolis algorithm to find $\min\{f(\mathbf{c}) : \mathbf{c} \in C\}$.

Problem 4.6 For the circular map of Section 4.4.4, show (i) that digest A can be rotated a distance a_1 in at most four reversals and (ii) that relative to one another the single digest maps can be rotated by any multiple of $g = \gcd\{a_1, a_2, \ldots, a_n, b_1, \ldots, b_m\}$.

Problem 4.7 A simple algorithm called bubble sort places $x_1, x_2, \ldots, x_n$ into order $x_{i_1} \leq x_{i_2} \leq \cdots \leq x_{i_n}$. The idea is to proceed through the list from start to end, exchanging x_i and k_{i+1} whenever $x_i > x_{i+1}$. When you pass through the list without an exchange, the list is sorted. (i) Outline the algorithm and (ii) analyze the worst case running time.

Chapter 5

Cloning and Clone Libraries

The word clone often conjures up a scene from a science fiction movie where some genetic material has been used to create an army of identical beings, often very strong and very beautiful. Although the results of cloning can often be marvelous, the clones of molecular biology are a good deal more mundane. This chapter is a natural extension of our earlier studies of restriction enzymes, which are essential tools in cloning. Cloning is a means of producing recombinant DNA molecules, new molecules which are formed from already existing ones. One common application is to use the bacterium *E.coli* to make nonbacterial proteins such as human insulin. See Table 5.1.

Table 5.1: *Some of the non-bacterial proteins produced by E. coli*

Protein	Use
Human insulin	Hormone; controls blood glucose levels
Human somatostatin	Hormone; regulates growth
Human somatotropin	Growth hormone; acts together with somatostatin
Human interferon	Anti-viral agent
Foot-and-mouth VP1 and VP3	Vaccines against foot-and-mouth virus
Hepatitis B core antigen	Diagnosis of hepatitis B

To begin with, we need a cloning vector. Often the vector is constructed from a virus that can infect a convenient host. In particular, bacteriophages are viruses that specifically infect bacteria. The virus can insert its chromosome into a bacteria. Lambda is the basis of a popular cloning vector. Other cloning vectors include plasmids, which exist in the bacterial cell and reproduce independently of the bacterial chromosome. Cosmids are a sophisticated cloning vector based on lambda. Cosmids combine elements of lambda with those of a bacterial plasmid.

The cloning vector is cut with a restriction enzyme, and a piece of DNA is inserted into the cut. Then the vector is transferred into the host where it can be replicated into an experimentally useful quantity of DNA. See Figure 5.1.

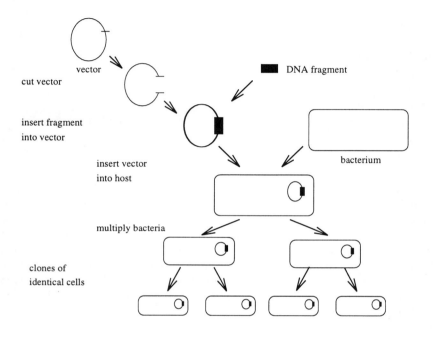

Figure 5.1: *Cloning DNA*

A collection of clones containing inserts that are DNA fragments from the genome of an organism is called a clone library. The vectors can only accept inserts within a certain size range dependent on the vector. This fact limits the DNA that can be put into a vector and, hence, the DNA that can be contained in a library.

We are going to assume that the DNA is from a genome of G bps, where G is large. The DNA fragments are produced by restriction digestion by one or more enzymes, and a cut site exists between any two base pairs with probability p. The cuts are made iid. In principle, this allows two adjacent bonds to be cut, and as sites are four or more base pairs in length, this is not realistic. However, p is small, such as $p = 1/5000$, and under the simple iid model, the cuts are adjacent, with vanishingly small probability.

Our numerical examples will deal with two cloning vectors. The first is a λ vector which is a modified version of the λ virus (whose *BamHI* and *EcoRI* maps appear in Chapter 4). For λ cloning vectors, the lower limit of clonable fragments is $L = 2$ kbps, and the upper limit is $U = 20$ kbps. The small fragments are discarded and a more realistic clonable range is 10 to 20 kbps. Cosmid vectors are also used where $L = 20$ and $U = 45$ kbps.

Our goal in this chapter is to see how representative clone libraries are. We will approximate the percentage of the genome that is in the library under various models.

5.1 A Finite Number of Random Clones

Here, we assume the clones are of fixed length L, randomly chosen from the genome. The question we study is: How much of the genome can be covered by N such random clones? Let b be an arbitrary fixed base in the genome.

$$\mathbb{P}(b \in \text{ random clone}) = \frac{L}{G},$$

and

$$\mathbb{P}(b \notin N \text{ random clones}) = \left(1 - \frac{L}{G}\right)^N = \left(1 - \frac{L}{G}\right)^{G\frac{N}{G}}$$

$$\approx e^{-\frac{NL}{G}}.$$

The approximation works because $L \ll G$ and $N \ll G$.

Therefore, a random base b belongs to a random clone of length L with probability $1 - e^{-N\frac{L}{G}}$. The result is formalized in

Proposition 5.1 (Clarke-Carbon) *If N random clones of length L are chosen from a genome of length G, the expected fraction f of the genome represented is approximately*

$$f \approx 1 - e^{-N\frac{L}{G}}.$$

Note that $N\frac{L}{G}$ equals the number of "genomes" included in the clones.

If 627 clones of length $L = 15$kbs are chosen in *E. coli* where $G = 5 \times 10^6$ bps, then $N\frac{L}{G} = 2$ and $1 - e^{-2} = 0.865$ is the fraction of the genome that is represented.

5.2 Libraries by Complete Digestion

Now consider a complete digestion of a genome of G bps. How much of the genome can be accommodated in a vector with lower limit L and upper limit U? The answer is an easy exercise in probability and calculus. Recall that the DNA has a restriction site with probability p.

Theorem 5.1 *Assume restriction sites are distributed along a genome of G bps according to a Bernoulli process with $p = \mathbb{P}(\text{restriction site})$. The expected* *fraction of the genome with fragment lengths $l \in [L, U]$ after complete digestion is approximately*

$$f = (pL + 1)e^{-pL} - (pU + 1)e^{-pU},$$

where $p > 0$ is assumed to be small and L to be large.

Proof. First note that fragments of length l are produced by the configuration cut–{no cut}$^{l-1}$–cut. This means that

$$\mathbb{P}(b \in \text{ fragment of length } l) = lp^2(1-p)^{l-1},$$

as each of l configurations containing b has probability $p^2(1-p)^{p-1}$. Now, with l large and p small,

$$\mathbb{P}(b \in \text{ fragment of length } l) \approx lp^2 e^{-p(l-1)}.$$

Therefore, b lies in a fragment with $L \leq l \leq U$ with probability

$$f = \sum_{l=L}^{U} lp^2 e^{-p(l-1)}.$$

Next we approximate this sum by an integral. This approximation is quite good for the small values of p we will consider. For example, $\max_l lp^2 e^{-p(l-1)}$ is at $l = \frac{1}{p}$ so that $\max_l lp^2 e^{-p(l-1)} = pe^{-1+p} \approx pe^{-1}$.

$$\sum_{l=L}^{U} lp^2 e^{-p(l-1)} \approx p^2 \int_{L}^{U} xe^{-p(x-1)} dx$$
$$= e^p \left\{ (pL + 1)e^{-pL} - (pU + 1)e^{-pU} \right\}.$$

∎

Alternative Proof. It is instructive to provide a continuous version of the proof. The geometric distribution of the number of basepairs in a restriction fragment is $\mathbb{P}(Z = m) = p(1-p)^m$. It is an exercise to show that when W is exponential with mean λ^{-1}, $\lambda = \log(1/(1-p))$ and $f_W(w) = \lambda e^{-\lambda w}$, $w > 0$, then $Z = \lfloor W \rfloor$. Therefore, we can consider restriction fragment lengths as continuous random variables. Let x be a point in the genome. Because the exponential is without memory, both the 5' and 3' lengths from x to the first restriction site is distributed as W. Let X and Y be iid where $X \stackrel{d}{=} W$ and $Y \stackrel{d}{=} W$. Then

$$F(X + Y \leq z) = \int_0^z \left\{ \int_0^{z-x} \lambda e^{-\lambda y} dy \right\} \lambda e^{-\lambda x} dx$$
$$= 1 - (1 + \lambda z)e^{-\lambda z}.$$

so that
$$f(z) = \lambda^2 z e^{-\lambda z}, \quad z \geq 0.$$
The probability that x is in a fragment that is clonable is
$$\int_L^U f(z)dz = \lambda^2 \int_L^U z e^{-\lambda z} dz$$
$$= (\lambda L + 1)e^{-\lambda L} - (\lambda U + 1)e^{-\lambda U}.$$
Of course, $\lambda = \log \frac{1}{1-p} \cong p$ and this result is consistent with Theorem 5.1. ∎

An interesting point arises in this alternate proof. Although restriction fragment lengths are exponentially distributed as W, the fragment containing a point x has distribution $X + Y$, where each is distributed as W. This is because the exponential is memoryless: $\mathbb{P}(W > x + y | W > x) = \mathbb{P}(W > y)$. An intuitive way of resolving this paradox (called the bus waiting paradox) is to realize that x is more likely to be in a long fragment than a short one. Sampling a fragment by choosing a basepair is very different from generating a fragment length according to W.

Consider two cloning vectors for *E. coli* where $G = 5 \times 10^6$. We digest *E. coli* with *EcoRI* which has $p = 1/5000$. For λ clones with $L = 2$ kbps, and $U = 20$ kbps,
$$f_\lambda = \left\{ \left(\frac{2}{5} + 1\right) e^{-2/5} - (4+1)e^{-4} \right\} = 0.845\ldots.$$
For cosmids with $L = 20$ kpbs and $U = 45$ kbps,
$$f_t = \left\{ (4+1)e^{-4} - (9+1)e^{-9} \right\} = 0.090\ldots.$$
Obviously we capture very little of the genome with cosmids. This is because $L = 20$ and $U = 45$ are far to the right of the mode of the fragment length distribution which is at $1/p = 5$ kpbs. Clearly, the values $L = 2$ kbps and $U = 20$ kbps for λ clones do bracket the mode, thereby increasing f.

Even $f_\lambda = 0.85$ is too small, leaving 15% of the genome out of the library. In addition there are very good reasons presented in the next section for making $U - L$ as large as possible. To increase f, molecular biologists have developed another strategy, discussed next.

5.3 Libraries by Partial Digestion

Partial restriction digests are performed by stopping the digest before all sites are cut. For our purposes, we will index a partial digest by the fraction μ of sites that are cut. It is necessary to get these two types of Bernoulli processes very clear. In the first case, sites are distributed in the genome, and each location has a site with probability p.

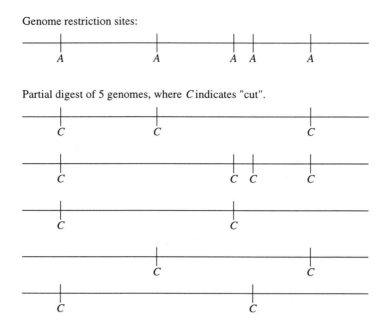

Figure 5.2: *Partial digestion of A sites*

The sites A are then fixed for the genome. When a partial digest is performed, each site is cut with probability μ. Remember that this process is taking place for many identical copies of the genome. For example, the cuts might occur at the locations indicated by C (for cut) in Figure 5.2.

There are two ways for a base b to fail to be in a partial digest library. One is that b belongs to no possible restriction fragment that can be cloned; denote this event by E. The other is that it does belong to a clonable fragment but that fragment is not incorporated into a clone; denote this event by F. We will prove in Section 5.3.2 that, for practical purposes, $\mathbb{P}(F) \cong 0$.

5.3.1 The Fraction of Clonable Bases

Note that $\mathbb{P}(E)$, the probability a base is unclonable, depends on the distribution of restriction sites, not on the partial digestion parameter μ. An obvious configuration that causes b to be unclonable is for the distance between the flanking restriction sites to be greater than U, the maximum clonable fragment size. By our last section, the probability of this event is

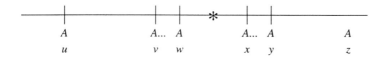

Figure 5.3: *Base b at position ∗ fails to be clonable.*

$$\sum_{l=U+1}^{\infty} lp^2 e^{-p(l-1)} \cong p^2 \int_{U+1}^{\infty} xe^{-p(x-1)}dx$$
$$= (p(U+1)+1)e^{-pU} \approx (pU+1)e^{-pU}.$$

However, the situation is much more complex than this. Although it might sound paradoxical, the base b can also belong only to fragments that are either smaller than L or larger than U. Figure 5.3 illustrates this possibility.

In Figure 5.3, assume that the base b is located at position ∗ and that the configuration of neighboring restriction sites is indicated by A's. The locations of these sites (bonds) are given by $u, v, \ldots, z$. Between positions v and w and between positions x and y any number of sites might occur; otherwise, all sites between u and z are shown. We impose the following conditions:

(i) $z - w > U$,

(ii) $x - u > U$,

(iii) $y - v < L$.

Condition (i) assures us that all fragments containing b and any site from z to the right are too long to be clonable. Condition (ii) similarly makes unclonable all fragments containing b and any site from u to the left. Because $y - v$ is too short to be clonable, the base b cannot be cloned. Now we compute the probability of this event.

A continuous model proves to be convenient. Because sites occur with probability p, with small p, we take a Poisson process on $(-\infty, \infty)$ with rate $\lambda = p$; that is, $\mathbb{P}(k \text{ sites in } [s, s+t)) = e^{-\lambda t}(\lambda t)^k/k!$. The distances between sites are independent exponential variables with mean $1/\lambda$. As p is small, this is an excellent model for restriction sites.

Our goal is to find the fraction f of a large genome that is unclonable. This is equivalent to the probability $\mathbb{P}(A_{t_0})$ of the event A_{t_0} that point t_0 is unclonable:

$$f = \lim_{G \to \infty} \frac{1}{G} \int_0^G \mathbb{I}(A_t)dt = \mathbb{P}(A_{t_0}).$$

Relabel t_0 by 0. The clones must have length in (L, U).

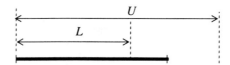

Shown in Figure 5.4 is a typical set of restriction sites at distances $X_1, X_2, \ldots$ to the left of 0. Clearly, 0 is unclonable if and only if there are no sites in $\bigcup_{i \geq 1}[\max\{0, L - X_i\}, \max\{0, U - X_i\}]$. Let the random variable W be defined by

$$W = \text{length}\left(\bigcup_{i \geq 1}[\max\{0, L - X_i\}, \max\{0, U - X_i\}]\right).$$

Then the probability that 0 is unclonable is

$$\mathbb{E}(e^{-\lambda W}).$$

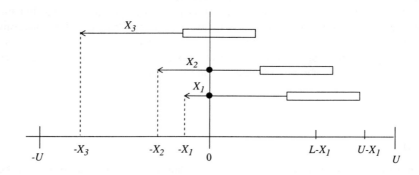

Figure 5.4: *Restriction sites to the left of 0*

Theorem 5.2 *Let $\{X_i\}_{i \geq 1}$ be iid exponentials with mean λ^{-1}. The probability that 0 is unclonable is $\mathbb{E}(e^{-\lambda W})$ where*

$$W = \text{length}\left(\bigcup_{i \geq 1}[\max\{0, L - X_i\}, \max\{0, U - X_i\}]\right)$$

and

$$\mathbb{E}(e^{-\lambda W}) \geq e^{-\lambda \mathbb{E}(W)},$$

where

$$\mathbb{E}(W) = U - Le^{-\lambda(U-L)} + (1 - e^{-\lambda(U-L)})/\lambda.$$

Cloning and Clone Libraries

Proof. $\mathbb{E}(e^{-\lambda W}) \geq e^{-\lambda \mathbb{E}(W)}$ by Jensen's inequality. To find $\mathbb{E}(W)$ we are interested in the coverage in $[0, U]$ by islands formed by the intervals of length $U - L$. See Figure 5.5. The right-hand end of these intervals is a Poisson process $\{U - X_i\}_{i \geq 1}$ of points at rate λ. Consider, first, points $t \in (0, L)$. t is not covered by an interval of length $U - L$ if and only if there are no events in $[t, t + U - L]$. Therefore, if $X_t = \mathbb{I}\{t \text{ uncovered}\}, \mathbb{E}(X_t) = e^{\lambda(U-L)}$, and

$$\mathbb{E}\left[\int_0^L X_t dt\right] = \int_0^L \mathbb{E}(X_t) dt = L e^{-\lambda(U-L)}.$$

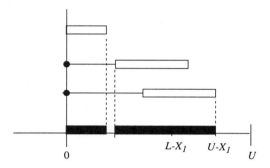

Figure 5.5: *Islands of intervals*

Now assume $t \in (L, U)$. t is not covered by an interval of length $U - L$ if and only if there is no event in $[t, U]$. Therefore, $\mathbb{E}(X_t) = e^{-\lambda(U-t)}$ and

$$\mathbb{E}\left[\int_L^U X_t dt\right] = \int_L^U \mathbb{E}(X_t) dt$$

$$= \int_L^U e^{-\lambda(U-t)} dt$$

$$= \frac{1 - e^{-\lambda(U-L)}}{\lambda}.$$

Finally $\mathbb{E}(W) = U - \mathbb{E}\left[\int_0^U X_t dt\right]$. ∎

5.3.2 Sampling, Approach 1

In this section we show that $\mathbb{P}(F)$, the probability that a base is clonable but is not incorporated into a clone, is practically zero if the number of DNA molecules is large enough. In other words, all nucleotides with clonable configurations can be considered as in some clones. The argument is rather crude and we will take another approach in Section 5.3.3.

Because the clonable range is between L and U, the expected number of restriction sites between the two ends of a clonable fragment is between pL and pU. Let μ be the fraction of restriction sites that are cut in a partial digest. The lower bound of the probability that any clonable fragment is obtained from one DNA molecule is approximately

$$\mu^2(1-\mu)^{pU}.$$

The logic for this is as above: The end sites must be cut, but no sites between them can be cut. Therefore, any clonable fragment will be obtained on the average once in not more than

$$\frac{1}{\mu^2(1-\mu)^{pU}}$$

DNA molecules.

Suppose every restriction site is the left end of some clonable fragment(s). There are approximately Gp restriction sites and each is approximately the left end of $p(U-L)$ distinct clonable fragments, so there are altogether $G(U-L)p^2$ clonable fragments. A very generous overestimate of the number of molecules required to yield all these fragments is

$$\frac{G(U-L)p^2}{\mu^2(1-\mu)^{pU}}.$$

This overestimate is obtained by multiplying the estimated number of clonable fragments by the estimated waiting time to observe each fragment. The next section takes a more sophisticated approach to this problem.

As an illustration, suppose *E. coli* is being digested with *EcoRI* with $\mu = 0.5$, and the cloning vector is pJC74, where $L = 19 \times 10^3$ and $U - L = 17 \times 10^3$. The number computed from the above formula is approximately 1.8×10^6, which is well within the limit of a typical sample size of about 2×10^9 molecules.

5.3.3 Designing Partial Digest Libraries

It is clear that the biologist can choose μ, the partial digestion parameter. Simply not adding enzyme makes $\mu = 0$; allowing the digestion to go to completion makes $\mu = 1$. Although it is harder to experimentally achieve a specific value of $\mu \in (0, 1]$, we will assume that is possible. It is not obvious what μ to choose or, to put it into other words, it is not obvious what the effects of μ are on a genomic library. Section 5.2 gave the representation when $\mu = 1$. Of course, $\mu = 0$ has no cuts at all and would be the worst possible choice. Here, we study $0 < \mu \leq 1$ in more detail.

A heuristic to guess the optimal μ is not too hard to come by. An average clonable fragment is $(U+L)/2$ bps in length. If p is the probability of a restriction

Cloning and Clone Libraries

site, there is, on the average, a restriction site every $1/p$ bps. Therefore there are approximately $\lfloor p(U+L)/2 \rfloor$ sites in an average clonable fragment. Because we must cut the fragment at its ends, we should average two cuts per $\lfloor p(U+L)/2 \rfloor + 1$ sites or about

$$\hat{\mu} = \frac{2}{\lfloor p(U+L)/2 \rfloor + 1}. \tag{5.1}$$

In this section, we show that this heuristic is not always correct, depending on the optimization problem being solved.

Poisson Approximation

We will have occasion in Chapter 11 to study the following problem. Let Z be a geometric random variable ($0 < \alpha < 1$):

$$\mathbb{P}(Z = k) = (1-\alpha)^k \alpha, \quad k \geq 0.$$

In our case we have n independent random variables $\{Z_i\}_{1 \leq i \leq n}$ where each Z_i is distributed as Z above. We will want the distribution of

$$Y = \max_{1 \leq i \leq n} Z_i.$$

To approach this problem, we define auxiliary random variables by multiplying independent Bernoullis:

$$X = B_1 B_2 \cdots B_t,$$

where $\mathbb{P}(B_i = 1) = 1 - \alpha = 1 - \mathbb{P}(B_i = 0)$. Note that $\mathbb{P}(Z \geq t) = \mathbb{P}(X = 1) = (1-d)^t$. Let $\{X_i\}_{1 \leq i \leq n}$ be n random variables distributed as X above. Then, if $I = \{1, \ldots, n\}$,

$$W = \sum_{i \in I} X_i$$

counts the number of Z_i that are at least t. We employ the following:

- Let $W = \sum_{i \in I} X_i$ and $\lambda = \mathbb{E}(W)$. *The Poisson approximation states that*

$$\mathbb{P}(W = k) \approx e^{-\lambda} \lambda^k / k.$$

This result is familiar in some cases. When $\mathbb{P}(X_i = 1) = r$ is small, $|I|$ is large and $|I|r$ is bounded, $W = \sum_{i \in I} X_i$ is binomial $(|I|, r)$ and can be well approximated by a Poisson with mean $\lambda = |I|r$. We will apply the approximation in this chapter and give a much more careful and rigorous statement of the approximation in Chapter 11, under the heading of Chen-Stein approximation, where the quality of the approximation is made explicit.

Returning to our problem of $Y = \max_{1 \leq i \leq n} Z_i$ and $X = B_1 B_2 \cdots B_t$, we have
$$W = \sum_{i \in I} X_i$$
and
$$\lambda = \mathbb{E}(W) = \sum_{i \in I} \mathbb{E}(X_i) = n(1-\alpha)^t.$$

We need to keep λ bounded away from 0 and ∞. Set $t = \log_{1/1-\alpha}(n) + c$; then $\lambda_n(t) = nn^{-1}(1-\alpha)^c = (1-\alpha)^c$ and

$$\begin{aligned}\mathbb{P}(Y = \max Z_i \leq t) &= \mathbb{P}(\text{all } X_i = 0) \\ &\cong \mathbb{P}(W = 0) = e^{-\lambda_n(t)} \\ &= e^{-(1-\alpha)^c}.\end{aligned}$$

Getting All Fragments

In Section 5.3.2, we gave a crude calculation to show that all clonable fragments should be in a standard sample of genomic DNA with $\mu = 0.5$. The model was to make a list of clonable fragments and to wait until fragment i appeared before waiting to see fragment $i + 1$. Obviously, many other clonable fragments j, with $j > i$, appear while we wait for fragment i. We now model this process in more detail.

As the genome has G bps, we expect M, the number of restriction sites, to be on the order of pG. We index partial digest fragment by k, where $k - 1$ is the number of uncut sites they possess. Because the terminal sites must be cuts, these k-fragments each have probability $\alpha = \mu^2(1-\mu)^{k-1}$. There are $M - (k+2) + 1 = M - k - 1$ such fragments in the genome. Our heuristic above suggests that $\mu = 2/(k+1)$ should optimize the number of k-fragments.

In the present model we will index the genomes in the sample by $1, 2, \ldots$ Let $W_{i,k} = 1$ if starting at site i we see a k-fragment in a random partial digest. Let

$$Z_{i,k} = \text{number of genomes until } \{W_{i,k} = 1\},$$

then
$$\mathbb{P}(Z_{i,k} = l) = (1-\alpha)^{l-1}\alpha,$$
where
$$\alpha = \mu^2(1-\mu)^{k-1}.$$

We, of course, are interested in the distribution of
$$Y_k = \max_{1 \leq i \leq M} Z_{i,k},$$

which is the minimum number of genomes until all k-fragments are observed. The smaller Y_k is, the more k-fragments are produced.

Section 5.3.3 gives some very useful insight into Y_k. The critical value of t was of the order
$$\log_{1/1-\alpha}(M) = \frac{\ln(M)}{\ln(1/(1-\alpha))}.$$
We are going to assume that $\mathbb{E}(Y_k) \approx \log_{1/1-\alpha}(M)$.

Theorem 5.3 *Restriction sites are distributed with probability p in a genome of length G. If the partial digestion parameter is $0 < \mu < 1$, the distribution of $Y_k =$ the waiting time in number of genomes until all k-fragments are observed satisfies*
$$\mathbb{P}(Y_k \le t = \frac{\ln(pG)}{\ln(1/(1-\alpha)) + c} \approx e^{-(1-\alpha)^c}.$$
The expectation $\mathbb{E}(Y_k)$ satisfies
$$\mathbb{E}(Y_k) \approx \mu^{-2}(1-\mu)^{-k+1}\ln(pG).$$
The value $\mu = 2/(2+k)$ minimizes $\mathbb{E}(Y_k)$.

Proof. As α is small,
$$\ln(1/(1-\alpha)) \approx \alpha = \mu^2(1-\mu)^{k-1}.$$
Therefore the central value of Y_k is of order
$$\mathbb{E}(Y_k) \approx \frac{\ln(M)}{\ln(1/(1-\alpha))}$$
$$\approx \mu^{-2}(1-\mu)^{-k+1}\ln(M).$$
Recall that we want μ to minimize Y_k. Thus, we should maximize $g(\mu) = \mu^2(1-\mu)^{k-1}$.
$$g'(\mu) = 2\mu(1-\mu)^{k-1} - (k-1)\mu^2(1-\mu)^{k-2}$$
and
$$g'(\mu) = 0 \text{ implies } \mu = 2/(1+k).$$
■

There are some important implications of this theorem. The behavior of $\mathbb{E}(Y_k)$ depends of the behavior of
$$g(\mu) = \mu^2(1-\mu)^{k-1}.$$
For fixed k, this function behaves like a quadratic near 0, and like $(1-\mu)^{k-1}$ for larger μ. Because k is usually larger than 3, the behavior for larger μ is $(1-\mu)^{k-1}$, which decreases much more rapidly than quadratic. Therefore, the effect of non-optimal μ is not symmetric about the mean. In fact, overdigestion

	$k = 5$			$k = 10$		
μ	α	$\frac{1}{\ln(1/(1-\alpha))}$	$\mathbb{E}(Y_k)$	α	$\frac{1}{\ln(1/(1-\alpha))}$	$\mathbb{E}(Y_k)$
2/11	0.0148	67.00	473.6	0.0054	183.6	1297.9
1/3	0.0219	45.06	318.5	0.0029	345.5	2442.3

Table 5.2: *Waiting times*

is much more harmful than underdigestion. Because clonable fragments have a variable number of k-fragments, we should aim at the smaller μ.

To see the effects of μ, consider cosmid vectors with $L = 20,000$ and $U = 40,000$. We take $p = 1/4000$. The k's corresponding to L and U are

$$k_L = 20000/4000 = 5$$

and

$$k_U = 40000/4000 = 10.$$

Therefore the optimal $\hat{\mu}$ values are

$$\hat{\mu}_L = 2/(1+5) = 1/3$$

and

$$\hat{\mu}_U = 2/(1+10) = 2/11.$$

Table 5.2 is for the *E. coli* genome where $G = 4.7 \times 10^6$ so that $M = G_p = 1175$. Note that using $\mu = 2/11$ for $k = 5$ where $\mu = 1/3$ is optimal increases the waiting time for all 4-fragments from 319 genomes to 474 genomes. However, using $\mu = 1/3$ for $k = 10$ where $\mu = 2/11$ is optimal increases the waiting time from 1298 genomes to 2442 genomes, almost doubling the waiting time. This illustrates the greater effects of overdigestion.

Maximum Representation

We now study the representation of the genome in clonable fragments using a more direct approach. The stochastic process that places sites at $X_1, X_1+X_2, X_1+X_2+X_3, \ldots$, where X_i are iid exponentials with mean λ, is called a Poisson process with rate λ. It is not hard to show that if sites are removed with probability

Cloning and Clone Libraries

$1 - \mu$, the result is a Poisson process with rate $\mu\lambda$. Therefore, if the partial digest parameter is μ, then Theorem 5.1 implies that

$$f = f(\mu) = (p\mu L + 1)e^{-p\mu L} - (p\mu U + 1)e^{-p\mu U} \qquad (5.2)$$

of the genome is clonable. When we consider N copies of the genome, dependence comes in from the potential sites fixed identically in all genomes. We now study this situation.

Let $\mathbf{X} = X_1, X_2, \ldots$ ($\mathbf{Y} = Y_1, Y_2, \ldots$) be iid fragment lengths from a point t moving to the left (right). These fragment lengths are fixed for all genomes. Then

$$\mathbb{P}(t \text{ is clonable}|\mathbf{X}, \mathbf{Y}) = \sum_{l=0}^{\infty}\sum_{k=0}^{\infty} \mu^2(1-\mu)^{k+l} \mathbb{I}\left\{\sum_{i=1}^{k+1} X_i + \sum_{j=1}^{l+1} Y_j \in (L, U)\right\}.$$

Because X_i and Y_j are iid, let $Z_i \stackrel{d}{=} X_i$, and

$$\mathbb{P}(t \text{ is clonable}|\mathbf{X}, \mathbf{Y}) = \sum_{n=0}^{\infty} \mu^2(1-\mu)^n (n+1) \mathbb{I}\left\{\sum_{i=1}^{n+2} Z_i \in (L, U)\right\}.$$

The conditional expectation of the number of clonable partial digest fragments containing t is $N\mathbb{P}(t \text{ is clonable}|\mathbf{X}, \mathbf{Y})$, so the remaining task is simply to uncondition. The sum of $n + 2$ exponentials has a gamma distribution, which is used next.

$$\mathbb{P}(t \text{ is clonable}) = \mathbb{E}(\mathbb{P}(t \text{ is clonable}|\mathbf{X}, \mathbf{Y}))$$
$$= \sum_{n=0}^{\infty} \int_L^U \mu^2(1-\mu)^n (n+1) \frac{p^{n+2} x^{n+1} e^{-px}}{\Gamma(n+2)} dx$$
$$= (p\mu)^2 \int_L^U xe^{-x} dx$$
$$= (p\mu L + 1)e^{-p\mu L} - (p\mu U + 1)e^{-p\mu U},$$

which is identical to $f(\mu)$ in Equation (5.2).

To maximize $f(\mu)$ for $\mu \geq 0$,

$$f'(\mu) = \mu(pU)^2 e^{-p\mu U} - \mu(pL)^2 e^{-p\mu L}$$

and $f'(\mu) = 0$ implies

$$\mu^* = \frac{2}{(U-L)p} \log \frac{U}{L}.$$

Because $f'(\mu)$ is positive for small positive μ and negative for large μ, μ^* gives the maximum of $f(\mu)$. Recalling $\mu \in [0, 1]$, the function $f(\mu)$ is maximized for

$$\hat{\mu} = \min\left\{\frac{2}{(U-L)p} \log \frac{U}{L}, 1\right\}.$$

This is very different from the heuristic value of $\hat{\mu}$ given in Equation (5.1).

5.4 Genomes per Microgram

Our results above were given in the number of genomes. Just how many genomes are in a normal sample of DNA? Normal samples are a few micrograms (μg), say 10 μg. Let us consider the E. coli genome of 5×10^6 bps. Some units are

$$1 \text{ bp} = 650 \text{ Daltons},$$

$$1 \text{ Dalton} = \text{molecular wt. of H},$$

$$1 \text{ mole H} = 1 \text{ g}.$$

Therefore,

$$10 \ \mu g = (10^{-5} \text{g}) \left(6 \times 10^{23} \frac{\text{Daltons}}{\text{g}} \right) = 6 \times 10^{18} \text{Daltons}$$

$$= 6 \times 10^{18} \text{Daltons} \times \frac{1 \text{bp}}{650 \text{Daltons}} \times \frac{\text{genome}}{5 \times 10^6 \text{bps}}$$

$$= 1.85 \times 10^9 \text{ genomes of } E.\ coli.$$

Problems

Problem 5.1 We replaced a sum by an integral. Compute the difference

$$p^2 \int_0^\infty x e^{-p(x-1)} dx - \sum_{l=0}^\infty l p^2 e^{-p(l-1)}.$$

Problem 5.2 We wish to clone a gene of g bps, where $0 < g < L$. Find the probability this gene can be cloned from a complete digest, where the probability of a restriction site is p. Hint: Theorem 5.1 solves this problem for $g = 1$.

Problem 5.3 In Problem 5.2, we have k independent enzymes with site probabilities p_i and we do k independent experiments. What is the probability the gene is cloned from at least one digest?

Problem 5.4 Distributed along an infinite genome according to a Bernoulli process are n restriction sites with $p = \mathbb{P}(\text{restriction site})$. Let F_i = length of the i-th fragment. Let $\mathbb{I}\{L \leq F_i \leq U\} = 1$ if the condition holds, 0 otherwise. Show that

$$\lim_{n \to \infty} \frac{\sum_{i=1}^n F_i \mathbb{I}\{L \leq F_i \leq U\}}{\sum_{i=1}^n F_i} = f,$$

where f is defined in Theorem 5.1. Hint: Apply the SLLN to the numerator and denominator.

Cloning and Clone Libraries

Problem 5.5 Make a table similar to Table 5.2 for cosmid vectors, $p = 1/5000$, and the human genome where $G = 3 \times 10^9$.

Problem 5.6 How many human genomes are there in $10\,\mu$g of DNA?

Problem 5.7 Consider two independent complete digests: Digest 1 (enzyme 1) has site probability p_1 and Digest 2 (enzyme 2) has site probability p_2. Digest 3 is the double digest of enzymes 1 and 2. Let f_i denote the fraction cloned from Digest i.

(i) Find f_3.

(ii) Find the fraction unclonable from Digest 1 which is clonable from Digest 2.

(iii) Find the fraction that cannot be cloned in either Digest 1 or Digest 2.

Chapter 6

Physical Genome Maps: Oceans, Islands and Anchors

In earlier chapters, we studied algorithms for the construction of restriction maps. Those maps are generally local maps, analogous to maps of a street in a town. Just as it is very useful to have these local restriction maps of unsequenced DNA, it is also very useful to construct maps of entire chromosomes or genomes. Such maps let us organize the unknown, analogous to an unexplored continent, into manageable units so that we can find our way around and locate more features of interest. In classical genetic mapping, the gene locations are given in units of genetic distance or recombination distance. Molecular biology has shown that changes in the DNA itself can be mapped. In this chapter, a different type of genomic map is studied, so-called physical maps where overlapping clones are used to span the genome.

To give an idea of the size of the problem, Table 6.1 shows four genomes and the number of clones required to cover each genome without overlap. The yeast clones are known as YACs, yeast artificial chromosomes, because the DNA is cloned into a genetically engineered yeast chromosome. The insert size is from 100 kb to 1 Mb and we take the most optimistic insert size.

	lambda (15 kb)	cosmid (40 kb)	yeast (1 Mb)
E. Coli	267	100	4
S. cerevisiae	1,333	500	20
C. elegans	5,667	2,125	85
Human	200,000	75,000	3,000

Table 6.1: *Genomes and vectors*

In this chapter, we consider the most straightforward method of physical mapping, overlapping random clones. To begin, a genomic library is constructed. Generally, we will assume here the library represents the genome and that all portions of the genome are equally likely to be cloned. We have seen in the last chapter that such assumptions can be flawed, but there are steps that can be taken to minimize these problems. For example, several enzymes can be used in library construction and the clones combined for a much more representative library. If α_i is the fraction uncloned by construction i, then $\prod_{i=1}^{k} \alpha_i$ is the fraction uncloned by all of k independent constructions.

The general assumption is that clones are drawn at random. Then each clone is characterized in a manner to be made specific later. Overlap is inferred from common characteristics. If the overlap can be detected with certainty, then the map of the genome proceeds at the rate predicted by the Clarke-Carbon formula (Proposition 5.1). There it is derived that if N random clones of length L are drawn from a genome of length G, then the expected fraction of the genome covered is approximately

$$1 - e^{-NL/G}.$$

Recall that the Clarke-Carbon formula was derived as a naive estimate of the fraction of a genome represented in a library, whereas here it is used as a naive estimate of the fraction of genome covered by a physical map. We will see below that this estimate is, in practice, very optimistic. This coverage could be approached if each clone were sequenced entirely, providing the most sensitive characterization possible. However, this would involve a vast amount of redundant sequencing, an expensive and time-consuming task. In fact, one of the uses of a physical map is to reduce the amount of genome sequencing required.

6.1 Mapping by Fingerprinting

The idea in mapping by random clone overlap is that clones are chosen at random and characterized in one of several ways. The characteristics are summarized in what biologists call a fingerprint. Many fingerprinting schemes involve digesting the cloned DNA with restriction enzymes. When two clones have enough fingerprint in common, they are declared to overlap. Clones that overlap are said to form *islands*. At the beginning of the mapping process, there are many isolated clones, which we refer to as *singleton* islands. As the clones begin to saturate the genome, the singleton clone islands are absorbed into fewer, longer islands. This is modeled here as a random process, and the mathematical results are useful in planning these large projects.

6.1.1 Oceans and Islands

For the purpose of analysis, we will make certain simplifying assumptions, which can be relaxed later. First, we will avoid the details of the particular fingerprinting scheme by simply considering an idealized scheme capable of detecting overlap between two clones whenever they share at least a fraction θ of their length. In reality, the minimum detectable overlap for most fingerprinting schemes will vary somewhat from clone to clone, for example depending on the number of restriction fragments in the clone. Nevertheless, we may think of θ as the *expected* minimum fraction, supposing that the criteria for overlap are sufficiently stringent that false positives and false negatives are rare.

Suppose that we have a perfectly representative genomic library, with all inserts of equal size. In fact, we model clone locations by a homogeneous Poisson process along the genome. Now we define the following symbols:

G = genome length in bp,
L = clone insert length,
N = number of clones,
$c = LN/G$ = expected number of clones covering a random point,
T = amount of overlap in bp needed to detect overlap,
$\theta = T/L$

Here we will rescale clone length to $1 = L/L$ and genome length to $g = G/L$. The expected number of clones covering a random point $c = LN/G = N/g$ remains the same. We assume the clones occur on the real line $(-\infty, \infty)$ according to a Poisson process with rate c. The genome corresponds to the interval $(0, g)$. The process of the location of right-hand ends of clones is a Poisson process $\{A_i\}$, $i \in \{\ldots, -1, 0, 1, 2, \ldots\}$, labeled so that $\cdots A_{-2} < A_{-1} < A_0 \leq 0 < A_1 < \cdots < A_N < g \leq A_{N+1} < \cdots$. Since clone ends occur with probability N/G, the Poisson approximation is valid when $N/G = c/L$ is small. Note that N is the random number of clones whose right ends belong to $(0, g)$. Because $\{A_i\}$ is a Poisson process,

$$\mathbb{P}(N = n) = \frac{e^{-cg}(cg)^n}{n!}, \ n \geq 0.$$

By the strong law of large numbers, with probability 1,

$$N/g \to c \text{ as } g \to \infty.$$

The interarrival times $A_i - A_{i-1}$ are distributed iid as exponentials with mean $1/c$ and density ce^{cx}, $x > 0$. A remarkable and useful fact about exponentials is the lack of memory property: If X is exponential,

$$\mathbb{P}(X > t + s | X > t) = \mathbb{P}(X > s).$$

Finally, we note the process of left endpoint locations is also a Poisson process with rate c.

Clones fall into "apparent" *islands* consisting of one or more members, based on overlaps detected by their fingerprints. The islands are only "apparent" because

Figure 6.1: *Oceans and islands*

some actual overlaps will go undetected. Islands with two or more members will be called *contigs*, and the gaps between islands will be called *oceans*. Islands of one member are called *singletons*. In Figure 6.1, $N = 6$. For $\theta = 0$, there are three islands; for $\theta = 0.2$, there are four apparent islands. In general, we neglect end effects, oceans and islands that overlap 0 and G.

The following results describe some expected properties of islands and oceans, both apparent and actual, as the mapping project proceeds.

Theorem 6.1 *Let θ be the fraction of length which two clones must share in order for the overlap to be detected, let N be the number of clones fingerprinted, and let c be the redundancy of coverage. With the above notation we have*

  (i) *The expected number of apparent islands is $Ne^{-c(1-\theta)}$.*

 (ii) *The expected number of apparent islands consisting of j clones $(j \geq 1)$ is*
$$Ne^{-2c(1-\theta)}(1 - e^{-c(1-\theta)})^{j-1}.$$

 (ii') *The expected number of apparent islands consisting of at least two clones (i.e., contigs) is*
$$Ne^{-c(1-\theta)} - Ne^{-2c(1-\theta)}.$$

(iii) *The expected number of clones in an apparent island is $e^{c(1-\theta)}$.*

(iv) *The expected length in base pairs of an apparent island is $L\lambda$, where*
$$\lambda = (e^{c(1-\theta)} - 1)/c + \theta.$$

 (v) *The corresponding results for the actual islands are obtained by setting $\theta = 1$. For example, the expected number of actual islands is Ne^{-c}.*

(vi) *The probability that an ocean of length at least xL occurs at the end of an apparent island is $e^{-c(x+\theta)}$. In particular, taking $x = 0$, the probability that an apparent ocean is real (as opposed to an undetected overlap occurring) is $e^{-c\theta}$.*

Proof. First we define an important quantity, $J(x)$ = the probability that two points x apart are not covered by a common clone. For $0 < x \leq 1$, this is equivalent to no arrivals in the interval of length $(1 - x)$ preceding the leftmost point or no events for a Poisson of mean $c(1 - x)$. Therefore,

$$J(x) = \begin{cases} e^{-c(1-x)} & \text{if } 0 \leq x \leq 1, \\ 1 & \text{if } x > 1. \end{cases}$$

Note that $J(x)$ was extended to $J(0) = \mathbb{P}(\text{no clone covers a point})$.

The number of islands is equal to the number of times we exit a clone without detecting overlap. Let E be the event that a given clone is the right-hand clone of an island. If the right-hand position is t, then this means the points $t - \theta$ and t are not covered by a common clone. Therefore, $\mathbb{P}(E) = J(\theta)$. Because there are N clones in $(0, g)$, the expected number of apparent islands is $NJ(\theta) = Ne^{-c(1-\theta)}$.

Now we consider the process of right-hand ends of islands. As above, we label them by

$$\cdots C_{-2} < C_{-1} < C_0 \leq 0 < C_1 < \cdots C_K < g \leq C_{K+1} < \cdots,$$

and K is the number of apparent islands with right-hand ends in $(0, g)$. The intensity of this process is $cJ(\theta)$, as $J(\theta)$ is the probability that a given clone is the right-hand end of an island.

To study the expected number of clones in an island, let M_j be the number of clones in the j-th island. $M_1, M_2, \ldots M_K$ are averaged to obtain $\mathbb{E}(M)$ by a limit argument. Define

$$\overline{M}_g = \frac{1}{K} \sum_{i=1}^{K} M_i.$$

At most $1 + \max\{k : k \text{ integer and } k < 1/(1 - \theta)\}$ islands overlap 0 or g (see Problem 6.5). Therefore, because $K \to \infty$, including or discarding the islands because overlapping 0 or g does not affect $\overline{M}_g$. Now $K/N \to J(\theta)$ and $\frac{1}{N} \sum_{i=1}^{K} M_i \to 1$ so that

$$\lim_{g \to \infty} \overline{M}_g = 1/J(\theta) = e^{c(1-\theta)}.$$

It is possible to obtain more information about the number of clones in an island. The event E that a given clone with $A_i = t$ is the right-hand end of an island has probability $\mathbb{P}(E) = J(\theta)$. Moving from t to the left, we ask whether we detect an overlapping clone or not. A clone has detectable overlap with $(t - 1, t)$ if it covers the interval $(t - 1, t - 1 + \theta)$. This event is independent of E and has probability $1 - J(\theta)$. Continuing in this fashion,

$$\mathbb{P}(M_i = j | E) = (1 - J(\theta))^{j-1} J(\theta),$$

or M_i has a geometric distribution with stopping probability $J(\theta)$. Multiplying this probability by the expected number of islands gives (ii), whereas the mean of the distribution is $J^{-1}(\theta)$, consistent with the limiting argument given above.

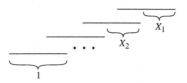

Figure 6.2: *Length of an island*

To prove (iv), consider an island consisting of M clones, where M has the geometric distribution described above. The length of an island is $X_1 + X_2 + \cdots + X_{M-1} + 1$, where each X_i is the distance until a new clone begins, $i < M$. (See Figure 6.2.) Define X_i to have density

$$f(x) = ce^{-cx}, \; 0 < x < 1 - \theta$$
$$\mathbb{P}(X_i = 1) = 1 - \int_0^{1-\theta} ce^{-cx} dx = e^{-c(1-\theta)}.$$

The expected length of an apparent island is then $\mathbb{E}(\Sigma_{1 \leq i \leq M} X_i)$. Evaluating this expectation requires some additional theory. The random variable M is a *stopping time* for $X_1, X_2, \ldots$ if the event $\{M = j\}$ is independent of $X_{j+1}, X_{j+2}, \ldots$. The expectation can then be evaluated by a result known as Wald's identity which states that $\mathbb{E}(\Sigma_{1 \leq i \leq M} X_i) = \mathbb{E}(X)\mathbb{E}(M)$ when M is a stopping time. This result is stated and proved as Theorem 6.2.

Now the stopping time M is defined by $M = \min\{i : X_i = 1\}$. We find

$$\mathbb{E}(X) = \int_0^{1-\theta} cxe^{-cx} dx + e^{-c(1-\theta)}$$
$$= c^{-1}(1 - e^{-c(1-\theta)}) + \theta e^{-c(1-\theta)}.$$

As above, $\mathbb{E}(M) = e^{c(1-\theta)}$ so that (iv) follows by $\mathbb{E}(M)\mathbb{E}(X)$.

Finally, for (vi) we require the probability that an ocean of length at least x occurs at the end of an apparent island at t means no arrivals in $(t - \theta, t + x)$, which has probability $e^{-c(x+\theta)}$. ∎

Theorem 6.2 (Wald's Identity) *If $X_1, X_2, \ldots$ are independent and identically distributed random variables with finite expectations and if M is a stopping time for $X_1, X_2, \ldots$, such that $\mathbb{E}M < \infty$, then*

$$\mathbb{E}\sum_{i=1}^M X_i = \mathbb{E}(M)\mathbb{E}(X).$$

Proof. Letting

$$Y_n = \begin{cases} 1 & \text{if } M \geq n, \\ 0 & \text{if } M < n, \end{cases}$$

Physical Genome Maps: Oceans, Islands and Anchors

we have that

$$\sum_{n=1}^{M} X_n = \sum_{n=1}^{\infty} X_n Y_n.$$

Thus,

$$\mathbb{E}\left(\sum_{n=1}^{M} X_n\right) = \mathbb{E}\left(\sum_{n=1}^{\infty} X_n Y_n\right) = \sum_{n=1}^{\infty} \mathbb{E}(X_n Y_n).$$

However, $Y_n = 1$ if and only if we have not stopped after successively observing $X_1, \ldots, X_{n-1}$. Therefore, Y_n is determined by $X_1, \ldots, X_{n-1}$ and is, thus, independent of X_n. From the last equation above we thus obtain

$$\mathbb{E}\left(\sum_{n=1}^{M} X_n\right) = \sum_{n=1}^{\infty} \mathbb{E}(X_n)\mathbb{E}(Y_n)$$

$$= \mathbb{E}(X)\sum_{n=1}^{\infty} \mathbb{E}(Y_n)$$

$$= \mathbb{E}(X)\sum_{n=1}^{\infty} \mathbb{P}\{M \geq n\}$$

$$= \mathbb{E}(X)\mathbb{E}(M). \qquad \blacksquare$$

The minimum detectable overlap θ clearly has a major effect on the progress of a mapping project. Figure 6.3 shows the expected number of islands as a function of the number c of genome equivalents of DNA fingerprinted, and Figure 6.4 shows expected average island length. In Figure 6.3, $Ne^{-c\sigma} = (G/L)ce^{-c\sigma}$ so that the expected number of islands is in units of G/L. This makes the figure independent of the details of the genome. Similarly, the average island length is in units of L. At the beginning of the project, the number of islands increases because new clones are unlikely to overlap others. The maximum number of islands occurs at $c = (1 - \theta)^{-1}$ and is equal to $(G/L)e^{-1}(1 - \theta)^{-1}$. After this point, the number of islands declines as gaps are closed. After some point, a directed strategy for bridging gaps must be employed, as it would require a huge amount of work to close all the gaps by fingerprinting random clones.

Note how decreasing the minimum detectable overlap from 50 to 25% greatly speeds the progress of the project. By contrast, the decrease from 25% minimum detectable overlap to the theoretical limit at 0% has relatively less effect. These results suggest that a fingerprinting scheme with $\theta \in (0.15, 0.20)$ may be a sensible goal, with further decrease being of limited value. Of course, the advantage of a smaller θ must be balanced with the increased effort to obtain a more sensitive fingerprint.

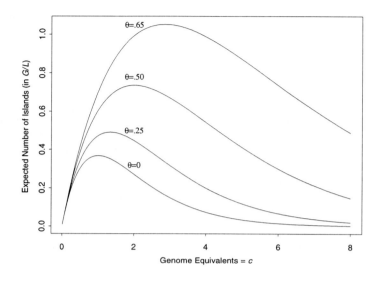

Figure 6.3: *Number of islands*

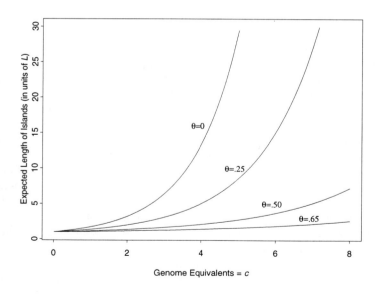

Figure 6.4: *Length of islands*

Physical Genome Maps: Oceans, Islands and Anchors

Next, we develop a more general version of Theorem 6.1(i) which relaxes the condition that L and θ are constant. Let the size L of the cloned insert be chosen according to some distribution with a density. The overlap necessary to detect overlap with a given clone will be an independent random variable Θ conditional on the length L of that clone. For example, $\Theta = \theta L$ would require fraction θ of the clone. The distribution of Θ is meant to reflect the fact that there are a variable number of restriction sites in clonable fragments.

We assume clone lengths L are iid with mean $\mathbb{E}(L)$. The density of $L/\mathbb{E}(L)$ is $f(l)$. We define a new version of $J(x)$ equal to the probability that two points a distance x apart are not covered by a common clone. Let

$$\mathcal{J}(x) = \mathbb{P}(L/\mathbb{E}(L) > x) = \int_x^\infty f(l)dl.$$

Then

$$J(x) = \exp\left\{-c\int_x^\infty \mathcal{J}(l)dl\right\}.$$

Below, we will write $\mathbb{E}(J(\Theta)) = \mathbb{E}(\mathbb{E}(J(\Theta)|L))$ because $J(\Theta)$ really is a function of (Θ, L).

Theorem 6.3 *With the assumptions and notation above including Θ and L as random variables:*

(i) The expected number of apparent islands is

$$N\mathbb{E}(J(\Theta)) = N\mathbb{E}(\mathbb{E}(J(\Theta)|L)).$$

(ii) The probability that an apparent island has j clones is at least

$$(1 - \mathbb{E}(J(\Theta)))^{j-1}\mathbb{E}(J\Theta).$$

Sketch of Proof. As the probability that a random clone is the right-hand clone of an island is $J(\Theta)$ and as there are N clones in total, the expected number of islands is $N\mathbb{E}(J(\Theta))$.

With variable θ or L, it is not possible to produce iid $X_1, X_2, \ldots$ for overlaps in a straightforward fashion as before. However, a "greedy" algorithm will produce lower bounds for the number of clones in an island, and for island lengths. See Figure 6.5 for an example of greedy island length versus apparent island length. When at a clone moving from right to left, take the clones in order of A_i, their right clone ends. This means sometimes leaving a clone for one of less coverage due to shorter L or larger θ, but it produces independent X_i. The number of clones in a greedy island is j with probability

$$\mathbb{E}\left(\prod_{i=1}^j (1 - J(\Theta_i))J(\Theta_1)\right). \qquad\blacksquare$$

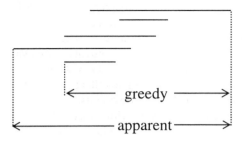

Figure 6.5: *Unequal length clone islands*

6.1.2 Divide and Conquer

In this section, we consider whether there is an advantage to constructing a physical map chromosome by chromosome, rather than all at once. At first glance, there appears to be no advantage: The formulas of the model are linear in N, the number of clones fingerprinted. In an organism with two chromosomes of equal length, one would expect the same number of islands if N clones were fingerprinted from a whole genomic library or if $N/2$ clones were fingerprinted from each of two chromosome-specific libraries–provided that, in each case, the fingerprinting scheme could detect matches between clones overlapping in a proportion θ of their length.

In fact, there are some second-order considerations that favor subdividing the project.

(i) If the rule for declaring overlaps were kept constant, the rate of false positives would be greater for genomes of larger size. In order to maintain the same rate of false positives, a greater proportion of overlap θ must be required for declaring overlap in a larger genome. However, the effect is not large (see Section 6.1.4).

(ii) If the genome were divided into two parts, an investigator might decide not to fingerprint an equal number of each half: As the project progressed, the investigator could fingerprint more clones from whichever half was progressing more slowly. However, the law of large numbers assures roughly comparable progress in each half (unless systematic cloning bias caused one half to be significantly less representative.) Only a slight increase in efficiency would result (unless the genome were decomposed into so many parts that the expected number of clones required to cover each part was small).

In addition to these mathematical considerations, there are various practical issues that might favor subdividing the project as well: For example, subdividing the project permits the adoption of improved fingerprinting strategies for later parts, as they become available. Also, the subdivided project can be pursued in different laboratories and the effect of parallel processing can be achieved.

6.1.3 Two Pioneering Experiments

The practical aspects of random clone mapping depends on the biology of the organism being mapped as well as the cloning vector. We have discussed lambda and cosmid cloning vectors. In the last few years, some innovative biologists have engineered yeast artificial chromosomes (YACs) so that a YAC can accept large inserts and be reproduced as yeast cells divide and multiply. The amount of DNA clonable into YACs varies but ranges from 10^5 bps to 10^6 bps. We have frequently used *E. coli* ($G = 5 \times 10^6$ bps) and human ($G = 3 \times 10^9$ bps) for examples. In this section, we add *S. cerevisiae* (yeast) with $G = 20 \times 10^6$ bps. Table 6.1 gave G and L for these organism and vector combinations.

The object of this section is to discuss two original and influential experiments. We will briefly describe the experiments and discuss the agreement of our results with the experimental data. One key consideration is whether the minimum detectable overlap may be taken to be a constant θ, independent of the clone under consideration. Theorem 6.2 gave some implications of nonconstant θ. This assumption will be satisfied by fingerprinting methods involving information derived from a large number of fragments (such as that used in the *E. coli*): In this case, clones will all have about the same density of fingerprint information and, therefore, will require roughly the same minimum overlap needed to recognize a match. By contrast, our assumption may not fit well enough to make it possible ever to declare overlap. For example, because the yeast project (*S. cerevisiae*, below) required that clones share five restriction fragment lengths in order to declare pairwise overlap, those clones containing fewer than six fragments could never be joined into islands. Thus, our formulas would be expected to agree closely with the data for the *E. coli* project but significantly to underestimate the number of single clone islands for the yeast project. This is indeed the case, as we shall see presently.

S. cerevisiae

Our first experiment was reported in 1986 by Olson et al. who studied 4946 lambda clones in yeast. Some relevant parameters for us are $G = 20 \times 10^6$ and $L = 15 \times 10^3$. A type a (see Section 6.1.4) fingerprint was used in which each clone was digested with *Eco*RI and *Hind*III and the fragment lengths measured.

Olson et al. arranged the 4946 clones into 1422 islands, consisting of 680 contigs and 742 singletons. The average number of fragments per clone was reported to be 8.36 and the criteria for declaring overlap included the requirement that a pair of clones share at least five fragments. The authors noted that most of the 742 singletons contained five or fewer fragments and, thus, even large overlaps involving such clones would be ignored in pairwise comparisons. Accounting for the problem of clones with five or fewer restriction fragments, what do our formulas predict for such a project?

A simple calculation shows that somewhat less than 14% of the clones would be expected to have five or fewer restriction fragments. Suppose that restriction sites are distributed according to a Poisson process with a mean of 8.36 fragments, or 7.36 restriction sites, per clone. Let K denote the actual number of restriction sites in a clone, then, for any integer k, $P(K = k) = \lambda^k e^{-\lambda}/k!$ with $\lambda = 7.36$. Accordingly, about 14% of all clones will have four or fewer sites. This estimate must be reduced slightly, as fragments smaller than 400 bps were not scored in deriving the estimate $\lambda = 7.36$. Taking this proportion at 13%, about 4300 clones would be expected to have 6 or more fragments and about 650 would have 5 or fewer.

With 5 out of an average of 8.36 common fragments required to declare overlap, we roughly estimate θ at $5/8.36 \approx 0.60$. Because a few other technical conditions were required, the estimate should be increased slightly. Here, we take $\theta = 0.63$.

With $N = 4303, \theta = 0.63$, and $c = 4.5$ (as reported by the authors), we expect 660 contigs and 154 single clone islands. Adding the 643 singletons expected from clones with 5 or fewer fragments yields a total of 660 contigs and 797 single clone islands. This agrees fairly well with the observed 680 contigs and 742 single clone islands, given the rough approximations involved.

As the yeast project was a first-generation mapping project, it is interesting to compare the results to the expected progress if one instead used the fingerprint subsequently developed for *E. coli*. With the same number of clones ($N = 4946, \theta = 0.20, c = 4.5$), one would expect only about 131 contigs and 4 singletons. Thus, as noted above, the use of a more informative fingerprint can greatly speed progress toward the completion of a physical map.

E. coli

The second experiment is a 1987 study of *E. coli* by Kohara et al. At $G = 4.7 \times 10^6$ bps, this is the smallest of the three genomes we consider. The study is distinguished by its experimental design and by its success. Kohara et al. did not make any startling technical advances; rather they very wisely applied known techniques to produce a major advance in knowledge of the *E. coli* genome. The result of their work was a restriction map of the entire genome for eight enzymes.

The fingerprint used by Kohara et al. is referred to below as a type c fingerprint— each clone was mapped for each of eight enzymes. The clones were overlapped by comparing the clone maps. See Figure 6.6 where an island of three clones is shown. The eight restriction enzymes are arranged from top to bottom in the following order: *Bam*HI (B), *Hind*III (D), *Eco*RI (E), *Eco*RV (F), *Bg*1II (G), *Kpn*I (Q), *Pst*I (S), and *Pvu*II (V). The final result of the Kohara et al. effort is an eight enzyme restriction map of the entire *E. coli* genome which is shown in Figure 6.7.

Physical Genome Maps: Oceans, Islands and Anchors   113

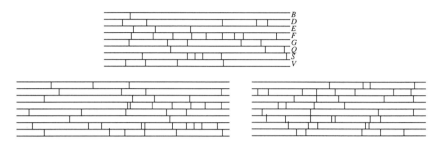

Figure 6.6: *Overlapping clones*

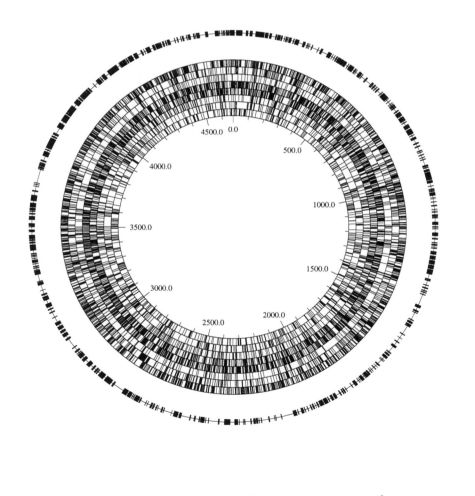

Figure 6.7: *Restriction map of the* E. coli *genome,* 4.7×10^6 *bps*

On the basis of the pairwise fingerprint comparison, Kohara et al. arranged 1025 clones into 70 islands, of which 7 were isolated single clone islands. A hybridization method was then used to find clones spanning the gaps, resulting in all but six gaps being closed.

The authors reported that the cloned inserts were 15.5 kb on average and that an overlap of about 3 kb could be detected. Taking the minimum detectable overlap to be $\theta = 3/15/5 \approx 0.19$, the genome size to be 4704 kb, and genome coverage $c = (1024)(15.5)/4704 \approx 3.38$, the formulas predict 67.16 islands of which 4.39 should be single clone islands. The agreement with the observed 70 and 7 is quite close. The small disagreement is even in accord with the results of Theorem 6.2 on variable θ and L.

Moreover, the formulas predict that most gaps will be small enough to be closed in one step via hybridization, but that a few should remain. Given an infinite library to screen via hybridization, Theorem 6.1(vi) predicts that about 2 of 70 gaps could not be closed because they would be longer than 15.5 kb. If a 1-kb overlap on each end were required to obtain a positive hybridization signal, about 4 of 70 gaps would be expected to remain because their lengths exceed 13.5 kb. Because a finite library containing only 2344 clones was screened by hybridization (including the clones that had been fingerprinted), a few additional gaps might remain simply because an optimally situated clone did not occur in the finite library. Thus, the six remaining gaps are well within expectation.

6.1.4 Evaluating a Fingerprinting Scheme

Many fingerprinting schemes involve digesting the clone DNA with other restriction enzymes. The simplest fingerprint associates a clone C with the set of fragment lengths of a single digestion. We can represent this by

$$f_a(C) = \{l_1, l_2, \ldots, l_k\}$$

if there are k fragments resulting from the digest.

There is a subtle variation on the above scheme. Recall that longer fragment lengths have larger errors associated with their length measurements. If we digested with an enzyme with a cutting probability of 4^{-4} instead of 4^{-6}, we could get more accurate measurements, but we would have many more (16 times) as many fragments. To circumvent this problem, the clone is first digested by the less frequently cutting enzyme ($p_1 = 4^{-6}$) and the fragments end labeled with radioactive material. Then a second restriction enzyme ($p_2 = 4^{-4}$) is added, producing many fragments, but a reduced number will be end labeled. See Figure 6.8 for illustration of this experiment. Although many small fragments are produced, only six are end labeled for careful measurement. We symbolize this procedure by

$$f_b(C) = \{l_{i_1}, l_{i_2}, \ldots, l_{i_k}\},$$

indicating that we have chosen a specific subset of the available fragments.

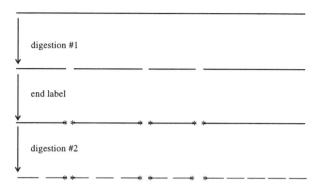

Figure 6.8: *Digestion*

The third method of fingerprinting we discuss is that of restriction mapping the clone. As we have discussed restriction maps extensively, there is no need to go into much detail here. The symbolic version of a map will be

$$f_c(C) = (l_1, l_2, \ldots, l_k).$$

In addition to specifying the data comprising the fingerprint, it is necessary to give a rule for deciding whether or not two clones overlap. Ideally, we would like

$$O(C_1, C_2) = \begin{cases} 1 \text{ if } C_1 \cap C_2 \neq \emptyset, \\ 0 \text{ if } C_1 \cap C_2 = \emptyset. \end{cases}$$

Naturally, it is impossible to achieve this ideal. We fingerprint clones to get a snapshot of their content without actually determining their DNA sequence. Of course, DNA sequence is an ultimate fingerprint, but determining sequence is very labor and material intensive. Therefore, we are subjected to the usual kinds of error in decision making. We formulate the function $O(C_1, C_2)$ as a hypothesis testing decision function.

$$H_0 : C_1 \cap C_2 = \emptyset,$$
$$H_1 : C_1 \cap C_2 \neq \emptyset.$$

To decide $H_1 : C_1 \cap C_2 \neq \emptyset$ we must have strong evidence that the data could not have arisen from disjoint clones. Biologists usually prefer to leave overlaps undetected rather than declare false overlaps. Given the magnitude of the problems this is quite difficult to avoid. Several thousand clones are often characterized in these experiments. With five thousand clones, there are

$$\binom{5000}{2} \approx 12.5 \times 10^6$$

clone pairs, so that even a Type I error probability of 10^{-6} leaves 12.5 expected false overlaps.

A good fingerprinting scheme should be able to detect relatively small overlaps between clones, while allowing an acceptable low rate of false positives given the size of the genome to be mapped. Here, we discuss how the choice of the fingerprint itself and the rule for declaring overlaps determine the minimum detectable overlap and the false positive rate, with the aim of providing general guidance for those designing a mapping project.

We consider two basic types of fingerprinting schemes. These two examples are meant to be illustrative, not exhaustive.

Type (a/b) The fingerprint consists of the lengths of the restriction fragment lengths produced following digestion by a single enzyme [type (a)] or a combination of enzymes used consecutively [type (b)] that produces an average of n fragments. The matching rule declares an overlap when two fingerprints share at least k fragment lengths. Because measurement error is roughly proportional to fragment length, two restriction fragments will be assumed to match if their lengths differ by at most $100\beta_1\%$. Typically, $0.01 \leq \beta_1 \leq 0.05$, depending on the gel system.

Type (c) The fingerprint consists of a restriction map for a single enzyme (or enzyme combination) that produces an average of n fragments. The matching rule declares an overlap when the lengths of k terminal fragments in the two maps agree. Fragments will be assumed to match if their lengths differ by at most $100\beta_2\%$, where β_2 may be larger than, smaller than, or equal to β_1, depending on how the restriction maps are made (partial digestion or double digestions). In our examples below, we will take $\beta_1 = \beta_2 = 0.03$.

Intuitively, it is clear that fingerprints of type (c) contain more information than fingerprints of type (a/b), provided that $\beta_1 = \beta_2$. The restriction map drastically restricts which fragment length matches are to be considered meaningful–namely, those corresponding to overlap between terminal segments of the two restriction maps.

The following result estimates the minimum detectable overlap and the chance of false positives for the two fingerprinting schemes specified above.

Theorem 6.4 *With the assumptions above, set the minimum detectable overlap for both fingerprinting methods to be approximately k fragments out of an expected n fragments per clone.*

(i) *For a fingerprint of type (a/b), the expected number of fragments shared by fingerprints of two nonoverlapping clones will be about $\lambda = \frac{1}{2}\beta_1 n^2$. Provided that λ is relatively small compared to n, the distribution of fragments shared by two nonoverlapping clones will be approximately Poisson; that* is, the probability that a declared overlap is a false positive is approximately

$$\mathbb{P}(X \geq k) = \sum_{i=k}^{\infty} e^{-\lambda} \frac{\lambda^i}{i!},$$

where X is a Poisson random variable with mean $\lambda = \frac{1}{2}\beta_1 n^2$.

(ii) For a fingerprint of type (c), the probability that a declared overlap is a false positive is approximately

$$4\left(\frac{1}{2}\beta_2\right)^k \left(1 + \frac{1}{2}\beta_2\right).$$

Proof. To compute the probability of a false match begin declared between two nonoverlapping clones, we first calculate the chance that two randomly chosen restriction fragments have matching lengths. If the restriction enzyme yields fragments with mean length λ^{-1}, then the restriction fragment lengths are well approximated by a continuous exponential distribution with density $f(x) = \lambda e^{-\lambda x}$, for $x > 0$. Suppose that we pick two fragments at random from such a distribution, that the first fragment has length x chosen from the distribution $f(x)$, and that the second fragment will match to within $100\beta\%$ provided that its length is between $x(1 - \beta)$ and $x(1 + \beta)$. Thus, the chance that two random fragments match is

$$\int_0^\infty \left[\int_{x(1-\beta)}^{x(1+\beta)} (\lambda e^{-\lambda y} dy)\right] \lambda e^{-\lambda x} dx = \frac{2\beta}{4 - \beta^2} \approx \frac{1}{2}\beta.$$

To be more precise, we could use the actual upper and lower limits for fragment sizes resolvable on the gel as the limits for the first integral. However, the simple approximation $\frac{1}{2}\beta$ is usually precise enough for use.

In comparing two fingerprints of type (a/b), there are n^2 ways to pick one fragment from each fingerprint, when each fingerprint consists of n fragments. Thus, the expected number of matching pairs will be $\frac{1}{2}\beta_1 n^2$. In fact, a proof can be obtained by using Poisson approximation (Problem 6.6).

In comparing two fingerprints of type (c), the situation is much more limited. The chance that two restriction maps will match in exactly k fragments is roughly $4\left(\frac{1}{2}\beta_2\right)^k$, as there are four orientations for the two maps and, once the orientations are fixed, at least the terminal k fragments in the two maps must match exactly. The chance that the maps will match in at least k fragments is then

$$4\left(\frac{1}{2}\beta_2\right)^k + 4\left(\frac{1}{2}\beta_2\right)^{k+1} + 4\left(\frac{1}{2}\beta_2\right)^{k+2} + \cdots = 4\left(\frac{1}{2}\beta_2\right)^k \left(1 - \frac{1}{2}\beta_2\right)^{-1}$$

$$\approx 4\left(\frac{1}{2}\beta_2\right)^k \left(1 + \frac{1}{2}\beta_2\right).$$

This completes the proof of Theorem 6.4. ∎

The two fingerprinting schemes produce roughly the same minimum detectable overlap $\theta \approx k/n$, but the rate of false positives is considerably higher for type (a/b) fingerprints than for type (c) fingerprints. [Indeed, the rate of false positives increases with n for type (a/b) fingerprints but is essentially independent of n for type (c).] In order to achieve a comparable rate of false positives, a larger value of k must be used for type (a/b) fingerprints, which would increase the overlap θ required for detection.

For example, suppose that we use an enzyme (or enzyme combination) that yields an average of 10 fragments per insert and gels that can resolve fragment lengths to within $\beta_1 = \beta_2 = 0.03$. If we simply measure fragments lengths and require that two such type (a) fingerprints match at 7 fragments, then the chance of a false positive will be about 0.0009 according to Theorem 6.4(i) and the minimum detectable overlap θ will be approximately $7/10 = 0.70$.

If we, instead, made a restriction map of these fragments, then the same false positive rate according to Theorem 6.4(ii) could be obtained by requiring that fingerprints share only two fragments in common. Thus, the minimum detectable overlap, yielding roughly the same false positive rate, would be $\theta \approx 2/10 = 0.20$, a substantial improvement in view of the results of Section 6.2. In general, the false positive rate when an overlap of two fragments is required is $4\left(\frac{1}{2}\beta_2\right)^2 = \beta_2^2$, which may be adequate for most purposes and yields $\theta = 2/n$.

One can clearly construct more detailed fingerprints by combining multiple fingerprints of type (a/b) or type (c) (as Kohara et al. did using eight different restriction maps), determined using separate gel lanes. If we require that each of the component fingerprints match in at least k fragments in order to declare an overlap, then the chance of a false positive is roughly the product of the false positive rates for each of the component fingerprints.

For example, if we wished to avoid constructing a restriction map in the situation described above, we could nevertheless achieve roughly the same false positive rate and the same minimum detectable overlap (of $\theta = 0.20$) by using multiple type (a) fingerprints involving a number of independent enzymes, each yielding about 10 fragments. Because each fingerprint would have a false positive rate of about 0.44 if overlaps were declared whenever fingerprints shared two bands, straightforward calculation shows that about nine independent enzymes are required. The choice of whether to construct a single restriction map (say, via partial digest in as in Kohara et al.) or to determine restriction fragment lengths for nine different enzymes would be governed in practice by an investigator's estimate of the work involved in each approach and the acceptable rate of false positives for the project. Also, note that the analysis above depends on the resolving power (β_1 and β_2) of the gels used.

Physical Genome Maps: Oceans, Islands and Anchors

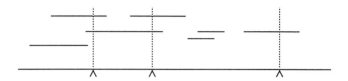

Figure 6.9: *Two anchored islands*

6.2 Mapping by Anchoring

There are alternate methods of physical mapping. Here, we consider linking random clones by a technique we will call anchoring. As before, we have a random genomic library of clones; the new feature is a random genomic library of so-called anchors. The anchor library contains very small genomic inserts that uniquely identify a genomic location which we model as points in the genome. Anchoring involves determining which clones contain a given anchor. The anchors link their associated clones into islands. Figure 6.9 shows a situation with seven clones and three anchors. There are three singleton unanchored islands, one singleton anchored island, and one island of three clones and two anchors.

6.2.1 Oceans, Islands and Anchors

We define the following symbols:

G = haploid genome length in bps,
L = clone insert length in bps,
N = number of clones,
M = number of anchors,
$c = NL/G$ = expected number of clones covering a random point,
$d = ML/G$ = expected number of anchors contained in a random clone.

Note that if we rescale clone length to $1 = L/L$ and genome length to $g = G/L$, then the rate of the Poisson processes remains the same, $c = NL/G = N/g$, and $d = ML/G = M/g$. As in the previous section, we model the process of right clone end locations by $\{A_i,\ i \in \{\ldots, -1, 0, +1, +2, \ldots\}\}$ so that $\cdots A_{-2} < A_{-1} < A_0 \leq 0 < A_1 < \cdots < A_N < g \leq A_{N+1} < \cdots$. As before, $N/g \to c$ as $g \to \infty$.

The process of anchors is described by another time homogeneous Poisson process with rate d, where the number of arrivals M in $(0, g)$ satisfies $M/g \to d$ as $g \to \infty$. We assume the anchor process and the clone process are independent.

Theorem 6.5 *With the above notation, we have the following:*

(i) *The probability q_1 that a clone contains no anchors is e^{-d}. The expected number of unanchored islands is Ne^{-d}.*

(ii) *The probability p_1 that a clone is the rightmost clone of an anchored island is*

$$p_1 = \begin{cases} \dfrac{d(e^{-c} - e^{-d})}{(d-c)} & \text{if } c \neq d, \\ ce^{-c} & \text{if } c = d. \end{cases}$$

The expected number of anchored islands is then Np_1.

(iii) *The expected number of clones in an anchored island is $(1 - q_1)/p_1$.*

(iv) *The probability p_2 that a clone is a singleton anchored island is*

$$p_2 = \begin{cases} \dfrac{d(c^2 - cd - d)}{(c-d)^2} e^{-(c+d)} + \dfrac{d^2}{(c-d)^2} e^{-2c} & \text{if } c \neq d, \\ \dfrac{2c + c^2}{2e^{2c}} & \text{if } c = d. \end{cases}$$

The expected number of singleton anchored islands is Np_2.

(v) *The expected length of an anchored island is λL, where*

$$\lambda = \begin{cases} \dfrac{(c-d)^2 e^{c+d} + c(cd - d^2 - c)e^c + d(2c - d)e^d}{cd(c-d)(e^c - e^d)} & \text{if } c \neq d, \\ \dfrac{2e^c + c^2 - 2c - 2}{2c^2} & \text{if } c = d. \end{cases}$$

(vi) *The expected proportion r_0 of the genome not covered by anchored islands is*

$$r_0 = \begin{cases} e^{-c} + \dfrac{c(d^2 - cd - c)}{(d-c)^2} e^{-(c+d)} + \dfrac{c^2}{(d-c)^2} e^{-2d} & \text{if } c \neq d, \\ \dfrac{2e^c + 2c + c^2}{2e^{2c}} & \text{if } c = d. \end{cases}$$

(vii) *The expected number of anchors in an anchored island is $d(1 - e^{-c})/cp_1$.*

(viii) *For any $x \geq 0$, the probability that an anchored island is followed by actual ocean of length at least xL is $e^{-c(x+1)}(1 - q_1)/p_1$. In particular, taking $x = 0$, the formula gives the probability that an anchored island is followed by an actual ocean rather than an undetected overlap.*

Proof. Now let N_u and N_a denote the number of unanchored (u) and the number of anchored (a) clones, respectively, in $(0, g)$ so that $N = N_u + N_a$. Define q_1 to be the probability a clone is unanchored. Now

$$q_1 = e^{-d}$$

because $q_1 = \mathbb{P}(\text{no anchor arrivals in } (0, 1))$. Therefore, the arrival intensity of unanchored clones is cq_1 and the arrival intensity of anchored clones is $c(1 - q_1)$. Also by the ergodic theorem, with probability 1

$$\frac{N_u}{g} \to cq_1 \quad \text{and} \quad \frac{N}{g} \to c, \quad \text{as } g \to \infty.$$

Therefore, $N_u \sim cgq_1 \sim Nq_1$ and (i) is shown.

As in the previous section, we now consider the process of right-hand ends of anchored islands. We label them in order by

$$\cdots C_{-2} < C_{-1} < C_0 \leq 0 < C_1 < \cdots < C_K < g \leq C_{K+1} < \cdots,$$

so that K is the number of anchored islands with right-hand ends in $(0, g)$.

Next, we calculate the intensity cp_1 of the process $\{C_j\}$. The value of p_1 is $\mathbb{P}(\text{clone ending at } t \text{ is the right-hand end of an anchored island} | \text{clone ends at } t)$. First, as in the earlier model, define $J(x) =$ the probability that two points x apart are not covered by a common clone. For $0 < x < 1$, this is equivalent to no arrivals in the interval of length $(1 - x)$ preceding the leftmost point, or no events in a Poisson of mean $c(1 - x)$. Therefore,

$$J(x) = \begin{cases} e^{-c(1-x)} & \text{if } 0 < x \leq 1, \\ 1 & \text{if } x > 1. \end{cases} \quad (6.1)$$

Returning to p_1, a given clone is rightmost in an anchored island if the distance V from the right-hand end to the nearest anchor to the left satisfies $V \leq 1$ and that no other clone spans this distance V. Therefore,

$$p_1 = \mathbb{E}(\mathbb{I}\{V \leq 1\} J(V))$$
$$= \int_0^1 de^{-dv} J(v) dv$$
$$= \begin{cases} \frac{d(e^{-c} - e^{-d})}{(d-c)} & \text{if } c \neq d, \\ ce^{-c} & \text{if } c = d, \end{cases}$$

and (ii) is established.

Now we turn to the expected number of clones in an anchored island. Let M_j be the number of clones in the j-th anchored island. $M_1, \ldots, M_K$ are defined for anchored islands with right-hand end in $(0, g)$. The average $\mathbb{E}(M)$ is established by a limit argument. First, define

$$\overline{M}_g = \frac{1}{K} \sum_{i=1}^K M_i.$$

There is a potential difficulty with islands that overlap 0 and g. Let $M'_g = \sum_{i=1}^{K} M_i - N_a$, where N_a is the number of clones in anchored islands entirely in $(0, g)$. Now $M'_g \leq M'' =$ number of clones in anchored islands that overlap 0 or g, so $\mathbb{E}(M'') < \infty$ and $M''/g \to 0$ as $g \to \infty$. Therefore,

$$\overline{M}_g = \frac{g}{K} \frac{1}{g} \sum_{i=1}^{K} M_i$$

$$= \frac{g}{K} \left\{ \frac{N_a}{g} + \frac{M'_g}{g} \right\}.$$

As $g \to \infty$, $K/g \to cp_1$ and $N_a/g \to c(1 - q_1) = c(1 - e^{-d})$, so that with probability 1,

$$\mathbb{E}(M) = \lim_{g \to \infty} \overline{M}_g = \frac{1 - q_1}{p_1}.$$

To be as clear as possible, M is a random variable with distribution function defined by the almost sure limit

$$F(t) = \lim_{g \to \infty} \frac{1}{K} \sum_{i=1}^{K} \mathbb{I}\{M_i < t\}.$$

This is part (iii) of our theorem.

Now we turn to the event E where a given clone C is a singleton anchored island. Let V be the distance from the left end to the first anchor in C and W be the distance from the right end to the first anchor in C. Now, if there is only one anchor, $W = 1 - V$. Given (V, W) and $V + W \leq 1$, we have E when no clones to the left of the leftmost end cover the anchor at distance V and no clones start within our clone to the left of the rightmost anchor. This conditional probability has value $J(V)J(W)$. The random variable (V, W) can be obtained from two iid exponentials (V', W') with parameter d by

$$(V, W) = \begin{cases} (V', W') & \text{if } V' + W' \leq 1, \\ (V', 1 - V') & \text{if } V' < 1 \text{ and } V' + W' > 1. \end{cases}$$

Therefore,

$$p_2 = \mathbb{E}\{\mathbb{I}\{V' < 1\} J(V') J(\min\{W', 1 - V'\})\}$$
$$= d^2 \int_0^1 \int_0^{1-v} e^{-d(u+v)} J(v) J(u) \, du \, dv$$
$$+ d \int_0^1 e^{-dv} J(v) J(1-v) e^{-d(1-v)} \, dv. \tag{6.2}$$

From Equation (6.2), part (iv) follows.

Physical Genome Maps: Oceans, Islands and Anchors

Next, we study the length of anchored islands. As before, set $S_i =$ length of the i-th anchored island and define $\mathbb{E}(S)$ to be the limiting average length of an island. Define

$$\overline{S}_g = \frac{1}{K}\sum_{i=1}^{K} S_i.$$

Note that a point in the genome can belong to at most two anchored islands, so

$$\frac{1}{g}\sum_{i=1}^{K} S_i \to r_1 + 2r_2, \qquad (6.3)$$

where r_i is the probability that a point is covered by exactly i anchored islands. As already remarked, $r_0 + r_1 + r_2 = 1$, so we calculate r_0 and r_2.

First, we calculate r_0. Take a point t in the genome and let W be the distance from t to the first anchor on the right and V the distance from t to the first anchor on the left. Let E be the event that t is not covered by any anchored island so that $\mathbb{P}(E) = r_0$. E occurs when clones starting to the left of $t - V$ end before t and when any clones starting in $(t - V, t)$ end before $t + W$. Given (V, W), then the first event has probability $J(V)$ and the second event has probability $J(W)/J(V + W)$. Clearly,

$$r_0 = \mathbb{E}\left(\frac{J(V)J(W)}{J(V+W)}\right). \qquad (6.4)$$

From Equation (6.4), part (vi) can be derived.

To obtain r_2, define E to be the event that t is covered by exactly two anchored islands. This event occurs when at least one clone starting to the left of $t - V$ ends in $(t, t + W)$, no clones starting to the left of $t - V$ end after $t + W$, and at least one clone starting in $(t - V, t)$ ends after $t + W$. Reasoning similar to the above gives

$$\mathbb{P}(E|V,W) = \left(1 - \frac{J(V)}{J(V+W)}\right) J(V+W) \left(1 - \frac{J(W)}{J(V+W)}\right)$$
$$= \frac{(J(V) - J(V+W))(J(W) - J(V+W))}{J(V+W)}.$$

and

$$r_2 = \mathbb{E}(\mathbb{P}(E|V,W)).$$

It follows from Equation (6.3) that

$$\overline{S}_g = \frac{g}{K}\frac{1}{g}\sum_{i=1}^{K} S_i \to \mathbb{E}(S) = \frac{r_1 + 2r_2}{cp_1} = \frac{1 - r_0 + r_2}{cp_1}$$

as $g \to \infty$. The equations for r_0, r_2 and $\mathbb{E}(S)$ show that

$$cp_1\mathbb{E}(S) = 1 - \mathbb{E}(J(V) + J(W) - J(V+W))$$
$$= 1 - 2\mathbb{E}(J(V)) + \mathbb{E}(J(V+W)) \qquad (6.5)$$

and now part (v) follows.

Now we look at $\mathbb{E}(H)$, the average number of anchors per anchored island. Let i index the i-th anchored island and R be the number of anchors in anchored islands $1, 2, \ldots, K$. Then

$$\mathbb{E}(H) = \lim_{g \to \infty} \frac{1}{K} \sum_{i=1}^{K} H_i = \lim_{g \to \infty} \frac{g}{K} \frac{R}{g}.$$

Recall the probability an anchor is not covered by an island is $J(0) = e^{-c}$, so that the intensity of anchors in islands is $d(1 - e^{-c})$. Therefore, $R/g \to d(1 - e^{-c})$ and part (vii) follows:

$$\mathbb{E}(H) = \frac{d(1 - e^{-c})}{cp_1}. \tag{6.6}$$

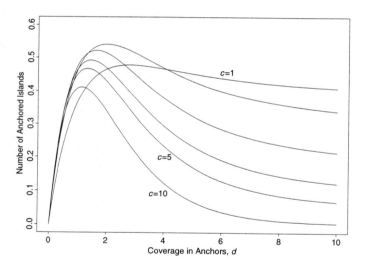

Figure 6.10: *Number of anchored islands for $c = 1, 2, 3, 4, 5, 10$*

Finally, we study ocean length. Suppose we have a clone with right-hand end located at t. Let V be the distance back from t to the first anchor. The conditional probability that t is at the right-hand end of an island, the island is anchored, and followed by an actual ocean of length at least k is equal to the probability that all clones that start to the left of t end to the left of t as long as $V < 1$ and is zero otherwise, multiplied by the probability of no clones starting in $(t, t + k)$. This probability, given V, equals

$$J(0)\mathbb{I}\{V < 1\}e^{-kc},$$

Physical Genome Maps: Oceans, Islands and Anchors

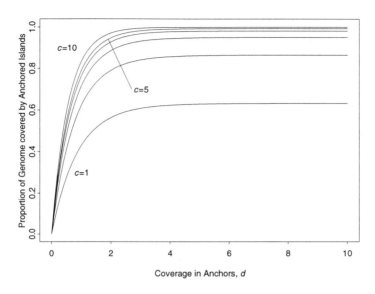

Figure 6.11: *Proportion of genome covered by anchored islands for $c = 1, 2, 3, 4, 5, 10$*

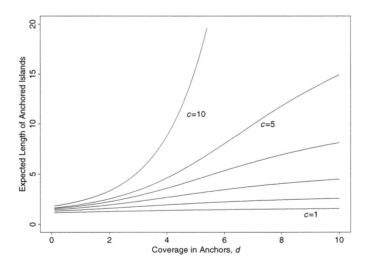

Figure 6.12: *Expected length of anchored islands for $c = 1, 2, 3, 4, 5, 10$*

so the desired probability is

$$\mathbb{E}(J(0)\mathbb{I}\{V \leq 1\}e^{-kc}) = e^{-c(k+1)}(1 - q_1) = e^{-c(k+1)}\mathbb{P}(V < 1).$$

We divide by the probability p_1 that a given clone is the end of an anchored island to obtain the conditional probability that an island is followed by an ocean of length k, given that the island is anchored. ∎

Corollary 6.1 *As $d \to \infty$, every island is anchored and all overlaps are detected.*

To illustrate numbers and lengths of anchored islands, along with coverage, See Figures 6.10–6.12.

6.2.2 Duality Between Clones and Anchors

There is a symmetry between clones and anchors when clones are fixed length (L). To see this, we represent both clones and anchors as points. Anchors are already points and we label a clone by its midpoint or center. Clearly, a clone overlaps an anchor if and only if the corresponding points are less than or equal to $L/2$ apart. The clone and anchor are then said to be *adjacent*. Adjacency then defines a bipartite graph where the connected components are

- isolated anchors
- isolated clones
- islands of at least one clone and one anchor.

Note that interchanging the labels of clone and anchor preserves the islands of anchored clones. There is some asymmetry between clones and anchors when we consider island length. Above, we measured length from the left end of the leftmost clone to the right end of the rightmost clone. In order to achieve symmetry, measure length from the center of the leftmost clone to the center of the rightmost clone. This is L smaller than the usual island length.

The next theorem formalizes some of the consequences of duality, where duality means interchanging the roles of clones and anchors.

Theorem 6.6 *With the above notation and definitions,*

(i) the expected number of clones in an anchored island is dual to the expected number of anchors in an anchored island;

(ii) the expected proportion of the genome not covered by anchored islands is dual to the probability that a clone lies in a singleton island.

Proof. To check (i), note that by Theorem 6.5(iii), the expected number of clones in an anchored island is
$$\frac{(1 - e^{-d})(d - c)}{d(e^{-c} - e^{-d})}.$$

By Theorem 6.5(vii), the expected number of anchors in an anchored island is

$$\frac{d(1 - e^{-c})(d - c)}{cd(e^{-c} - e^{-d})}.$$

These quantities are, indeed, equal by an interchange of c and d.

To show (ii), the probability of singleton islands equals the probability of an unanchored island (q_1) plus the probability of a singleton anchored island (p_2) or ($q_1 + p_2$). This is dual to the formula for r_0 in Theorem 6.5(vi).   ∎

6.3   An Overview of Clone Overlap

In this short section, we give a survey of techniques used to infer clone overlap. This is not an attempt to exhaust all possible methods but to give a more general picture of some methods already in use.

We began this chapter with a description of clone overlap by restriction fragment information. Each clone is digested and some subset of the fragment length data is obtained for what we called type (a/b) fingerprints. In addition, ordered fragment lengths – a restriction map – were used in type (c) fingerprints.

The next clone overlap technique we studied was anchoring, where the presence of a given anchor sequence could be determined for each clone. This is done by DNA-DNA hybridization and we now will call this method probe-clone hybridization. There are two subcases for probe-clone hybridization. The model in Section 6.2 was for the situation where the probes or anchors are unique in the genome, sequences that biologists call sequence tagged sites (STS). The presence of an STS in two clones unambiguously implies that the clones overlap in the genome.

The other case of probe-clone hybridization is to take probes that are not unique in the genome, that is, where a random clone C contains probe A with probability p. If there are M probes and N clones, then the number of probes hybridizing to clone C_i is a binomial random variable $Y_i = \text{Bin}(M, p)$. We can control the number of false overlaps in our model when we only declare overlap when $Z_{i,j} > t$, where

$$Z_{i,j} = \text{number of probes common to } C_i \text{ and } C_j.$$

When $C_i \cap C_j = \emptyset$, $Z_{i,j}$ is $\text{Bin}(M, p^2)$. From this it follows from results on large deviations (Section 11.1.6) that the expected number of (false) declared clone overlaps will be bounded above by

$$\binom{N}{2} e^{-M\mathcal{H}},$$

where $\mathcal{H} = \mathcal{H}\left(\frac{t}{M}, p^2\right) = \frac{t}{M} \log \frac{t/M}{p^2} + \left(1 - \frac{t}{M}\right) \log \frac{(1-t/M)}{(1-p^2)}$. Here, of course, $t > Mp^2$ is required.

For our last method of clone overlap we consider determining clone-clone overlap by hybridization. This sounds as if all $\binom{N}{2}$ clone-clone hybridization experiments are required but that can be avoided in the following way. First sequence the ends of each clone, 100–200 bases say. This gives N probes (by using the "ends" together) specific for the individual clones. Then each clone must be analyzed for the presence or absence of these probes. This can be accomplished by a clever technique known as *pooling*. Let A_i, $i = 1, \ldots, M_A$, be a set of distinct clones and $\cup_{i=1}^{M_A} A_i$ consists of all clones. Suppose B_j, $j = 1, \ldots, M_B$, is a different disjoint covering of the set of clones.

Each pool A_i and B_j can be analyzed in the following way. Arrange the N probes in a $\sqrt{N} \times \sqrt{N}$ grid, so that each probe has an (i, j) location. Form clone pools by pooling each row (A_i) and each column (B_j). Then hybridization of a clone pool to the grid will have the effect of being positive on the row or column with which the pool is associated. If we look at the grid of positive hybridization values for A_i and B_j, those (k, l) where positive hybridization occurs include (i, j) as well as (i', j') where clone (i', j') contains overlap with clone (i, j). See Figure 6.13.

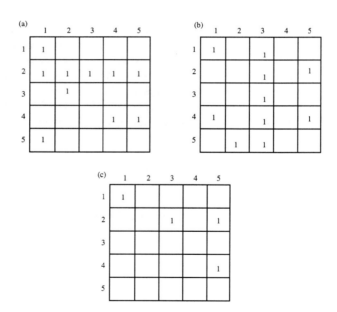

Figure 6.13: *Hybridization for (a) row 2, (b) column 3, and (c) the clones overlapping (2,3)*

Let us look at the design a little more closely. If $\mathcal{A} = \{A_1, \ldots, A_{M_A}\}$ and $\mathcal{B} = \{B_1, \ldots, B_{M_B}\}$ are disjoint covers of a set of N clones as described above,

we now require
$$|A_i \cap B_j| \leq 1 \quad \text{for all } i, j,$$
so that the overlap inference is guaranteed to be unique. The work of hybridization is $O(M_A + M_B)$.

Proposition 6.1 *Assume $\mathcal{A}$ and $\mathcal{B}$ are disjoint covers of sizes M_A and M_B with $|A_i \cap B_j| \leq 1$ for all i and j. Then $\min\{M_A + M_B\} = 2\sqrt{N}$.*

Proof. Note that $\cup_{i,j}(A_i \cap B_j)$ is the set of all N clones so that $|\mathcal{A}| \cdot |\mathcal{B}| = M_A \cdot M_B \geq N$. Therefore, $M_B \geq N/M_A$ and

$$M_A + M_B \geq M_A + \frac{N}{M_A},$$

and calculus shows the lower bound is minimized for $M_A = \sqrt{N} = M_B$. ∎

In fact, extending this strategy to $\sqrt[k]{N}$ clones per dimension in k dimensions suggests the question of the dimension that minimizes the number of pooled hybridizations:
$$\min_k k N^{1/k}$$
which occurs for $k = \ln(N)$.

6.4 Putting It Together

The previous sections of this chapter discussed expected map coverage and techniques for detecting clone overlap under some idealized conditions. What about determining the map from overlap data? The graphs for clone overlap maps are just interval graphs, so most of Chapter 2 applies here. If the overlap data are unambiguous and come from a linear (or circular) genome, then map assembly is just an interval graph problem and is easily solved. Of course, real data are almost never unambiguous! Even when the uniqueness assumptions of the STS/anchor sites are satisfied, errors in hybridization experiments will result in data that are not consistent with interval graphs. We will not pursue it here, but it should come as no surprise that while data can be checked to see if it is consistent with an interval graph in linear time, most generalizations of the assembly problem are NP-complete.

In this section, we will present some results for probe-clone mapping when probes are not unique in the genome. We will give some results characterizing clone layouts where we assume there are no errors in hybridization data.

The goal is to find the layout of clones on the real line that maximizes some objective function. First, we will discuss layouts without specifying an objective function. For a given layout, define an *atomic interval* as a maximal interval that contains no clone endpoint in its interior. The *height* of an atomic interval is the number of clones containing the interval. When an atomic interval has height 0 it is called a *gap*.

In what follows, we assume N fixed length clones have been placed in a layout in order (by left endpoints) $\pi_1, \pi_2, \ldots, \pi_N$. A layout is associated with a path through the $N \times N$ lattice $\{(i, j) : 1 \leq i, j \leq N\}$ in the following way. A cell (i, j) with $i \leq j$ corresponds to an atomic interval with clones $\pi_i \pi_{i+1} \cdots \pi_j$ and no others present. We connect (i, j) with $(i, j + 1)$ when we encounter the left end of clone π_{j+1} and we enter an interval with clones $\pi_i \pi_{i+1} \cdots \pi_{j+1}$ present. We connect (i, j) with $(i + 1, j)$ when we encounter the right end of clone π_i and we enter an interval with only clones $\pi_{i+1} \cdots \pi_j$ present. This means that gaps correspond to cells like $(j + 1, j)$. Figure 6.14 gives a simple illustration of a layout and its associated lattice path. We define a *lattice path* as a path from $(1, 1)$ to (n, n) where each edge is from a cell (i, j) to its right neighbor $(i, j + 1)$ or its lower neighbor $(i - 1, j)$ and where every cell (i, j) on the path satisfies $j \geq i - 1$.

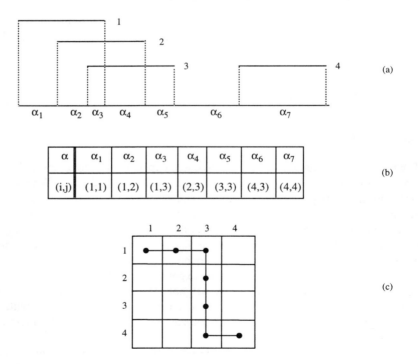

Figure 6.14: *A layout (a) along with the lattice path (b) and (c) associated with the atomic intervals $\alpha_1, \alpha_2, \ldots$ for the permutation $\pi = (1, 2, 3, 4)$*

Not only are there $N!$ orders of left endpoints, but for each ordering of left endpoints there are many layouts with different clone overlaps. Our probe data

Physical Genome Maps: Oceans, Islands and Anchors

$(d_{ij}) = D$ where $d_{ij} = \mathbb{I}\{\text{probe } j \text{ is contained in clone } i\}$ eliminates some of these layouts. Suppose that probe $j \notin C_{\pi(k)}$ and $j \notin C_{\pi(m)}$, but $j \in C_{\pi(l)}$ for $\pi(k) < \pi(l) < \pi(m)$. Then $C_{\pi(k)}$ and $C_{\pi(m)}$ cannot overlap in any layout consistent with π and D, because if they did, $C_{\pi(l)} \subset C_{\pi(k)} \cup C_{\pi(m)}$ and probe j would belong to at least one of $C_{\pi(k)}$ or $C_{\pi(m)}$. Therefore, all lattice paths passing through (k', m') with $k' \leq k < m \leq m'$ are inconsistent with D and such cells are called *excluded cells* under the data D. A layout is *consistent with the data* D if for each atomic interval each probe j is in all or none of all clones containing that interval. Implicit in our argument is that we have assumed probes to be points. We are not guaranteed to detect overlaps in this scheme that are smaller than the probe length.

Theorem 6.7 *A layout is consistent with the data D if and only if its lattice path does not pass through any cells excluded under the data D.*

Proof. The discussion preceding the theorem implies that lattice paths corresponding to a layout consistent with the data D do not pass through any excluded cells.

For the converse, we must show that any lattice path that does not pass through any excluded cells under the data D corresponds to a layout and probe placement consistent with D. Let $d_{ij} = 1$ and (k, l) be on the path with $k \leq i \leq l$, where by assumption, (k, l) is not an excluded cell. Suppose some clone in $\pi_k \cdots \pi_l$ does not contain probe j. All clones in this atomic interval that do not contain probe j must have right endpoints to the right or to the left of clone i, as otherwise (k, l) is excluded. Without loss of generality, we assume all are to the right of clone i.

Let clone k' be the rightmost clone in the interval (k, l) that does not contain probe j. Then, the first cell in row $k' + 1$ on the lattice path corresponds to an atomic interval where all clones do not contain probe j. All $d_{ij} = 1$ can be satisfied by placing probe j in this manner. See Figure 6.15. ∎

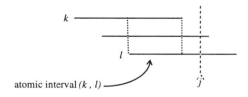

Figure 6.15: *Illustration for Theorem 6.7*

Theorem 6.7 tells us that any path that does not pass through a cell excluded under the data D corresponds to a layout consistent with the data D. In fact, it is possible for given π to find the layout and probe placement that gives the data D. Without loss of generality, $\pi = (1, 2, \ldots, N)$.

First of all we find all excluded cells. The data $d_{ij} = 1$ if and only if probe j is in clone i. When $d_{k-1,j} = 0$, $d_{k,j} = d_{k+1,j} = \cdots = d_{l,j} = 1$, and $d_{l+1,j} = 0$, cells (k', l') are excluded when $k' < k \leq l < l'$. To find all excluded cells, we first find such runs and then these "local" exclusions, which are determined by looking at each probe.

In Figure 6.16, an example with seven probes and seven clones is presented. The clones are indexed by left endpoints, A to G, while the probes are indexed by 1 to 7. In Figure 6.16(a) is the data D. From these data we can find the lower-leftmost excluded point for each run of 1's in the columns of D. These are marked by "x" in Figure 6.16(b). Finally, in Figure 6.16(c) is the entire excluded region marked by "x"'s along with a path corresponding to a consistent layout.

The layout in Figure 6.16(c) is special. It corresponds to the layout that has the maximum clone overlaps or minimum number of probes that are consistent with the data. See Problem 6.14.

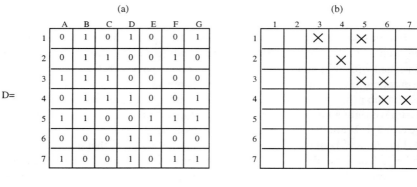

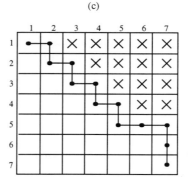

Figure 6.16: *Example of layout*

Problems

Problem 6.1 Find the value of c at which the maximum expected number of islands occurs (Theorem 6.1) and determine the value of the maximum.

Problem 6.2 Use Theorem 6.1 to derive the fraction f of clonable DNA in the Carbon-Clarke formula (Proposition 5.1). Carefully explain your derivation.

Problem 6.3 Suppose each of N clones of length L is sequenced and that the genome map proceeds according to Carbon-Clarke. Estimate the ratio of DNA sequenced to DNA mapped for large N.

Problem 6.4 Suppose for a clone mapping experiment of coverage c, that l bps are sequenced on each end of the input, with $2l \leq L$. What expected fraction of the genome will be sequenced in this way? If $l \ll L$, estimate the length of sequence islands and the size of the oceans between islands.

Problem 6.5 Assume we are mapping by fingerprinting with clones of length L requiring θ or more overlap. Show that a point in the genome can belong to as many as $1 + \max\{k : k \text{ integer}, k < 1/(1-\theta)\}$ islands.

Problem 6.6 Apply the Chen-Stein method of Poisson approximation from Chapter 11 to establish Theorem 6.4(i) with an explicit bound.

Problem 6.7 Two independent, homogeneous Poisson processes have rates λ_1 and λ_2, $\lambda_2 < \lambda_1$. Clones of length L are characterized by first cutting into fragments with process #1 and then with process #2. Two clones are declared to overlap if they share matching **landmark fragments**. A landmark fragment occurs when a process #1 fragment is cut by process #2. If both these fragments are cut by process #2 and the resulting restriction maps match fragment by fragment within $100 \times \beta\%$, clone overlap is declared. Find the expected number of landmark fragments that match between two nonoverlapping clones.

Problem 6.8 With $J(x)$ from Equation (6.1), use Equation (6.2) to establish Theorem 6.5(iv).

Problem 6.9 Use Equation (6.5) to establish Theorem 6.5(v).

Problem 6.10 Use Equation (6.4) to establish Theorem 6.5(vi).

Problem 6.11 Use Equation (6.6) to establish Theorem 6.5(vii).

Problem 6.12 In the discussion on clone pooling for an $\sqrt{N} \times \sqrt{N}$ grid of clones, rows and columns were pooled. Describe two other easily accomplished poolings, each with $(\sqrt{N} - 1)$ pools, where each pool has $(\sqrt{N} + 1)$ clones.

Problem 6.13 If in the proof of Theorem 6.7, all clones not containing probe j in the atomic interval (k, l) are to have right endpoints to the left of clone i, let k' be the rightmost such clone. Describe the location of the first atomic interval where all clones contain probe j.

Problem 6.14 For the example of Figure 6.16 (i), find the maximum overlap layout and place labeled probes on the layout. (ii) Find also the minimal overlap layout and place probes on that layout.

Problem 6.15 State and prove the results of Theorem 6.5 for the case where clone lengths L are iid random variables with mean $\mathbb{E}(L)$.

Chapter 7

Sequence Assembly

The fundamental value and role of DNA sequencing has frequently been emphasized in this book. The rapidly growing nucleic acid and protein sequence databases depend on the ability to read DNA. A Nobel prize was given to Gilbert and to Sanger in 1980 for their 1976 inventions of two methods of reading DNA sequences. At present, using the Gilbert or the Sanger method, it is routine to read a contiguous interval or string of DNA up to 450 basepairs in length. In Chapter 8 we will see that there are many sequenced DNAs of length 50,000 to over 300,000. The process of using the ability to read short substrings to determine strings that are 100 to 1000 times longer is quite involved. In Section 7.1, we will study the so-called shotgun sequencing method to accomplish this. Shotgun sequencing has been employed since the invention of rapid sequencing in 1976 and, although there are many methods that modify this essentially random approach, it is still widely used. Recently, there has been a proposal to develop sequencing by hybridization, a method that essentially uses the k-tuple content of a sequence as data for sequence determination. The computer science aspects of sequencing by hybridization which we discuss in Section 7.2 are quite distinct from that for shotgun sequencing. It is certain that other new techniques will be developed or significantly refined to make genome sequencing routine. It is less certain what role mathematics or computer science will play, but the excitement of this important enterprise is contagious.

7.1 Shotgun Sequencing

In this section we will use the ability to read a strand of DNA of length l, where usually $l \in [350, 1000]$. All procedures are based on the following. Identical copies of the same single stranded DNA are subject to four different reactions, one reaction for each base. Each reaction results in a collection of single stranded molecules, beginning at 5′ and ending at one of the bases in the sequence specific

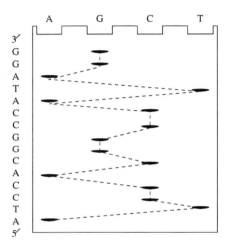

Figure 7.1: *Sequencing data for* ATCCACGGCCATAGG

to the reaction. Each molecule is labeled, sometimes at the 5' end and sometimes at the terminating base. The DNA from each reaction is put in a separate lane and the four lanes analyzed by electrophoresis. See Figure 7.1 and Section 2.4.

One of the complicating subtleties of the sequence assembly problem is that the orientation of a sequenced single stranded fragment f is unknown; that is, f could be either strand of the molecule we wish to sequence. Therefore, when comparing two fragments f_1 and f_2, they could be on the same strand, or it might be necessary to take the reverse complement of one of them, f_1^r, to have them on the same strand.

The basic shotgun sequencing problem has data generated from the molecule $\mathbf{a} = a_1 a_2 \cdots a_L$ to be sequenced. The data are the sequences of fragments $f_1, f_2, \ldots, f_N$ at random, unknown locations in $\mathbf{a}$. Our problem is to infer $\mathbf{a}$ from $\mathcal{F} = \{f_1, f_2, \ldots, f_N\}$.

In Section 7.1.2, we discuss the computational complexity of shotgun sequencing. Then, in Section 7.1.3, a greedy algorithm is shown to give a sequence at

Sequence Assembly

most four times the shortest possible sequence that contains all $f_i \in \mathcal{F}$. The usual approach to sequencing, allowing errors in f_i, is outlined in Section 7.1.3. In Section 7.1.4, we discuss how to estimate errors in the assembled sequence.

7.1.1 SSP is NP-complete

First we abstract and idealize our problem into the shortest common superstring problem (SSP). Here we have the set $\mathcal{F} = \{f_1, \ldots, f_N\}$ of fragments or strings and we wish to find a string S of minimal length (a shortest common superstring) such that f_i is a substring of S for all i. Here we do not have fragment reversals or errors in reading fragments.

For an example of SSP, consider the fragment set,

$$f_1 = \text{ATAT},$$
$$f_2 = \text{TATT},$$
$$f_3 = \text{TTAT},$$
$$f_4 = \text{TATA},$$
$$f_5 = \text{TAAT},$$
$$f_6 = \text{AATA}.$$

The shortest superstring S containing these f_i follows:

$$S = \text{TAATATTATA},$$

$$
\begin{array}{ll}
f_1 = & \text{ATAT}, \\
f_2 = & \text{TATT}, \\
f_3 = & \text{TTAT}, \\
f_4 = & \text{TATA}, \\
f_5 = \text{TAAT}, & \\
f_6 = & \text{AATA}.
\end{array}
$$

Theorem 7.1 (Gallant et al. 1980) *SSP is NP-complete.*

The proof that SSP is NP-complete does not give much insight into sequence assembly, and is not given here.

It is possible to introduce error into SSP in many ways. One way is by the *sequence reconstruction problem* (SRP) which is, given the fragment set $\mathcal{F}$ and an error rate $\epsilon \in [0, 1)$, to find a sequence S such that for all $f_i \in \mathcal{F}$ there is a substring $\mathbf{a}$ of S such that

$$\max_i \{\min\{d(\mathbf{a}, f_i), d(\mathbf{a}, f_i^r)\}\} \leq \epsilon |\mathbf{a}|.$$

In SRP, we have both reversible fragments and error. Of course, problem SRP is still NP-complete.

Theorem 7.2 *SRP is NP-complete.*

Proof. We restrict ourselves to the alphabet {A,C,G,T}. Start with an example SSP with $\mathcal{F}_{SSP} = \{s_1, s_2, \ldots, s_n\}$. We construct a SRP problem as follows. For each s_i, we create f_i by replacing every letter x by AAxCC. For example, if $s_i = $ ATG, then $f_i = $ AAACCAATCCAAGCC.

Note that $f = $ AAx_1CC$\cdots$AAx_nCC has $f^r = $ GGx_n^cTT$\cdots$GGx_1^cTT. Therefore f_i cannot overlap any f_j^r. Thus, the solution to SRP with $\epsilon = 0$ gives a superstring for $\mathcal{F}_{SSP}\{SSP}$. If any GGx^{c}TT occur in the SRP solution, simply replace x^c by x. Conversely any superstring of $\mathcal{F}_{SSP}$ gives a reconstruction for SRP with $\mathcal{F} = \{f\}$ defined above. The reconstruction has five times the length of the superstring. A polynomial time algorithm for SRP thus gives a polynomial time algorithm for SSP. ∎

7.1.2 Greedy is at most Four Times Optimal

In this section, we will study a greedy algorithm for superstring assembly. The greedy algorithm sequentially merges fragments, taking the pair with the most overlap first. It is possible to prove that greedy is at most 2.75 times optimal, although we do not present that proof here, and it is conjectured that greedy is at most 2 times optimal. Because many DNA sequence assembly algorithms use a greedy algorithm, it is interesting to study greedy algorithm behavior in the idealized setting of SSP.

Our set of strings is $\mathcal{F} = \{f_1, f_2, \ldots, f_N\}$. We assume $f_i \neq f_j$, $i \neq j$, and that no string f_i is a substring of f_j, all i, j. For strings f_i and f_j, let v be the longest string such that $f_i = uv$ and $f_j = vw$. $v = \emptyset$ is possible, whereas $u = \emptyset$ is not allowed. Then define the *overlap* $\text{ov}(i, j) = |v|$ and the prefix length $pf(i, j) = |u|$. For example, if $f_1 = $ ATAT and $f_4 = $ TATA, then $\text{ov}(1, 4) = 3$ and $pf(1, 4) = 1$, as

$$f_1 = \text{ATAT},$$
$$f_4 = \phantom{\text{AT}}\text{TATA}.$$

Although in this example $\text{ov}(1, 4) = \text{ov}(4, 1)$ this is not generally true.

The *prefix graph* $G = \{V, E, pf\}$ for the fragment $\mathcal{F}$ is an edge weighted, directed graph with N vertices, $V = \mathcal{F}$, N^2 edges $E = \{(f_i, f_j) : 1 \leq i, j \leq N\}$ and each edge (s_i, s_j) has weight $pf(i, j)$.

For illustration, we will use the set from Section 7.1.1 when $\mathcal{F} = \{f_1, \ldots, f_6\}$ with

$$f_1 = \text{ATAT},$$
$$f_2 = \text{TATT},$$
$$f_3 = \text{TTAT},$$
$$f_4 = \text{TATA},$$
$$f_5 = \text{TAAT},$$
$$f_6 = \text{AATA}.$$

Sequence Assembly

It will turn out that superstrings will correspond to cycles in the prefix graph. A cycle in G is a path with initial and terminal vertices equal. The cycle of Figure 7.2 corresponds to the substrings of Figure 7.3 or

$$
\begin{aligned}
f_1 &= \text{ATAT} \\
f_4 &= \text{TATA} \\
f_3 &= \text{TTAT} \\
f_2 &= \text{TATT} \\
f_1 &= \text{ATAT}
\end{aligned}
$$

Superstring $S = $ ATATTATAT

We have $|S| = 9$.

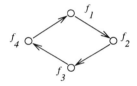

Figure 7.2: *Cycle of f_1, f_2, f_3, f_4*

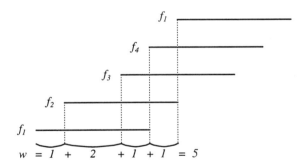

Figure 7.3: *Substrings of $f_1 f_2 f_3 f_4$*

Recall that a Hamiltonian cycle visits each vertex exactly once and that finding minimum weight Hamiltonian cycles is *NP*-complete. Define MWHC(G) to be the sum of the weights in a minimum weight cycle. This is a traveling salesman problem (TSP); see Section 4.4.2. OPT(s) is the minimum length superstring.

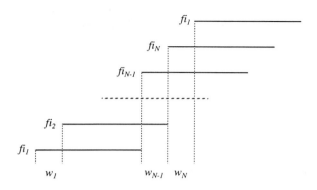

Figure 7.4: *Hamiltonian cycle*

Lemma 7.1 MWHC(G) $\leq \min\{|S| : S \text{ is a superstring for } \mathcal{F}\} = OPT(S)$.

Proof. The cycle achieving $MWHC(G)$ has sum of cycle weights less than or equal to that of the weighted Hamiltonian cycle for S which has weight $W = \sum_{j=1}^{N} w_i$ and $w_i=$ prefix weight (See Figure 7.4.) Each sequence is contained in this configuration of overlapped strings $f_{i_1}, f_{i_2}, \ldots, f_{i_N}$ (we do not include the second occurrence of f_{i_1}). Therefore, where $i_{N+1} \equiv i_1$,

$$\text{MWHC}(G) \leq W = \sum_{j=1}^{N} pf(i_j, i_{j+1}) = \sum_{j=1}^{N-1} pf(i_j, i_{j+1}) + pf(i_N, i_1)$$

$$\leq \sum_{j=1}^{N-1} pf(i_j, i_{j+1}) + pf(i_N, i_1) + \text{ov}(i_N, i_1)$$

$$= \sum_{j=1}^{N-1} pf(i_j, i_{j+1}) + |f_{i_N}| = |S|$$

∎

Now assume that we allow any number of cycles where each vertex is assigned to exactly one cycle and the total weight of cycles is minimized. Let the cycle cover (CYC) weight of G be defined by

$$\text{CYC}(G) = \sum_{i} \text{weight of cycle } i = \sum_{i} W_i.$$

Lemma 7.2 CYC(G) $\leq$ MWHC(G) $\leq$ OPT(S)

Computing CYC(G) is much easier than MWHC(G). In fact, the following greedy algorithm does this. The idea is that the two strings with maximum overlap are merged.

Algorithm 7.1 (Greedy)

1. $T \leftarrow \{f_1, f_2, \ldots, f_N\} S \leftarrow \emptyset$
2. while $T \neq \emptyset$, do
    for $s, t \in T$ with $\max\{ov(s,t)\}$ ($s = t$ possible)
    (a) if $s \neq t$ merge s and t to uvw;
    (b) if $s = t$ remove s from T; add s to S.
3. when $T = \emptyset$ output T the concatenation of strings in S.

Lemma 7.3 *Overlap between merged strings in algorithm greedy can be determined from overlap between the original strings.*

Proof. The lemma fails if and only if merging f_1 and f_2 (say) into $f = uvw$ has some f_i as a substring of f. As f is not a substring of f_1 or f_2 by assumption, it must include the overlap v between f_1 and f_2 and some of the prefix u. Therefore, $ov(f_1, f) > ov(f_1, f_2)$, which is a contradiction. ∎

We now apply the algorithm greedy to our example data $\mathcal{F} = \{f_1, \ldots, f_6\}$. Note that the merged strings can be indexed by the sequence of original strings. To begin, we compute the data $ov(f_i, f_j)$.

	1	2	3	4	5	6
1	2	3	1	3	1	0
2	0	1	2	1	1	0
3	2	3	1	3	1	0
4	3	2	0	2	2	1
5	2	1	1	1	1	3
6	3	2	0	2	2	1

1. There are several i, j with $ov(f_1, f_j) = 3$. We arbitrarily choose f_1 and f_4. Merge f_1 and f_4 to obtain ATATA.

2. Because $ov(f_6, f_1) = 3$, merge f_6 and $f_1 f_4$ to obtain $f_6 f_1 f_4 =$ AATATA.

3. Because $ov(f_5, f_6) = 3$, merge f_5 and $f_6 f_1 f_4$ to obtain $f_5 f_6 f_1 f_4 =$ TAATATA.

4. The last (accessible) length 3 overlap is $f_3 f_2$, which we merge to obtain $f_3 f_2 =$ TTATT.

5. The overlaps come from the ends: $ov(f_2, f_5) = 1$, $ov(f_4, f_3) = 0$, and the self overlaps: $ov(f_4, f_5) = 2$ and $ov(f_2, f_3) = 2$.
   Merge f_4 and f_5 which are two ends of the same string:
   $$S \leftarrow \{f_5 f_6 f_1 f_4\} \bigcup \emptyset.$$

6. Merge f_2 and f_3 which are two ends of the same string:

$$S \leftarrow \{f_3 f_2\} \bigcup \{f_5 f_6 f_1 f_4\},$$
$$T = \emptyset.$$

7. Concatenate strings in T:

$$T = \text{TTATTTAATATA}$$

The idea is that each superstring in T is a cycle in the graph G. When we form our superstring for the output, we break cycles and concatenate.

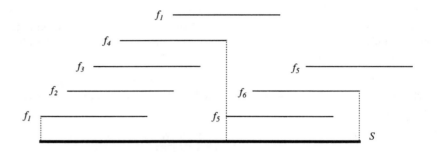

Figure 7.5: *Concatenating cycles C_1 and C_2*

In Figure 7.5 the cycles are $C_1 = f_1 f_2 f_3 f_4 f_1$ and $C_2 = f_5 f_6 f_5$. If we set $l_i = \max\{|f_j| : f_j \in \text{cycle } C_i\}$, and w_i is the weight of cycle i, then

$$|S| \leq \sum_i (l_i + w_i)$$

A string $s = s_1 s_2 \cdots s_n$ has *period* p if s is a substring of $s_1 s_2 \cdots s_p s_1 s_2 \cdots s_p \cdots s_1 s_2 \cdots s_p = (s_1 s_2 \cdots s_p)^k$ for some k. The following lemma is useful in the proof of Theorem 7.3.

Lemma 7.4 *If a string s has two periods of length p_1 and p_2 and $|s| \geq p_1 + p_2$, then s has a period of length $\gcd(p_1, p_2)$.*

Proof. It can be shown that, when $p_1 < p_2$, the string has a period of length $p_2 - p_1$. Applying Euclid's algorithm gives us that s has a period of length $\gcd(p_1, p_2)$. ∎

Theorem 7.3 *If the greedy algorithm produces superstring T, then*

$$|T| \leq 4 \, \text{OPT}(S).$$

Proof. To begin we note that for every cycle in our graph G, the cycle weight w is equal to the period length p of a superstring associated with the cycle. Clearly the cycle weight w which is the number of letters until the repetition of the first string is a period of T. If there were a smaller period, that would contradict the construction of T.

Let C_1 and C_2 be two disjoint cycles in G with $f_1 \in C_1$ and $f_2 \in C_2$. Then the overlap u of f_1 and f_2 has length $|u| \leq w_1 + w_2$. To see this, note that f_i has period w_i. If $|u| > w_1 + w_2$, then f_1 and f_2 and, hence, all strings in C_1 and C_2 should be in the same cycle of weight $\gcd(w_1, w_2)$, which contradicts the minimality of the cycles.

Take the longest string from each cycle and merge them optimally, say in order $l_1 l_2 \cdots l_k$. Each adjacent pair l_i, l_{i+1} cannot overlap by more than the cycle weights $w_i + w_{i+1}$, so total overlap is bounded by $2 \sum_i w_i$. The merged string has length L. Then

$$L \geq \sum_i l_i - 2 \sum_i w_i = \sum_i (l_i - 2w_i),$$

so

$$\text{OPT}(S) \geq L \geq \sum_i (l_i - 2w_i).$$

Finally,

$$|T| \leq \sum_i (w_i + l_i)$$
$$= \sum_i (l_i - 2w_i) + \sum_i 3w_i$$
$$\leq \text{OPT}(S) + 3\,\text{OPT}(S) = 4\,\text{OPT}(S).$$

■

7.1.3 Assembly in Practice

Real DNA sequence assembly has several problems that the idealized superstring problems do not. There is the issue about whether f_i or f_i^r is the proper choice. More problematic is the fact that real sequence data comes with errors, mismatches, and insertions or deletions (indels). Several strategies have been developed for sequence assembly, but all have the following outline:

- fragment overlap statistics
- fragment layout or approximate alignment
- final multiple alignment

For the first step, fragment overlap statistics, we must calculate how a fragment f_i overlaps f_j (or f_j^r). This is to put the fragment pair into an alignment and calculate an alignment score or likelihood for the alignment. The computational cost is $\binom{N}{2}$ times the cost of an individual alignment. Holding f_i fixed in orientation, f_j and f_j^r can align with f_i in the following ways:

```
     fi ─────────              fi       ─────────
     fj       ─────────        fj ─────────

         fi ─────────              fi ─────────
     fj ─────────                       fj ─────────
```

There is also the possibility of no overlap

```
     fi ─────────              fj ─────────
```

Of course, f_j^r can replace f_j, doubling the above possible relationships between two fragments.

Overcap can be evaluated by an alignment score (See Chapter 9). Equal aligned letters receive positive scores while unequal aligned letters receive negative score. Inserted or deleted letters (indels) receive negative score. $S(\mathbf{a}, \mathbf{b})$ is the maximum score over all alignments of $\mathbf{a}$ and $\mathbf{b}$. We let $s(\mathbf{a}, \mathbf{b})$ be a similarity measure on $\{A,C,G,T\}$, and $g(k) = k\delta$ to be the indel penalty. Define

$$A(\mathbf{a}, \mathbf{b}) = \max \left\{ S(a_k a_{k+1} \cdots a_i, b_l b_{l+1} \cdots b_j) : \begin{cases} 1 \leq k \leq i \leq n, \\ 1 \leq l \leq j \leq m, \\ \text{and at least one of} \\ j = n \text{ or } l = m \text{ holds} \end{cases} \right\}.$$

A dynamic programming algorithm finds $A(\mathbf{a}, \mathbf{b})$. An array of related algorithms is presented in Chapter 9. We only present the algorithm here and refer the reader to Chapter 9 for a more complete treatment. Define

$$A(i, j) = \max\{S(a_k a_{k+1} \cdots b_l b_{l+1} \cdots b_j) : 1 \leq k \leq i \text{ and } 1 \leq l \leq j\}.$$

Algorithm 7.2 (Overlap)
```
input:   a, b, s(.,.), δ
output:  A(i, j)
         A(0, j) = A(i, 0) ← 0 for i = 1, ..., n; j = 1, ..., m
         for i = 1, ..., n
           for j = 1, ..., m
                            ⎧ A(i − 1, j) − δ,
             A(i, j) ← max ⎨ A(i, j − 1) − δ,
                            ⎩ A(i − 1, j − 1) + s(aᵢ, bⱼ) ⎭
         end
     end
```

Sequence Assembly

This is like the local alignment algorithm but does not have a 0 in the recursion. (See Chapter 9.) The score for the best alignment for all four overlap possibilities is obtained by looking on the borders:

$$A(f_i, f_j) \leftarrow \max\{A(i, |f_j|), A(|f_i|, j); 1 \leq i \leq |f_i|, 1 \leq j \leq |f_j|\}.$$

These alignments correspond to the overlaps of f_i and f_j represented above.

For the second step, fragment layout, we must arrange the fragments into an approximate alignment. We describe a greedy algorithm to do this. Fragments contained in others are first identified and associated with the fragment into which they best fit. For each remaining (noncontained) pair we must choose an orientation from

$$\max\{A(f_i, f_j), A(f_i, f_j^r)\}.$$

Requiring $A(f_i, f_j) \geq C$ ensures high quality overlaps. The first pair we layout is

$$\max_{i,j}\{\max\{A(f_i, f_j), A(f_i, f_j^r)\}\}.$$

As the best overlaps are found, that pair of fragments are put into contigs or islands. The location or actual overlap of the fragments is **not** the score we use. Note that relative orientation is set at this point. When two contigs are put into a large contig, the overall relative orientation is easy to find.

It is possible to make errors in a greedy approach and schemes such as stochastic annealing can be used to probe other alternate assemblies. Clearly this is a version of the Hamiltonian path problem from Sections 7.1.1 and 7.1.2 where we want to visit all vertices at minimum cost.

Although we have given a brief sketch of practical sequence assembly, anyone who has read the multiple alignment chapter (Chapter 10) will realize that this is a very complex problem. Sequence assembly is multiple alignment with new degrees of freedom as to where the fragments belong. The several packages to accomplish this problem have not yet proved to be robust for the size and complexity of data coming from sequencing laboratories.

7.1.4 Sequence Accuracy

It is natural to ask how accurate assembled DNA sequences are. Most of the information about this important topic comes from sequence that is determined in different laboratories and then compared. Still, sequence assembly is basically a random process and we should be able to make a statistical statement about the finished product. What we present in this section is an estimate of the probability that the i-th letter of an assembled sequence is A, C, G, T or $-$. It is necessary to make a number of assumptions.

The key assumption is that the fragment assembly is correct. This is unrealistic in practice, but the reader should view this section as giving a first analysis of a difficult problem. It is also assumed that all fragments and indeed all portions of each fragment are equally reliable. As we know that sequence is frequently less reliable as we move along the gel, this, too, is a simplifying assumption. Also, we assume sequencing errors are independent of their local context and that they occur at constant rate across the entire sequence. Finally, we assume the sequence is composed of iid letters.

Now we set the notation for our problem. As usual, $\mathcal{F} = \{f_1, f_2, \ldots, f_N\}$ are the set of fragments. These fragments are aligned into an assembled sequence which corresponds to columns i in a matrix and the fragments corresponding to rows j. The matrix is conveniently filled in by $\emptyset$ where the fragments do not align; that is, each element x_{ij} in our matrix is a member of $\mathcal{B} = \{A, C, G, T, -, \emptyset\}$, where $\{A,C,G,T,-\}$ indicates an aligned fragment and $\emptyset$ is to fill in the beginning and end of the row. The true sequence corresponding to the n columns is $s = \{s_1, s_2, \ldots, s_n\}$, where $s_i \in \{A, C, G, T, -\} = \mathcal{A}$. It is necessary to keep track of fragment orientation because specific sequencing errors can depend on orientation. This is done by

$$r_j = \begin{cases} 0 & \text{fragment } j \text{ as is,} \\ 1 & \text{fragment } j \text{ is reverse complemented.} \end{cases}$$

Our goal is to estimate

$$\pi_i(a) = \mathbb{P}\left(s_i = a | x_{i,j}, j = 1, \ldots, N\right).$$

We first assume knowledge of the sequencing error rates:

$$p(b|a) = \mathbb{P}(x_{ij} = b | s_i = a), a \in \mathcal{A}, b \in \mathcal{B}.$$

A simple application of Bayes rule then gives the next equation. With the above notation,

$$\pi_i(a) = \frac{p(a) \prod_{j=1}^{N}[(1 - r_j)p(x_{ij}|a) + r_j p(x_{ij}^c | a^c)]}{\sum_{b \in \mathcal{A}} p(b) \prod_{j=1}^{N}[(1 - r_j)p(x_{ij}|b) + (r_j p(x_{ij}^c | b^c)]}. \quad (7.1)$$

As we might not have good estimates of $p(b|a)$, we now give an algorithm to estimate $\pi_i(a)$ when $p(b|a)$ is unknown. The algorithm is a special case of the EM (expectation maximization) algorithm. If the true DNA sequence were known, it is a simple matter to estimate composition and error rates. The base a occurs

$$n_a = \sum_{i=1}^{n} \mathbb{I}(s_i = a)$$

times in s. The number of times a was recorded as b in a fragment is

$$n_{ab} = \sum_{i=1}^{n} \sum_{j=1}^{N} [(1 - r_j)\mathbb{I}(x_{ij} = b)\mathbb{I}(s_i = a)\mathbb{I} + r_j \mathbb{I}(x_{ij}^c = b)\mathbb{I}(s_i^c = a)] \quad (7.2)$$

for all $a \in \mathcal{A}$ and $b \in \mathcal{B}$. Maximum likelihood estimates of the base composition and error rate parameters are given by

$$\hat{p}(a) = \frac{n_a}{n}, \tag{7.3}$$

$$\hat{p}(b|a) = \frac{n_{ab}}{n_a}. \tag{7.4}$$

This situation suggests the following algorithm for the simultaneous estimation of the error rates and the distribution π.

Algorithm 7.3 (Accuracy)

1. Initialize the consensus distribution. Set $\pi_i(x) = 1.0$, where x is the most frequently occurring letter at column i.
2. Estimate $p(a)$ and $p(b|a)$ for all a, b. Set the counts n_a and n_{ab} equal to their conditional expected values

$$\hat{n}_a = \sum_{i=1}^{N} \pi_i(a),$$

$$\hat{n}_{ab} = \sum_{i=1}^{n} \sum_{j=1}^{N} \left[(1 - r_j) \mathbb{I}(x_{ij} = b) \pi_i(a) + r_j \mathbb{I}(x_{ij}^c = b) \pi_i(a^c) \right]$$

and estimate $p(a)$ and $p(b|a)$ as before [Equations (7.3) and (7.4)].
3. Recompute $\pi_i(a)$ for all i and a according to Equation (7.1), with $p(a)$ and $p(b|a)$ replaced by their current estimates.
4. Continue. If the changes in $\hat{p}(b|a)$ and $\hat{p}(a)$ are less than ϵ, for all a and b, stop, otherwise go to step 2.

7.1.5 Expected Progress

A reasonable stochastic model for sequence assembly is closely related to the ocean and islands model for physical mapping by fingerprinting random clones (Chapter 6). Here we have N fragments of length l located at random in a longer sequence of length L. The model is that left ends of fragments occur according to a Poisson process at rate N/L. The fraction overlap required to establish overlap is θ. This fraction is small. For example, 20 bps might be determined adequate to establish overlap, so $\theta = \frac{20}{l} = \frac{20}{350}$ is a typical value of θ.

Clearly, Theorem 6.1 applies directly. Relevant earlier results are collected in the next theorem. The fraction of the $[0, L]$ covered by at least k fragments is useful for estimating the underlying sequence and for understanding the rate of progress. Note that we assume $l \ll L$ and that effects from the two sequence ends are assumed to be negligible.

Theorem 7.4 *For a shotgun sequencing project using the above notation and assumptions, let $c = Nl/L$. Then:*

(i) *The fraction of $[0, L]$ covered by k fragments is*
$$\frac{e^{-c}c^k}{k!}.$$

(ii) *The expected number of apparent sequence islands is $Ne^{-c(1-\theta)}$.*

(iii) *The expected length in bps of an apparent island is*
$$l\left\{\left(e^{c(1-\theta)} - 1\right)/c + \theta\right\}.$$

(iv) *The probability that there is at least xl bps between apparent sequence islands is $e^{-c(x+\theta)}$.*

Proof. Only (i) is not already stated in Theorem 6.1. Consider $t \in (0, L)$. Then t is covered by k fragments if and only if $K = k$ left clone ends occur in $[t - l, t]$. Therefore the depth K is $\mathcal{P}\left(\frac{N}{L}l\right) = \mathcal{P}(c)$. If $X_k(t) = \mathbb{I}(t \text{ is covered by } k \text{ fragments})$,

$$\frac{1}{G}\int_0^G X_k(t)dt \to \mathbb{E}(X_k) = \frac{e^{-c}c^k}{k!}.$$

∎

It is of interest to know the largest distance between sequence islands ($\theta = 0$) because the experimentalist will switch from shotgun sequencing when that distance is small enough. There are Ne^{-c} oceans, each ocean length Y_i is exponential with distribution function $1 - e^{-cx}$ and mean $1/c$. As will be shown in Chapter 11, $\mathbb{E}\left(\max_{1 \leq i \leq Ne^{-c}} Y_i\right) \approx \frac{1}{c}\log(Ne^{-c}) = \frac{\log N}{c} - 1 = \frac{L \log N}{lN} - 1$. More precise details about this distribution appear in Chapter 11.

7.2 Sequencing by Hybridization

Recently, a new approach to sequencing DNA was proposed, sequencing by hybridization (SBH). In the language of this book, the method might also be called sequencing by k-tuple composition. The idea is to build a two-dimensional grid or matrix of all k-tuples. At each (i, j) is attached a distinct k-tuple or probe. This matrix of probes will be referred to as a sequencing chip. Then a sample of the single stranded DNA to be sequenced is presented to the matrix. This DNA is labeled with a radioactive or fluorescent material. Each k-tuple present

Sequence Assembly

AAA	ACA	AGA	ATA
AAC	ACC	AGC	ATC
AAG	ACG	AGG	[ATG]
AAT	ACT	AGT	ATT
CAA	CCA	CGA	CTA
CAC	CCC	[CGC]	CTC
CAG	[CCG]	CGG	CTG
CAT	CCT	CGT	CTT
GAA	[GCA]	GGA	GTA
GAC	[GCC]	GGC	GTC
GAG	GCG	GGG	[GTG]
GAT	GCT	GGT	GTT
TAA	TCA	TGA	TTA
TAC	TCC	[TGC]	TTC
TAG	TCG	TGG	TTG
TAT	TCT	[TGT]	TTT

Figure 7.6: *The matrix of 3-tuples and hybridization results*

in the sample is hybridized with its reverse complement in the matrix. Then, when unhybridized DNA is removed from the matrix, the hybridized k-tuples can be determined with a device detecting the labeled DNA. In Figure 7.6, we present a grid of the $4^3 = 64$, 3-tuples along with the hybridization results from a = ATGTGCCGCA.

There are some technical difficulties with this method, both experimental and mathematical. A major experimental difficulty results from the hybridization between k-tuples. Some k-tuples, such as those with high G-C content, hybridize more strongly than others. In addition, there can be nonperfect hybridization involving mismatches. All this contributes to errors in the SBH data, where k-tuples in the sequence are not hybridized. For this section, we assume perfect hybridization data. Another experimental implication for SBH data is that multiplicities are ignored. When a k-tuple occurs more than once, we can only learn that it occurred at least once. So far, SBH is not a practical method to sequence DNA.

The mathematical aspects of SBH are also nontrivial. Note that if k is too small, for example, the SBH data is too clumped to find the sequence. It is, for example, not useful to learn that all sixteen 2-tuples are present in a DNA sequence 100 letters in length. Usually in these cases almost all k-tuples will have a hybridization signal, but the data will not allow us to resolve the sequence. Obviously, it is an advantage to take k as large as possible. In fact, if the sequence to be determined is of length n, it would be ideal to have a grid of all 4^n n-tuples.

Then the sequence could be read from one hybridization on the matrix. Obviously, this is experimentally impractical, and our goal is to resolve as much sequence as we can from our k-tuple matrix. Biologists work on increasing the feasible size of k; currently, $k = 8$ has been constructed and perhaps $k = 10$ can be obtained.

To summarize, we have data $\mathbb{I}_w = 1$ if w is a substring of a and $\mathbb{I}_w = 0$ if w is not, for each k-tuple w. Hybridization data is nonrandom and of quite distinct character. We build a graph from the set of k-tuples of a defined to be the *spectrum* $S(\mathbf{a}) = \{w : w = s_i s_{i+1} \cdots s_{i+k-1}, 1 \leq i \leq n + 1 - k\}$. Here, we make $S(\mathbf{a})$ a multiset so that $|S(\mathbf{a})| = n - k + 1$. The first graph we consider is H with vertex set $S = V_H$ and directed edges between $u, v \in S$ if the last $k - 1$ letters of u are equal to the first $k - 1$ letters of v. For example, the sequence $\mathbf{a} = $ ATGCAGGTCC has vertex set ATG, TGC, GCA, CAG, AGG, GGT, GTC, TCC, and the graph H in Figure 7.7. The motivation for the graph is that an edge between u and v accounts for the adjacent overlapping occurrence of u and v in the DNA sequence. A Hamiltonian path visiting all vertices gives a solution to the SBH assembly problem.

In Figure 7.7, the Hamiltonian path is unique. However, in Figure 7.8, we show another graph for the sequence $\mathbf{a} = $ ATGTGCCGCA. There are several alternative branchings and we are faced with the *NP*-complete problem of finding Hamiltonian paths.

An alternate data structure changes the basic graph problem from a search for Hamiltonian paths to a search for Eulerian paths. There are efficient algorithms to find Eulerian paths and necessary and sufficient conditions for the existence of Eulerian paths. The directed graph G has as vertex set V_G, the set of $(k-1)$-tuples from the spectrum, where each k-tuple of the spectrum contains two $(k-1)$-tuples. The $(k - 1)$-tuple u is joined by a directed edge to v if the spectrum S contains a k-tuple whose first $(k - 1)$-tuple is u and second $(k - 1)$-tuple is v. There can be

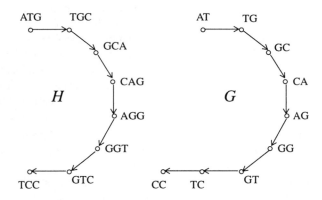

Figure 7.7: *Graphs H and G for* ATGCAGGTCC

Sequence Assembly

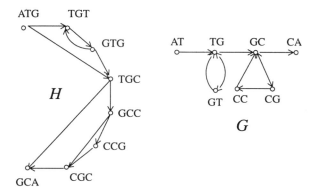

Figure 7.8: *Graphs H and G for* ATGTGCCGCA

multiple edges in this graph and V_G is **not** a multiset. Examples of the directed graph G is shown in Figure 7.7 and Figure 7.8. Note that the graph G in Figure 7.8 is much simpler than graph H.

Euler's theorem for directed graphs gives conditions for the existence of an Eulerian path. Define for vertex v

$$\text{in}(v) = \text{indegree}(v)$$
$$\text{out}(v) = \text{outdegree}(v)$$
$$\text{d}(v) = \text{in}(v) - \text{out}(v)$$

Label the starting vertex s and the terminal vertex t. There is an Eulerian path if and only if

$$\text{in}(v) = \text{out}(v) \quad \text{for } v \neq s, t,$$
$$\text{out}(s) - \text{in}(s) = 1,$$
$$\text{out}(t) - \text{in}(t) = -1.$$

The key turns out to be the intersection graph of cycles in G. We now define this graph. First, add an arc from t to s, so that our interest is now in Eulerian cycles. Then decompose G into simple cycles: $v_{i_1} \to v_{i_2} \ldots v_{i_k} = v_{i_1}$, where no $v_i = v_j$ except for $v_{i_k} = v_{i_1}$. An edge can be used in at most one cycle C, but vertices can be used arbitrarily many times. For these cycles, define the intersection graph G_I of the cycles $C_1, C_2, \ldots, C_l$ where if cycles C_i and C_j have l vertices in common, connect them by l edges in G_I.

Returning to graph G in Figure 7.8; the simple cycles look like

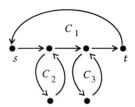

so the graph G_I is

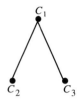

which is a tree; that is, G_I has no cycles. It turns out that this property holds in general. The next theorem holds for general graphs G.

Theorem 7.5 *The intersection graph G_I of simple cycles from G is a tree if and only if there is a unique Eulerian cycle in G.*

Proof. Let G_I be a tree with n vertices. The proof is by induction. If G_I contains one vertex, then G has one cycle and the theorem is true. Assume the statement is true for all trees up to $n - 1$ vertices. Consider a leaf corresponding to a cycle C in the n vertex tree G_I. C has only one vertex v in common with the graph G' with the cycle C removed. Clearly, the graph G' is a tree with $n - 1$ vertices and by induction has a unique Eulerian cycle E passing through v. This implies that the unique Eulerian cycle in G begins at v, passes through E, back to v, through C, and ends at v.

Let G have a unique Eulerian cycle. We prove G_I is a tree. Assume G_I is not a tree and, therefore, contains a cycle. The cycle with k vertices in G_I corresponds to k cycles in G. See Figure 7.9. It is easy to see that combining these cycles in G we have at least two Eulerian cycles, obtaining a contradiction. ∎

To apply Theorem 7.5 to SBH data, we need to transform the SBH graph G in a simple way. To see the necessity for this, consider the case where a particular k-tuple is repeated twice (and no other k-tuple repeats exist). As the first repeat can be uniquely identified with s and the second with t, there is a unique Eulerian cycle. However, the cycle graph is

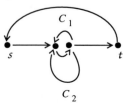

Sequence Assembly

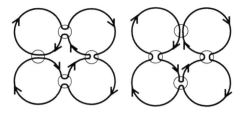

Figure 7.9: G where G_I has $k = 4$ cycles

where two vertices with $\text{in}(v) = \text{out}(v) = 2$ are the repeated $(k-1)$-tuples, and the lower arc of C_2 represents the path between the k-tuple repeats. The intersection graph is

which is not a tree. To fix this difficulty we recursively merge all vertices v_i and v_j into v_i^* where all arcs entering v_j are those leaving from v_i and $\text{in}(v_j) = \text{in}(v_i)$. If the new graph is G^*, we note that G^* has an Eulerian path from s to t (or whatever vertex s and t are mapped to) if and only if there is an Eulerian path from s to t in the original SBH graph G. Our example becomes

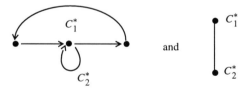

If the k-tuple repeat occurs three times, then the graph G^* has cycles

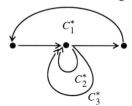

and G_I^* is not a tree:

7.2.1 Other SBH Designs

The matrix of probes is called a sequencing chip. Thus far, we have studied the classical chip $C(k)$ of all k-tuples. It is, of course, possible to look at other possibilities than all k-tuples and ask what is the best design. First, we set a criterion for the design.

For the sequence $\mathbf{s} = s_1 s_2 \cdots s_m s_{m+1} \cdots s_n$, we assume the first m letters have been determined. We will estimate the probability $\mathbb{P}(U(m,n))$ of unambiguously extending the sequence $s_1 s_2 \cdots s_m$ in one letter to the right. In typical statistical style, we will require for a small probability α that

$$1 - \mathbb{P}(U(m,n)) \leq \alpha.$$

Our criteria of goodness will be

$$n_{\max} = \max\{n : 1 - \mathbb{P}(U(m,n)) \leq \alpha\}.$$

This is just the largest n such that the extension probability is small. Below we assume the sequence letters are iid uniform.

We will consider four different sequencing chips. The chip $C(k)$ has already been described. The *capacity* $\|C\|$ of a chip is the number of probes. $\|C(k)\| = 4^k$. Recall that R denotes purines A,G and Y denotes pyrimidines T,C. In addition W (weak) denotes A,T whereas S (strong) denotes C,G. X denotes an arbitrary letter A, C, G, or T. Our sequencing chips will now have a pool of probes at each position.

• The *binary* chip $C_{\text{bin}}(k)$ is composed of all probes in the sets

$$\{\text{W},\text{S}\}\{\text{W},\text{S}\}\cdots\{\text{W},\text{S}\}\{\text{A},\text{C},\text{G},\text{T}\} = \{\text{W},\text{S}\}^k\{\text{A},\text{C},\text{G},\text{T}\}$$

and

$$\{\text{R},\text{Y}\}^k\{\text{A},\text{C},\text{G},\text{T}\}.$$

Clearly, $\|C_{\text{bin}}(k)\| = 2 \times 2^k \times 4$ and each probe is of length $k+1$.

• The *gapped* chip $C_{\text{gap}}(k)$ is composed of all probes in the sets

$$\{\text{A},\text{C},\text{G},\text{T}\}^k$$

and

$$\{\text{A},\text{C},\text{G},\text{T}\}^{k-1}\{\text{X}\}^{k-1}\{\text{A},\text{C},\text{G},\text{T}\}.$$

Here, the second $k-1$ letters are just place holders without sequence information. $\|C_{gap}(k)\| = 2 \bullet 4^k$

- The *alternating* chip $C_{alt}(k)$ is composed of all probes in the sets

$$(\{A, C, G, T\})^{k-1} \{X\} \{A, C, G, T\}$$

and

$$(\{A, C, G, T\})^{k-2} \{X\} \{A, C, G, T\}^2.$$

Here, alternating letters have positional information. $\|C_{alt}(k)\| = 2 \times 4^k$.

It might not seem like these designs will differ substantially in resolution power from the classical or *"uniform"* chip $C(k) = C_{unif}(k)$ discussed earlier. It will turn out otherwise. For ease of calculation, assume all k-tuples have probability 4^{-k}.

For the uniform chip,

$$\mathbb{P}_{unif}(U(m,n))$$
$$= \mathbb{P}(s_{m-k+2} \cdots s_m x \notin S(\mathbf{s}) \text{ for } x \neq s_{m+1}) \cong \left((1 - 4^{-k})^{n-k+1}\right)^3.$$

We have neglected self-overlapping words in this calculation and we have assumed the independence of $s_{m-k+2} \cdots s_m x$ and $s_{m-k+1} \cdots s_n y$ for $x \neq y$. Therefore,

$$1 - \mathbb{P}_{unif}(U(m,n)) \cong \frac{3(n-k+1)}{4^k} \cong \frac{3n}{4^k} = \frac{3n}{\|C(k)\|}.$$

The solution for $n_{\max}$ is then

$$n_{\max} = \frac{\|C(k)\|\alpha}{3}.$$

For the binary chips, the ambiguity results from cases where the spectrum $S_{bin}(\mathbf{s})$ contains both a $v'y$ probe and a $v''y$ probe, where $y \neq s_{m+1}$ and $s_{m-k+1} \cdots s_m$ written in the $W-S$ alphabet is v'; $s_{m-k+1} \cdots s_m$ written in the $R-Y$ alphabet is v''. The probability that $v'y$ is not in $S_{bin}(\mathbf{s})$ is approximately $\left(1 - \frac{1}{2^k 4}\right)^{n-k}$ so

$$\mathbb{P}_{bin}(U(m,n)) = \mathbb{P}(\text{both } v'y \text{ and } v''y \text{ not in } S_{bin}(s))$$
$$\cong 1 - \left(\left(1 - \frac{1}{2^k 4}\right)^{n-k}\right)^2 \cong \left(\frac{n}{2^k 4}\right)^2 = \frac{n^2}{2^{2(k+2)}}.$$

Note that $\mathbb{P}(v''|v') = \mathbb{P}(v'')$ for a uniform alphabet. We have assumed that $v'y$ and $v''y$ are approximately independent. Therefore,

$$1 - \mathbb{P}_{bin}(U(m,n)) \cong 1 - \left(1 - \frac{n^2}{2^{2(k+2)}}\right)^3 \cong \frac{3n^2}{2^{2(k+2)}} = \frac{12n^2}{\|C_{bin}(k)\|^2}.$$

Therefore, for binary chips,

$$n_{\max} = \frac{1}{\sqrt{12}} \|C_{\text{bin}}(k)\| \sqrt{\alpha}.$$

For $k = 8$ and $\alpha = 0.01$, we get for the classical chip $C(8) = C_{\text{unif}}(8)$ the value of $n_{\max} \cong 210$. For binary chips $n_{\max} = \frac{1}{\sqrt{12}} \|C\| \sqrt{\alpha}$, and with $\alpha = 0.01$ and $\|C\| = 4^8$, we get $n_{\max} \approx 1800$, a huge gain. This result also holds for alternating and gapped chips.

7.3 Shotgun Sequencing Revisited

In Section 7.1, we learned that quite accurate fragment sequences are relatively easily obtained, and that the sequence a is determined by depth of coverage. The disadvantage is that the problem is computationally hard and the fragment errors, although not great in number, can mislead the analyst. In addition, the relative fragment orientations are unknown. The conventional method of shotgun sequence assembly, outlined in Section 7.1.3, is based on pairwise overlaps, not the multiple overlaps actually in the data.

In Section 7.2, we saw that SBH sequencing by k-tuple content can be recast as an Eulerian graph problem. However, since experimental reality keeps k small and implies that the data will be error prone, most sequencing is still done using variations of the method of shotgun sequencing.

Next, we will sketch a way to apply the ideas of SBH sequencing to shotgun sequencing. Recall that we have a set of fragments $\mathcal{F} = \{f_1, f_2, \ldots, f_N\}$. The basic idea of the new algorithm is to apply the mathematical ideas of SBH to shotgun sequencing. We are given fragments of approximate length $l \in [350, 1000]$ and can just read along a fragment and determine all k-tuples where k is chosen by the person setting up the algorithm. If there are 2% errors, on the average there will be $50 - k$ correct k-tuples in every 50 bps of fragment data. With $k \in [10, 20]$, this gives a substantial fraction of correct data. The data determined from $f_1, f_2, \ldots, f_N$ will look like

$$\cup_{w \in \cup_i (f_i \cup f_i^r)} \left(w; \cup_{\alpha=1}^{n(w)} (i_\alpha, j_\alpha) \right),$$

where w is a k-tuple occurring in $\cup f_i$ in fragment f_i at location j_α, or at that location in the reversed fragment f_i^r. The k-tuple w occurs $n(w)$ times.

To avoid the troublesome feature of fragment orientation, the fragment data is simply enlarged by a factor of 2 to include f^r as well as f. The algorithm will then produce two identical sequences of which we only report one. The speed of the method makes this a minor inefficiency and allows us to sidestep the usual fragment orientation pitfalls.

Next is a sketch of the procedure. First take fragment data and produce k-tuple data. Then construct the Euler graph on the $(k - 1)$-tuples (as in SBH). An edge

Sequence Assembly

has a weight associated with the number of associated fragments. Then collapse all edges, edge$_1$ → vertex → edge$_2$, where the data on the two edges is completely consecutive without contradiction. Call these collapsed edges *super edges*. Then perform a greedy Eulerian tour beginning with the heaviest edge(s). As the tour proceeds, a sequence is produced without (direct) reference to a multiple alignment. After all contigs or islands of sequence have been produced, eliminate the duplicates.

Biologists require a multiple alignment to check the algorithm and to check their basic fragment data. This is easily done after the Euler sequences have been produced. Simply apply the hashing methods from Chapter 8 to see where a fragment might align well to the Euler sequence. This gives candidate alignment diagonals. Then apply dynamic programming restricted to a narrow band along these diagonals. Most fragments fit almost perfectly with the Euler sequence, and a multiple alignment that the experimentalist can examine is rapidly produced.

Algorithm 7.4 (Assembly)
```
input:    N, k; f₁, f₂, ..., f_N
output:   Sequence assembly
```

1. Convert $f_1, f_2, \ldots, f_N, f_1^r, f_2^r, \ldots, f_N^r$ into $(w_\alpha, i_\alpha, j_\alpha)$ for all α occurrences of w, when $|w| = k$.

2. Construct the Euler graph on $(k-1)$-tuples for the k-tuples from 1. Each edge w has $\alpha = 1$ to $n(w)$ pairs of (fragment=i_α, position=j_α).

3. Collapse edges into super edges.

4. Perform greedy Eulerian tour(s).

5. Align fragment to sequence produced by (4).

If the reads are 100% accurate and no k-tuple repeats exist, then this algorithm is guaranteed to give the correct sequence. It actually does very well in the presence of errors and repeats.

Next, we give a simple example in which the complication of f and f^r is ignored for ease of presentation. The sequence a = ATGTGCCGCA from Section 7.2 and Figure 7.8 is used. Our project has four fragments as shown:

$$\begin{array}{ll} a = & ATGTGCCGCA \\ \hline f_1 = & GTGCCG \\ f_2 = & GCCGCA \\ f_3 = & ATGTG \\ f_4 = & TGTGCC. \end{array}$$

The data for $k = 3$ becomes

ATG : (3,1)
TGT : (3,2),(4,1)
GTG : (1,1),(3,3),(4,2)
TGC : (1,2),(4,3)
GCC : (1,3),(2,1)
CCG : (1,4),(2,2)
CGC : (2,3)
GCA : (2,4).

The graph G is now

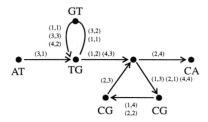

The sequence **a** is easily read begining from a heavy edge and moving forward and backward.

Problems

Problem 7.1 In SSP, prove that

$$\max_i |f_i| \leq |S| \leq \sum_{i=1}^{N} |f_i|.$$

Give an example with $N = 5$, $|f_i| \neq |f_j|$, $i \neq j$, where $\max_i |f_i| = |S|$. Give an example where $|S| = \sum_{i=1}^{N} |f_i|$, for arbitrary N. ($|\mathcal{A}| = 4$.)

Problem 7.2 For the stochastic model for shotgun sequencing in Section 7.1.5, let the orientation of a fragment be 5' to 3' (+) or 3' to 5' (−) iid each with probability 1/2. Find the fraction of $[0, L]$ covered by at least one + fragment and one − fragment.

Problem 7.3 For the sequence **a** = AATGATAGGCAGCCAC, (i) find the graph G and (ii) find all sequence reconstructions consistent with G using $k = 3$.

Problem 7.4 Take a word $x_0 x_1 \cdots x_{r-1}$ and define $\mathbf{a} = a_1 \cdots a_n$ by $a_i = x_l$, where $l = i \bmod r$, where $n \geq 2r$. Suppose $S(\mathbf{a})$ is the spectrum where the SBH k-tuples satisfy $1 \leq r \leq k$. Further assume that $\mathbf{x} = x_0 x_1 \cdots x_{r-1}$ has no self-overlaps. (i) Find $S(\mathbf{a})$ and (ii) find all sequences $\mathbf{b} = b_1 \cdots b_n$ such that $S(\mathbf{a}) = S(\mathbf{b})$.

Problem 7.5 Show that for gapped chips, $n_{\max} = \frac{1}{\sqrt{12}}\|C_{\text{gap}}(k)\|\sqrt{\alpha}$.

Problem 7.6 Suppose there are 4^k distinct elements in $S(\mathbf{a})$. How many reconstructions have spectrum $S(\mathbf{a})$? How long are these (shortest) reconstructions?

Problem 7.7 Suppose our sequences are for a two letter alphabet $\{R, Y\}$. It is desirable to assign probes so that they differ as little as possible from their neighbors. (i) Consider a one-dimensional array. Prove by induction that 2^l words can be arranged so that each word differs from each adjacent neighbor in exactly one position. (ii) Use (i) to construct a two-dimensional array of all 2^{2l} tuples of length $2l$ so that each word differs from each of its four adjacent neighbors in exactly one position.

Problem 7.8 In Section 7.3, change f_2 in the example to $f_2 = \text{CCCGCA}$, introducing and error. Execute the algorithm assembly.

Chapter 8

Databases and Rapid Sequence Analysis

Rapid DNA sequencing has radically altered biology as we have discussed in Chapter 1. It is difficult to grasp how quickly this transformation has taken place. The vast amount of data has caused the growth of databases that collect and distribute the sequence data as well as important related data. The major DNA databases are the EMBL Data Library which is run by European countries, GenBank which is the U.S. database and DDBJ, the DNA database of Japan. Today, these databases are essentially identical in content. The growth of the international nucleic acid database is shown in Figure 8.1.

In the 1980s, molecular biology became dependent on these databases for a current view of known sequences. There was some difficulty achieving timely entry of the sequences, but, today, almost all sequences are entered within a month of publication. The reason rapid entry is important is so that biologists can determine the relationship of their sequences with other sequences that have been determined. In Section 8.5, we will present a widely used technique for screening a database for sequence relationships.

Today, some sequences are published in a journal as well as entered into a database. Increasingly, journals will not accept more than a small fraction of the DNA or protein sequence that is determined. Genome projects promise to greatly increase the rate and volume of DNA sequence. Therefore, databases will become even more central to biology. Section 8.1 discusses a release of the nucleic acid database.

What is the utility of being able to look up a DNA or protein sequence? As described in Chapter 1, a DNA sequence is a blueprint for an organism's structure and function. Clearly, it is important to know the coding DNA, the DNA regulating the gene, and so on. This information can be stored along with the sequence. Protein sequences, of course, represent the working parts of an

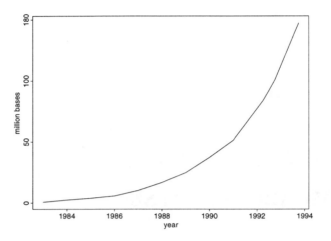

Figure 8.1: *DNA database size*

organism and are of much biological interest. Some proteins have a structural role, such as those in skin, hair, and bone whereas others have a role as components of molecular machines. Some amino acids are essential to the working of the protein whereas others are much less important. In Sections 8.2, 8.3, and 8.4, we will study some elementary computer science techniques to summarize or represent sequences. First, we put the sequence into a compact or suffix tree form, where much information about the sequence can be easily read. Then we give an easy algorithm for hashing and chaining.

Another reason for organizing biological sequences into databases is to learn new biology. Evolution conserves useful sequence patterns over great amounts of evolutionary time. When a new sequence has a great deal of similarity with a sequence already in a database, there is a good chance that the biological functions might also be similar. In this way, new and useful biological hypotheses are formed by sequence comparison. Without being much more specific, in Section 8.5 we use the ideas of this chapter to motivate a heuristic sequence comparison method. In Section 8.6, we give a rigorous sequence comparison method related to these ideas.

8.1 DNA and Protein Sequence Databases

GenBank is indexed by release number. Release 82.0 appeared in Spring 1994 and has 169,896 loci (separate sequence segments) representing 180,589,455 bases. Release 76.0 (April 1993) had 111,911 loci with 129,968,355 bases. Release 82.0, thus, is 39% larger in number of bases than Release 76.0. A statistical summary of Release 82.0 is presented in Section 8.1.3.

8.1.1 Description of the Entries in a Sequence Data File

LOCUS — A short unique name for the entry, chosen to suggest the sequence's definition. Mandatory.

DEFINITION — A concise description of the sequence.

ACCESSION — The primary accession number is a unique, unchanging code assigned to each entry. This code should be used when citing information from GenBank. Mandatory.

KEYWORDS — Short phrases describing gene products and other information about an entry. Mandatory

SEGMENT — Information on the order in which this entry appears in a series of discontinuous sequences from the same molecule. Optional.

SOURCE — Common name of the organism or the name most frequently used in the literature. Mandatory.

ORGANISM — Formal scientific name of the organism (first line) and taxonomic classification levels (second and subsequent lines). Mandatory.

REFERENCE — Citations for all articles containing data reported in this entry. Includes four subkeywords and may repeat. Mandatory.

AUTHORS — Lists the authors of the citation. Mandatory.

TITLE — Full title of citation. Optional.

JOURNAL — Lists the journal name, volume, year, and page numbers of the citation. Mandatory.

STANDARD — Lists information about the degree to which the entry has been annotated and the level of review to which it has been subjected. Mandatory.

COMMENT   Cross-references to other sequence entries, comparisons to other collections, notes of changes in LOCUS names, and other remarks. Optional.

FEATURES   Table containing information on portions of the sequence that code for proteins and RNA molecules and information on experimentally determined sites of biological significance. Optional.

BASE COUNT   Summary of the number of occurrences of each base code in the sequence. Mandatory.

ORIGIN   Specification of how the first base of the reported sequence is operationally located within the genome. Where possible, this includes its location within a larger genetic map. Mandatory.

                 The ORIGIN line is followed by sequence data (multiple records).

//   Entry termination symbol. Mandatory at the end of an entry/exactly one record.

8.1.2 Sample Sequence Data File

An example of a complete sequence entry file follows. (This example has only two entries.) Note that in this example, as throughout the data bank, numbers in square brackets indicate items in the REFERENCE list. For example, in AAURRA, [1] would refer to the paper by Huysmans et al.

GBSMP.SEQ

                        Sample Sequence Data File

           2 loci,   280 bases, from   2 reported sequences

    LOCUS   AAURRA   118 bp ss-rRNA   RNA   16-JUN-1986
    DEFINITION   A.auricula-judae (mushroom) 5S ribosomal RNA.
    ACCESSION   K03160
    KEYWORDS   5S ribosomal RNA; ribosomal RNA.
    SOURCE   A.auricula-judae (mushroom) ribosomal RNA.
    ORGANISM   Auricularia auricula-judae Eukaryota; Planta; Mycophyta; Basidiomycotina; Hymenomycetes; Russulaceae.

REFERENCE	1 (bases 1 to 118)
AUTHORS	Huysmans, E., Darris, E., Vandenberghe, A. and De Wachter, R.
TITLE	The nucleotide sequences of the 5S rRNAs of four mushrooms and their use in studying the phylogenetic position of basidiomycetes among the eukaryotes
JOURNAL	Nucl Acid Res 11, 2871-2880 (1983)
STANDARD	full staff review
FEATURES	from to/span description
rRNA	1 118 5S ribosomal RNA
BASE COUNT	27 a 34 c 34 g 23 t
ORIGIN	5' end of mature rRNA.
	1 atccacggcc ataggactct gaaagcactg catcccgtcc gatctg-caaa gttaaccaga
	61 gtaccgccca gttagtacca cggtggggga ccacgcggga atc-ctgggtg ctgtggtt

//

LOCUS	ACARR58S 162 bp ss-rRNA RNA 15-MAR-1989
DEFINITION	A.castellanii (amoeba) 5.8S ribosomal RNA.
ACCESSION	K00471
KEYWORDS	5.8S ribosomal RNA; ribosomal RNA.
SOURCE	A.castellani (amoeba; strain ATCC 30010) rRNA.
ORGANISM	Acanthamoeba castellanii
	Eukaryota; Animalia; Protozoa; Sarcomastigophora; Sarcodina; Rhizopoda; Lobosa; Gymnamoeba; Amoe-bida; Acanthopodina; Acanthamoebidae.
REFERENCE	1 (bases 1 to 162)
AUTHORS	Mackay,R.M. and Doolittle,W.F.
TITLE	Nucleotide sequences of AcanthamoebA.castellanii 5S and 5.8S ribosomal ribonucleic acids: Phylogenetic and comparative structural analyses
JOURNAL	Nucleic Acids Res. 9, 3321-3334 (1981)
STANDARD	full automatic
COMMENT	[1] also sequenced A.castellanii 5S rRNA <K03160>. NCBI gi: 173608
FEATURES	Location/Qualifiers
rRNA	1..162
	/note="5.8S rRNA"
SOURCE	1..162
	/organism="Acanthamoeba castellanii"
BASE COUNT	40 a 39 c 44 g 39 t

ORIGIN 5' end of mature rRNA.
1 aactcctaac aacggatatc ttggttctcg cgaggatgaa gaacgcagcg aaatgcgata
61 cgtagtgtga atcgcaggga tcagtgaatc atcgaatctt tgaacgcaag ttgcgctctc
121 gtggtttaac cccccgggag cacgttcgct tgagtgccgc tt
//

8.1.3 Statistical Summary

The database is divided into divisions determined by biology; plants sequences, for example, form one division. First, we show some sequences greater than 100,000 bps in length:

Locus	Length (bp)	Division	Accession Number
CHMPXX	121024	PLN	X04465
CHNTXX	155844	PLN	Z00044
CHOSXX	134525	PLN	X15901
CLEGCGA	143172	PLN	X70810
D26185	180136	BCT	D26185
EBV	172281	VRL	V01555
ECO110K	111401	BCT	D10483
ECOUW76	225419	BCT	U00039
ECOUW82	136254	BCT	L10328
ECOUW89	176195	BCT	U00006
HE1CG	152260	VRL	X14112
HEHCMVCG	229354	VRL	X17403
HEVZVXX	124884	VRL	X04370
HS1ULR	108360	VRL	D10879
HS4B958RAJ	184113	VRL	M80517
HSECOMGEN	150223	VRL	M86664
HSGEND	112930	VRL	X64346
HUMNEUROF	100849	PRI	L05367
HUMRETBLAS	180388	PRI	L11910
IH1CG	134226	VRL	M75136
MPOMTCG	186608	PLN	M68929
MTPACG	100314	PLN	X55026
PANMTPACGA	100314	PLN	M61734
SCCHRIII	315338	PLN	X59720
VACCG	191737	VRL	M35027
VARCG	186102	VRL	L22579
VVCGAA	185578	VRL	X69198

Databases and Rapid Sequence Analysis

The next table gives the number of entries and bases by division of the database.

Division	Entries	Bases
PRIMATE	31972	30328835
MAMMALIAN	5628	6183786
VERTEBRATE	6558	7270430
INVERTEBRATE	11234	18729402
RODENT	20581	22836624
PLANT	16154	27150929
BACTERIAL	15107	27433286
PHAGE	968	1414274
RNA	3603	2176197
VIRAL	15876	20597295
UNANNOTATED	1490	1391910
SYNTHETIC	1717	2572139
EST SEQUENCES	33727	10672722
PATENT	5281	1831626

8.2 A Tree Representation of a Sequence

A *suffix tree* is very useful for locating repeats within or between sequences. This is very valuable to a biologist to find exact repeats between his sequence and another in the database. Although we do not give a formal definition, the concept is most easily illustrated by an example. Although we will give a short DNA sequence, the idea works with any finite sequence.

To illustrate the concept of suffix trees, let a =AATAATGC$, where $ signals the end of the sequence. For each $i, i = 1$ to 9, let the substring S be the shortest substring beginning at i which does not occur elsewhere in a. This substring is said to identify i. For example, position $i = 4$ is identified by AATG. These identifying substrings are organized into a suffix tree which represents the information:

Position	Identifying Substring
1	AATA
2	ATA
3	TA
4	AATG
5	ATG
6	TG
7	G
8	C
9	$

The n terminal notes of the suffix tree for $\mathbf{a} = a_1 a_2 \cdots a_n$ consist of $1, 2, \ldots, n$. The sequence of labels on the edges from the root to terminal node i is the

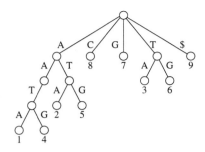

Figure 8.2: *Suffix tree for AATAATGC$*

identifying substring for position i. The suffix tree for the length 9 sequence from above is given in Figure 8.2. Two sequences (or more) can be processed simultaneously to give a suffix tree where the longest matching regions can be easily found.

Figure 8.2 is a very simplified version of a suffix tree. In more formal presentations, edges are labeled with strings rather than single characters. The strings for leaves begin with the current character continuing to the end of the sequence. Also, whenever an internal node has only one child, it is removed and the two character strings merged. The concatenation of letters from root to leaf will give the associated suffix.

Although it should now be clear what a suffix tree is, no algorithm has been presented for its construction. In Section 8.4, we will give an algorithm which is adequate for the task, but, first, we will consider the information contained in a suffix tree.

The longest repeat in the sequence is AAT which begins at positions 1 and 4. This information is obtained simply from reading the labels at the tips of the longest branches of the tree. Moving up one level, we find the length two repeats: AT beginning at positions 2 and 5 and AA beginning at positions 1 and 4. (Of course, these are contained within the length 3 repeats.) Finally, we can read the length 1 repeats: A begins at positions 1, 4, 2, and 5; T begins at positions 3 and 6; G begins at position 7; and C begins at position 8. Contained in this economical structure is all the repeat information of the sequence **a**, if we are interested in repeats of frequency at least 2.

8.3 Hashing a Sequence

In this section, we study the problem of finding all repeats of length k in our sequence $\mathbf{a} = a_1 a_2 \ldots a_n$. Recall that $a_i \in \mathcal{A}$ where $|\mathcal{A}| = d$. A word of length k such as $a_i a_{i+1} \ldots a_{i+k-1}$ is also called a k-*tuple*, and there are d^k

possible k-tuples. Although the location of the repeats of length k is less detailed information than a suffix tree, we will see in Section 8.4 that it is very useful in sequence comparison. In what follows, we assume that $d^k \ll n$.

The first task is to associate each k-tuple with an integer. Let $e : \mathcal{A} \to \{0, 1, \ldots, d-1\}$ be a one-to-one mapping. Define

$$s_i = \sum_{j=0}^{k-1} e(a_{i+j}) d^{k-1-j}.$$

Note that
$$s_{i+1} = s_i d + e(a_{i+k}) \bmod(d^k).$$

Each k-tuple, $a_i a_{i+1} \cdots a_{i+k-1}$, is therefore associated with a unique integer in $\{0, 1, \ldots, d^k - 1\}$. Therefore, we have a one-to-one mapping between $\mathbf{a} = a_1 \cdots a_n$ and $\mathbf{s} = s_1 s_2 \cdots s_{n-k+1}$. Below, we assume $\mathbf{s}$ has been determined.

8.3.1 A Hash Table

Efficient methods of ordering a list of numbers have been developed. The process is called sorting. One of the most elementary is *bubble sort* in which the input data $s_1, s_2, \ldots, s_m$ is put into numerical order with $s_{i_1} \le s_{i_2} \le \cdots \le s_{i_m}$.

Algorithm 8.1 (Bubble Sort)
```
input:    s_1, s_2, ..., s_m
output:   s_{i_1} ≤ s_{i_2} ≤ ... ≤ s_{i_m}
  1. bound ← m
  2. l ← 0
       for i = 1 to bound -1
           if s_i > s_{i+1}
               interchange s_i and s_{i+1}
               l ← i
  3. if l = 0, end
     if l ≠ 0, bound ← l
           return to step 2.
```

Proof. The first iteration of step 2 considers all pairs beginning with s_1 and s_2. Therefore, all pairs after the last exchange (l) are in order:

$$s_{l+1} \le s_{l+2} \le \cdots \le s_m.$$

In the new list we need only order the list to the l-th entry ending with s_l. The list decreases by at least 1 in length at every iteration of step 2 and, when $l = 0$, the list is ordered. ∎

We see from the proof that the worst case running time for this algorithm occurs when $s_1 > s_2 > \cdots > s_m$. The first pass of the algorithm produces the order $s_2, s_3, \ldots, s_m, s_1$ and each succeeding pass puts the leading member of the list into its correct order. The number of steps required is bounded by $(m-1) + (m-2) + \cdots + 1 = m(m-1)/2 = O(m^2)$. Another algorithm known as *quicksort* has time requirement $O(m \log m)$ in the expected case and is known to be one of the fastest possible sorting algorithms.

We now assume that the sequence $\{s_i\}$ has been ordered: $s_{i_1} \leq s_{i_2} \leq \cdots \leq s_{i_m}$. It is easy to make a hash table (w, b_w, e_w), $w = 0$ to $d^k - 1$ which indicates where each k-tuple ($= w$) in the sequence begins (b_w) and ends (e_w) in the ordered sequence $s_{i_1}, s_{i_2}, \ldots, s_{i_m}$, and this table can be made in linear time. For example, $b_w = 17$ and $e_w = 23$ means that k-tuple w occurs at $i_{17}, i_{18}, \ldots$ and i_{23} in the original sequence. $b_w = 0$ means that w does not occur anywhere in the sequence. Essentially, this table tells us where each w occurs in the original sequence.

8.3.2 Hashing in Linear Time

With this approach, we define an array A of dimension n by d^k. $A(i, j)$ is defined to be the location of the j-th occurrence of k-tuple i in the sequence $s_1 s_2 \cdots$. The array is constructed by moving down $s_1 s_2 \cdots$ and, when k-tuple i is encountered, adding the position to the i-th row of the array.

Although this procedure takes linear time $O(n)$, it takes storage $O(nd^k)$. This should be compared to Section 8.3.1 where the time required in $O(n \log n)$ and storage is $O(n)$.

8.3.3 Hashing and Chaining

We begin with $\mathbf{s} = s_1 s_2 \cdots$. The chain will allow us to find all occurrences of k-tuples in the sequence. Recall that $s_i \in \{0, 1, \ldots, d^{k-1} - 1\}$. The sequences α and γ will be explained after the algorithm.

Algorithm 8.2 (Chain)
```
input:    s₁s₂···sₘ
output:   α(0), α(1), ..., α(dᵏ − 1)  and  γ(1), γ(2), ..., γ(m)
  1.  α = β ← (−1, −1, ...)
  2.  for i = 1 to m
            if β(sᵢ) = −1
                α(sᵢ) ← i,  β(sᵢ) ← i,  γ(i) ← end
            otherwise β(sᵢ) > 0
                γ(β(sᵢ)) ← i
                γ(i) ← end
                β(sᵢ) ← i
```

The output of this algorithm allows us to find the first location of i by reading $\alpha(i)$ and then the succeeding locations can be found by $\gamma(\alpha(i))$, $\gamma(\gamma(\alpha(i)))$, $\cdots$. Thus, $\{\gamma^k(\alpha(i))\}_{k \geq 0}$ gives the locations of i. Therefore, all length k repeats can be found in time $O(m)$ for making the sequence numerical plus $O(m)$ for chaining. Storage is, of course, $O(m)$. Actually, instead of a new sequence γ, we can write over s itself. For example consider s $= (1, 0, 1, 1)$ where we compute $\alpha = (\alpha(0), \alpha(1)) = (2, 1)$ and $\gamma = (3, \text{end}, 4, \text{end})$.

8.4 Repeats in a Sequence

To give the essential idea of the algorithm, consider a $=$ AATAATGC\$. At the first step, collect all positions with identical letters into groups. At each additional step, repeat the procedure for the successive letters that follow within each group. See the tree in Figure 8.2.

In this section, we will give an algorithm which finds all repeats with frequency at least 2 in a sequence. To be more explicit, at stage i the algorithm can produce all length i repeats. A by-product of the algorithm is all information necessary to construct a suffix tree.

The suffix tree simply enumerates the patterns beginning at each position until that pattern is unique. At each node, the "next" letters are sorted, becoming the sets of positions for the "daughter" nodes. Below **Suffix**$(B, depth)$ has as input a set of positions and depth. It sorts the letters at depth from the positions in B. The algorithm repeats uses this function to traverse the suffix tree until each position has unique suffix.

Algorithm 8.3 (Suffix)
```
input:   (B, depth)
output:  (list, depth) for α ∈ A
    for all α ∈ A
        list(α) = ∅
    for all i ∈ B
        list(a_{i+depth}) ← list(a_{i+depth}) ⋃ {i}
```

Algorithm 8.4 (Repeats)
```
input:   a_1, a_2, ..., a_n
    for node = top
        list(node) ← {1, 2, ..., n}
    for all nodes with |(list(node)| > 1
        suffix(list(node), depth)
```

To illustrate this algorithm, recall our example with **a** =AATAATGC$. We initialize with
$$\text{list}(top) \leftarrow \{1, 2, \ldots, 9\}.$$
Calling suffix(list(top), 1), we return

$$\text{list}(A) = \{1, 2, 4, 5\},$$
$$\text{list}(C) = \{8\},$$
$$\text{list}(G) = \{7\},$$
$$\text{list}(T) = \{3, 6\},$$
$$\text{list}(\$) = \{9\}.$$

For illustration, we only continue with two of these lists. Calling **Suffix**(list(T), 2), we return

$$\text{list}(TA) = \{3\},$$
and
$$\text{list}(TG) = \{6\}.$$

The algorithm continues until the tree of Figure 8.2 is produced.

8.5 Sequence Comparison by Hashing

The goal of this section is to describe an algorithm to locate rapidly unusual similarity between sequences. To gain an intuitive idea of sequence similarity, we present the example with **a** = CTAATCC and **b** = AATAATGC. A standard method of visualizing the sequences is to put them into a *dot matrix* form, where the • indicates that $a_i = b_j$. See Figure 8.3. We will also be interested in only writing • at (i, j) when $a_i a_{i+1} = b_j b_{j+1}$. Then we obtain the matrix in Figure 8.4 [Note that the matrix in Figure 8.4 is much more sparse than that in Figure 8.3]

Sequences that have a good deal of similarity will have diagonal regions that have many •'s. We will give several more precise definitions, but in this section we will simply study the diagonal sums, that is, the number of matches on each diagonal. Let $\mathbf{a} = a_1 \cdots a_n$ and $\mathbf{b} = b_1 \cdots b_m$. For counting k-tuple matches, define for $-m + k \leq l \leq n - k$

$$S_l = \sum_{i=1}^{n-k+1} \mathbb{I}\{a_i a_{i+1} \cdots a_{i+k-1} = b_{i-l} b_{i-l+1} \cdots b_{i-l+k-1}\}.$$

Here, $j + l = i$, and the count of matching words on a diagonal is S_l. The diagonals are indexed by $l = i - j$. Figure 8.5 shows $S_l, -7 \leq l \leq 6$.

Databases and Rapid Sequence Analysis

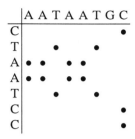

Figure 8.3: *Matching 1-tuples*

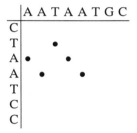

Figure 8.4: *Matching 2-tuples*

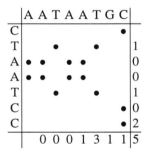

Figure 8.5: *1-tuple diagonal sums*

It is quite easy to compute the diagonal sums. A direct algorithm is given in the formula for S_l and requires $O(nm)$ time and space. To improve on this obvious method, consider hashing b with k-tuples. Then for each k-tuple in a (beginning at $i = 1$) the matching k-tuples in b can be found by look-up. Then the offset $l = i - j$ can be calculated and the sums S_l easily computed.

Next, we make a hash table for 2-tuples in b = AATAATGC:

2-tuple	positions in b
AA	1,4
AT	2,5
GC	7
TA	3
TG	6

Recall that we must compute S_l for $k - m \leq l \leq n - k$ or $-6 \leq l \leq 5$. For $i = 1$ to $n - k = 5$, we find the matching positions in **b**, compute the offset $l = i - j$, and increment S_l. For example, $i = 1$ has $w = $ CT, which is not in the hash table for **b**. $i = 2$ has $w = $ TA which appears in **b** at position 3. Therefore $S_{2-3} = S_{-1} = 0 + 1$. Continuing this process;

i	1	2	3	4	5	6
w	CT	TA	AA	AT	TC	CC
S_{-1}		1	2	3	3	3
S_2			1	2	2	2

The result of this computation is $S_{-1} = 3$ and $S_2 = 2$. Note that this is consistent with Figure 8.4.

The following algorithm, **Fast**, computes these diagonal sums following this hashing method. The method first hashes sequence **b** and then proceeds from $i = 1$ to $i = n - k + 1$, incrementing the diagonal sums. The smallest diagonal index is $1 - (m - k + 1) = k - m$ whereas the largest is $(n - k + 1) - 1 = n - k$.

Algorithm 8.5 (Fast)
```
input:    a_1,...,a_n; b_1,...b_m; k
output:   S_ν, k - m ≤ ν ≤ n - k
```
1. $S_\nu \leftarrow 0$, $k - m \leq \nu \leq n - k$

2. chain **b**

    for each $w \in \{0, 1, \ldots, d^k - 1\}$

    $\gamma^l(\alpha(w)), l \geq 0$ gives (successive) locations of $w \in$ **b**.

3. for $i = 1$ to $n - k + 1$

    $w \leftarrow a_i a_{i+1} \cdots a_{i+k-1}$
    until $\gamma^l(\alpha(w)) = $ end
    $S_{i-\gamma^l(\alpha(w))} \leftarrow S_{i-\gamma^l(\alpha(w))} + 1$

Actually, we are interested in regions of the diagonal where the sums are large. Thus far, we have computed

$$S_\nu = \#k\text{-tuple matches} - 0 \times (\#\text{nonmatches}).$$

Databases and Rapid Sequence Analysis

Now we introduce a penalty g for nonmatches and set

$$S_\nu = \text{\#}k\text{-tuple matches} - g \times (\text{\#nonmatches}).$$

Of course, $g = 0$ in the first formula.

Algorithm 8.6 (Fastgap)
```
input:    a₁,...,aₙ; b₁,...,bₘ; k
output:   Sᵥ, k - m ≤ ν ≤ n - k
```
1. $S_\nu \leftarrow 0$ $\text{loc}_\nu \leftarrow 0$, $k - m \leq \nu \leq n - k$
2. chain **b**

   for each $w \in \{0, 1, \ldots, d^k - 1\}$

   $\gamma^l(\alpha(w)), l \geq 0$ gives (successive) locations of $w \in \mathbf{b}$.
3. for $i = 1$ to $n - k + 1$

   $w \leftarrow a_i a_{i+1} \cdots a_{i+k-1}$
   until $\gamma^l(\alpha(w)) = \text{end}$
   $\nu = i - \gamma^l(\alpha(w))$
   $S_\nu \leftarrow S_\nu + 1 - g \times (i - \text{loc}_\nu - 1)$
   $\text{loc}_\nu \leftarrow i$

For our final algorithm, we compute the interval of each diagonal that gives the largest score. This allows us to locate intervals of the diagonal that give high scores, without accounting for the other parts of the diagonal. Algorithms that find matching intervals or substrings of the sequences are called local algorithms.

Algorithm 8.7 (lfast)
```
input:    a₁,...,aₙ; b₁,...,bₘ; k
output:   Sᵥ, k - m ≤ ν ≤ n - k
```
1. $M_\nu \leftarrow 0$ $\text{loc}_\nu \leftarrow 0$, $k - m \leq \nu \leq n - k$
2. chain **b**

   for each $w \in \{0, 1, \ldots, d^k - 1\}$

   $\gamma^l(\alpha(w)), l \geq 0$, gives (successive) locations of $w \in \mathbf{b}$.
3. for $i = 1$ to $n - k + 1$

   $w \leftarrow a_i a_{i+1} \cdots a_{i+k-1}$
   until $\gamma^l(\alpha(w) = \text{end}$
   $\nu \leftarrow i - \gamma^l(\alpha(w))$
   $S_\nu \leftarrow 1 + \max\{S_\nu - g(i - \text{loc}_\nu + 1), 0\}$
   $\text{loc}_\nu \leftarrow i$
   $M_\nu \leftarrow \max\{M_\nu, S_\nu\}$

Because the object of computing S_l is to locate regions of similarity between sequences, a program must monitor the sums S_l for those which are unusually large. This is the topic of Chapter 11, but there exist statistical tests very useful for this purpose.

As for issues of computational efficiency, the direct "compute the sums" algorithm takes time $O(nm)$. The hashing-based algorithm takes time $O(\#\bullet\text{'s})$.

Theorem 8.1 *If the sequences* $\mathbf{a} = a_1 \cdots a_n$ *and* $\mathbf{b} = b_1 \cdots b_m$ *have iid letters with* $p_\alpha = \mathbb{P}(\text{letter } \alpha)$, *then the hashing algorithm to compute* S_l, $k - m \leq l \leq n - k$, *takes expected time* $O((\sum_{\alpha \in \mathcal{A}} p_\alpha^2)^k nm)$ *under the assumption that* $k \ll \min\{n, m\}$.

Proof. Neglecting the time to hash $\mathbf{b}$, it is clear that the hashing-based algorithm takes computation time T proportional to the number of $\bullet$'s. Thus,

$$\mathbb{E}(T) = \mathbb{E}\left[\sum_{i=1}^{n-k+1}\left[\sum_{j=1}^{m-k+1} \mathbb{I}\{a_i \cdots a_{i+k-1} = b_j \cdots b_{j+k-1}\}\right]\right]$$

$$= \sum_{i=1}^{n-k+1}\sum_{j=1}^{m-k+1} \mathbb{P}(a_i \cdots a_{i+k-1} = b_j \cdots b_{j-k+1})$$

$$= (n - k + 1)(m - k + 1)(\mathbb{P}(a_i = b_j))^k$$

$$= (n - k + 1)(m - k + 1)\left(\Sigma_{\alpha \in \mathcal{A}} p_\alpha^2\right)^k.$$

The hashing algorithm requires $\mathbf{b} = b_1 \cdots b_m$ to be put into a hash table. Although $O(m)$ is a realistic time to hash $\mathbf{b}$, the maximum time this should take is $O(m \log m)$, and the running time of the algorithm is

$$O((\Sigma_{\alpha \in \mathcal{A}} p_\alpha^2)^k nm + cm \log m) = O((\Sigma_{\alpha \in \mathcal{A}} p_\alpha^2)^k nm).$$

∎

Obviously $O((\Sigma_{\alpha \in \mathcal{A}} p_\alpha^2)^k nm)$ is just order nm. But when, for example, $\mathcal{A} = \{A, C, T, G\}$, $p_\alpha = 1/4$, and $k = 6$, we have $(\Sigma_{\alpha \in \mathcal{A}} p_\alpha^2)^k = 1/4^6 \approx 0.0002$. This is an enormous reduction in computing time. Some widely used programs for sequence comparison are database searches based on this method, known as FASTN, FASTA, and so on. (N = nucleotide, A = amino acid).

8.6 Sequence Comparison with at most l Mismatches

In the last section, we gave an algorithm that maximized the number of k-tuple matches on a diagonal or interval. We were able to penalize for nonmatches. These

Databases and Rapid Sequence Analysis

nonmatches, which are nonidentical aligned letters (one from each sequence), are now called *mismatches*. To make a better defined problem, we study the problem of finding matching strings (contiguous subsequences) with at most l mismatches. The ideas of hashing can be used to advantage here. The first result characterizes matchings with at most l mismatches, although in Lemma 8.1 it appears for boolean words. See Figure 8.6 for an example.

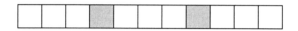

Figure 8.6: *Worst case mismatch distribution for $t = 11, l = 2$, and $k = \lfloor \frac{11}{3} \rfloor = 3$.*

Lemma 8.1 *Let $\mathbf{c} = c_1 c_2 \cdots c_t$ with $c_i \in \{0, 1\}$ have at most l zeros. Then*

(i) $\mathbf{c}$ *contains at least $t - (l+1)k + 1$ k-tuples of 1's, and*

(ii) $\mathbf{c}$ *contains at least one k-tuple of 1's with $k = \lfloor \frac{t}{l+1} \rfloor$.*

Proof. The word $\mathbf{c}$ has $t - k + 1$, k-tuples. Each 0 in $\mathbf{c}$ is in at most k, k-tuples. As there are at most l zeros, the zeros belong to at most $l \times k$, k-tuples. Therefore $\mathbf{c}$ has at least $t - k + 1 - lk = t - (l+1)k + 1$, k-tuples of 1's. Note that

$$t - (l+1)\lfloor \frac{t}{l+1} \rfloor + 1 \geq 1,$$

proving (ii). ∎

Clearly, every match between $a_1 \cdots a_t$ and $b_1 \cdots b_t$ corresponds to a boolean word $\mathbf{c}$:

$$c_i = \begin{cases} 0 & \text{if } a_i \neq b_i, \\ 1 & \text{if } a_i = b_i. \end{cases}$$

This gives the next easy result.

Theorem 8.2 *Let $a_1 a_2 \cdots a_t$ and $b_1 b_2 \cdots b_t$ match with at most l mismatches.*

(i) For $k \leq \lfloor \frac{t}{l+1} \rfloor$, $a_1 \cdots a_t$ and $b_1 \cdots b_t$ share at least $t - (l+1)k + 1$, k-tuples.

(ii) For $k = \lfloor \frac{t}{l+1} \rfloor$, $a_1 \cdots a_t$ and $b_1 \cdots b_t$ share a k-tuple.

This immediately gives a rigorous algorithm for approximate matching. The idea is to use k-tuples to filter out nonmatching regions.

Algorithm 8.8 (Filtration)
```
input:   a, b
output:  matches with l mismatches
```
1. set $k = \lfloor \frac{t}{l+1} \rfloor$
2. find all locations (i, j) of shared k-tuples
$$a_i a_{i+1} \cdots a_{i+k-1} = b_j b_{j+1} b_{j+k-1}$$
3. extend (i, j) to left and right until $\begin{cases} l+1 \text{ mismatches} \\ \text{end of } \mathbf{a} \text{ or } \mathbf{b} \end{cases}$

Let the letters of **a** and **b** be iid. Then it is easy to show that if X = number of potential matches, then
$$\mathbb{E}(X) = (n - k + 1)(m - k + 1)p^k,$$
where $p = \mathbb{P}(a = b)$.

When the algorithm is run on DNA sequences, often there are many potential matches that are not of length t with less than or equal to l mismatches and must be rejected. Next, we show how to improve the rejection rate by applying the $\lfloor \frac{t}{l+1} \rfloor$ idea twice.

To apply our idea twice, we define a *gapped k-tuple* beginning at i with *gap size* s to be a set of positions $i, i+s, i+2s, \ldots, i+(k-1)s$. See Figure 8.7. These gapped k-tuples can also serve as an additional filter to detect potential matches.

Lemma 8.2 *Let* $\mathbf{c} = c_1 c_2 \cdots c_t$ *have at most l 0's. Then* $\mathbf{c}$ *has at least one gapped* $\lfloor \frac{t}{l+1} \rfloor$*-tuple with gap size* $l+1$ *with all 1's.*

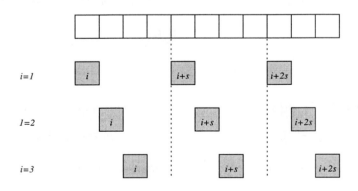

Figure 8.7: *The set of gapped* $\lfloor \frac{11}{3} \rfloor = 3$ *tuples with gap size* $s = 2 + 1 = 3$.

Proof. Consider $l + 1$ $\lfloor \frac{t}{l+1} \rfloor$-tuples with gap size $l + 1$ starting at $1, 2, \ldots, l + 1$ of **c**. These $l + 1$ gapped $\lfloor \frac{t}{l+1} \rfloor$-tuples are nonoverlapping. In addition, because

$$(l+1) + (\lfloor \frac{t}{l+1} \rfloor - 1)(l+1) \le t,$$

each $\lfloor \frac{t}{l+1} \rfloor$ tuple fits into **c**. At most l of them contain 0's, so at least one must have all 1's. ■

Combining our results, we have the following "double filtration."

Theorem 8.3 *Let $a_1 a_2 \cdots a_t$ and $b_1 b_2 \cdots b_t$ match with at most l mismatches. With $k = \lfloor \frac{t}{l+1} \rfloor$, the strings $a_1 a_2 \cdots a_t$ and $b_1 b_2 \cdots b_t$ share a (continuous) k-tuple and a gapped k-tuple of size $l + 1$.*

Before we write a sketch of a double filtration algorithm, we need to define a distance between k-tuples. Let the coordinate of a k-tuple be (i, j) if the **a** tuple begins at a_i and the **b** tuple begins at b_j. The distance between tuples v_1 and v_2 with coordinates at (i_1, j_1) and (i_2, j_2) is defined to be

$$d(v_1, v_2) = \begin{cases} i_1 - i_2 & \text{if } i_1 - i_2 = j_1 - j_2, \\ \infty & \text{otherwise.} \end{cases}$$

Algorithm 8.9 (Double Filtration)
```
input:    a, b
output:   length l mismatches
```
(i) set $k = \lfloor \frac{t}{l+1} \rfloor$

- find all locations (i, j) of continuous k-tuples where there exists a gapped k-tuple with gap size $k + 1$ of distance $d \in [-l, t - l]$

(i) extend (i, j) to left and right until $\begin{cases} l+1 \text{ mismatches or} \\ \text{end of } \mathbf{a} \text{ or } \mathbf{b} \end{cases}$

To conclude, we estimate the advantage of double filtration.

Theorem 8.4 *Suppose $\mathbf{a} = a_1 a_2 \cdots a_n$ and $\mathbf{b} = b_1 b_2 \cdots b_m$ are iid sequences with $p = P(a_i = b_j)$. Let X be the number of potential length t matches with up to l mismatches.*

(i) *For continuous k-tuple filtration with $k = \lfloor \frac{t}{l+1} \rfloor$,*

$$\mathbb{E}(X) = (n - k + 1)(m - k + 1) p^k.$$

(ii) *For double filtration with a continuous $k = \lfloor \frac{t}{l+1} \rfloor$-tuple and gapped k-tuple of gap size $l + 1$,*

$$\mathbb{E}(X) \le (n - k + 1)(m - k + 1)(t + 1) p^{2k - \delta},$$

where $\delta = \lceil \frac{k}{l+1} \rceil$.

Proof. As earlier noted (i) is straightforward. To prove (ii), note that a gapped k-tuple with gap size $l+1$ has at most $\lceil \frac{k}{l+1} \rceil$ letters in common with a continuous k-tuple. With a fixed continuous k-tuple, there are no more than k such intersecting gapped k-tuples.

$$\mathbb{E}(X) = \sum_{(i,j)} \sum_{\substack{\text{gapped} \\ k\text{-tuples}}} \mathbb{P}(\text{continuous } k\text{-tuple at}(i,j)) \times \mathbb{P}(\text{gapped } k\text{-tuple}|\text{continuous } k\text{-tuple})$$

Obviously,
$$\mathbb{P}(\text{continuous } k\text{-tuple at}(i,j)) = p^k.$$

There are no more than $t - l - (-l) + 1 = t + 1$ gapped k-tuples of gap size $l+1$ of distance $d \in [-l, t-l]$. Each can intersect no more than $\delta = \lceil \frac{k}{l+1} \rceil$ letters with the continuous k-tuple. Therefore

$$\sum_{\substack{\text{gapped} \\ k\text{-tuples}}} \mathbb{P}(\text{gapped } k\text{-tuple}|\text{continuous } k\text{-tuple}) \leq (t+1) p^{k-\delta}$$

∎

8.7 Sequence Comparison by Statistical Content

The techniques of this chapter can be used to find the statistical content of a sequence. The simplest statistics that are used to summarize a sequence are the single letter counts or the 1-tuple counts. Usually, k-tuple counts for all k up to some fixed value comprise the count statistics. For $|\mathcal{A}| = d$, the k-tuple counts can obviously be obtained in time $O(n \times d^k)$ and storage $O(d^k)$ for $\mathbf{a} = a_1 a_2 \cdots a_n$. It is common practice to find these statistics for genes, introns, and even genomes.

When comparing two sequences, a question asked is whether the underlying probability distributions of the two sequences are themselves different. We will not present the detailed results here, but in Chapter 12.1 we give a multivariate central limit theorem for the count statistics from a sequence. It is not difficult to test whether two independent observations from multivariate normals are identically distributed. This gives us a way to compare two sequences.

Another problem of interest is to use statistical content to locate possible similarities with $\mathbf{a} = a_1 a_2 \cdots a_n$ in a database. For this problem, it is easy to find the content of all intervals of length n as we move along the database and test each interval for identity with $\mathbf{a}$. Issues of multiple hypothesis testing come up here as well as dependence between overlapping intervals.

Problems

Problem 8.1 Make a suffix tree for the first 20 letters of mushroom 5S ribosomal RNA.

Problem 8.2 Suppose that gaps are weighted by the function $w(i)$ where, in Algorithm 8.6, $w(i) = gi$. Modify Algorithm 8.6 to accommodate this change.

Problem 8.3 In some cases, matches are weighted according to the matching letters by $s(\cdot, \cdot)$ defined on $\mathcal{A} \times \mathcal{A}$. For example, $s(A, A) = 2$ and $s(T, T) = 3$ might be required. Of course, we have used $s(\alpha, \alpha) = 1$ for all $\alpha \in \mathcal{A}$. Modify Algorithm 8.7 to accommodate this change.

Problem 8.4 At each potential match in Theorem 8.4(i) what is the probability distribution of the number of additional positions outside the k-tuple must be checked in the "extend to the left and right" until $l + 1$ mismatches are located? You can neglect the effects of the sequence ends.

Problem 8.5 Show that in Algorithm 8.9 the distance between the continuous and gapped k-tuples must be in $[-l, t - l]$.

Chapter 9

Dynamic Programming Alignment of Two Sequences

The methods of sequence comparison in the previous chapter have the advantage of being simple to understand and rapid to execute on a computer. The basic problem being solved was left vague, however. An entire class of problems is most easily motivated by considering evolution at the molecular level. Chimpanzee and man had a recent common ancestor; the wings of birds and those of bats were evolved independently. How do we distinguish between common ancestry and common appearances? The notion that genomic DNA is the blueprint for a living organism leads to the idea that evolution must be directly related to changes in DNA. The study of history of these changes is called molecular evolution.

Molecular evolution began to be studied in the 1960s when a few protein sequences were available, notably cytochrome c (which assists in electron transport) and hemoglobin. These last sequences were known for a variety of organisms and, under the assumption that closely related organisms have similar sequences, family trees were constructed for these sequences. The broad details are usually clear–chimpanzee is more closely related to man than to rattlesnake–but some of the close relationships are difficult to deduce from the data sets. Eventually, there will be enough data to settle some of the currently controversial issues, but deeper and more subtle questions will continue to come up. Molecular evolution is a new topic that is beginning to come into its own. For our purposes, we will use evolution at the molecular level to motivate the formulation of our sequence comparison problems. In Chapter 14, we will study the problem of inferring evolutionary trees from sequences.

The simplest events that occur during the course of molecular evolution are substitution of one base for another and the insertion or deletion of a basepair. You might think of these as typing errors made by the replication machinery, although other events such as radiation can cause these changes. These events are easily indicated in our linear representation of DNA or protein sequences. For example, if $\mathbf{a}$ = ACTGC undergoes a substitution of C for T = a_3, then $\mathbf{a} \to \mathbf{b}$ = ACCGC. Biologists usually represent the transformation by an alignment where $\mathbf{a}$ is written over $\mathbf{b}$ with letters appropriately aligned. For example,

$$\mathbf{a} = \text{ACTGC},$$
$$\mathbf{b} = \text{ACCGC}.$$

If a_2 is deleted, we maintain the alignment by replacing a_2 = C with $-$, a null character:
$$\mathbf{b} = \text{ACCGC},$$
$$\mathbf{c} = \text{A--CGC}.$$

Here, $\mathbf{a}$ has become $\mathbf{c}$ = ATGC. Next, an insertion of T between b_3 = G and b_4 = C is shown by
$$\mathbf{c} = \text{ACCG--C},$$
$$\mathbf{d} = \text{A--CGTC}.$$

Note that it is not possible to distinguish insertions from deletions in an alignment. From the last alignment alone, the insertion of T in $\mathbf{d}$ might have been a deletion of T in $\mathbf{c}$.

As more time separates sequences, changes accumulate that obscure their relationships. In the above example, the relationship between $\mathbf{a}$ and $\mathbf{d}$ is

$$\mathbf{a} = \text{ACTG--C},$$
$$\mathbf{d} = \text{A--CGTC}. \qquad (9.1)$$

The goal of sequence alignment is to infer the true evolutionary relationships between two sequences without knowledge of the evolutionary events themselves. Without knowing $\mathbf{a} \to \mathbf{b} \to \mathbf{c} \to \mathbf{d}$ we surely would prefer the next alignment to (9.1):

$$\text{ACTG--C},$$
$$\text{AC--GTC},$$

because alignment (9.1) has three identities, one mismatch and two indels, whereas this alignment has four identities and two indels.

It is important to clearly set the definitions for describing sequence alignments. Two letters arranged over one another are called *matched*. If two matched letters are equal, the match is called an *identity*; otherwise the match is called a substitution or mismatch. An insertion or deletion (*indel*) is one or more letters aligned against "$-$." When only the matches and not the details of the indels are specified, the resulting arrangement is called a *trace*.

Let us look further at our simple example where $\mathbf{a}$ = ACTGC and $\mathbf{d}$ = ACGTC. The alignment in (9.1) has three identities, one mismatch, and two indels. This alignment represents a specific hypothesis about the evolution of the sequences; three of the nucleotides have not changed since the common ancestor, there has been (at least) one substitution, and two nucleotides have been either inserted or deleted. Sequences are aligned by computer to find good alignments. A computer is necessary because of the exponential number of possible alignments. For example, as we will see below, two sequences of length 1000 have over 10^{600} possible alignments.

To compute "best" alignments, we will need a way to score an alignment. There remains the question of how to score an alignment. We present a simple heuristic derivation of "alignment score." If p is the probability of an identity, q the probability of a substitution, and r the probability of an indel of a single letter, the last alignment above has probability Pr, where

$$\Pr = p^3 q r^2.$$

Define score S by the log likelihood:

$$S' = \log \Pr = 3(\log p) + (\log q) + 2(\log r).$$

Define $S = S' - 5 \log s$, where s is a constant satisfying $\log(p/s) = 1$. We have simply subtracted a constant from S'. S becomes

$$S = 3 - \mu - 2\delta,$$

where $\mu = \log(s/q)$ and $\delta = \log(\sqrt{s}/r)$. Fortunately, score, defined by

$$S = \max \{\#\text{identities} - \mu \#\text{substitutions} - \delta \#\text{indels}\},$$

can be efficiently computed as we will see in this chapter. It also has the simple maximum likelihood interpretation we have presented. We caution the reader that this simple reasoning is only a heuristic. The heuristic will be extended in Section 11.2.1, regarding statistics and alignment scores.

Needleman and Wunsch (1970) wrote a paper titled "A general method applicable to the search for similarities in the amino acid sequence of two proteins." It was surely unknown to the authors that their method fit into a broad class of algorithms introduced by Richard Bellman under the name dynamic programming. Their paper has had a great deal of influence in biological sequence alignment. Its great advantage is that an explicit criterion for optimality of alignment is stated as well as an efficient method of solution given. Insertions, deletions, mismatches (negative similarity), and matches (positive similarity) were allowed in the alignments.

During early 1970s, Stan Ulam and other mathematicians became interested in defining a distance $D(\mathbf{a}, \mathbf{b})$ on sequences. The minimum distance alignment was defined to be the one with the smallest weighted sum of mismatches, insertions, and deletions. The advantage of a distance was the construction of a metric space on the space of sequences:

1. $D(\mathbf{a}, \mathbf{b}) = 0$ if and only if $\mathbf{a} = \mathbf{b}$.

2. $D(\mathbf{a}, \mathbf{b}) = D(\mathbf{b}, \mathbf{a})$ (symmetry).

3. $D(\mathbf{a}, \mathbf{b}) \leq D(\mathbf{a}, \mathbf{c}) + D(\mathbf{c}, \mathbf{b})$ for any $\mathbf{c}$ (triangle inequality).

The emphasis on sequence metrics came from the fact that a matrix of sequence distances was often used to construct an evolutionary tree. An algorithm, very similar to that of Needleman and Wunsch, can be used to calculate the distance between sequences.

The historical order is reversed here. Distance methods are described in Section 9.3 with similarity methods in Section 9.4. We find similarity to be the most satisfactory since all problems known to be solvable with distance methods can be solved with similarity methods. However in Section 9.6 a similarity solution is given that has no distance counterpart. Still the metric space associated with a distance makes it worthwhile to present distance. Section 9.5 shows several simple modifications to solve a related problem of best fit of a short sequence into a long one. Section 9.6 studies the important problem of locating segments of two sequences which are unexpectedly similar, although the full sequences might not have a good alignment. Variations of these themes allow a variety of problems to be solved within one framework.

9.1 The Number of Alignments

In this section, a brief combinatorial treatment of sequence alignments is given. Biology provides the motivation for aligning sequences and for considering how difficult alignment is. It is then a mathematical task to estimate the number of sequence alignments. The results are applicable to biology in a negative sense; they assure one that a huge number of possible alignments exist and that direct enumeration is hopeless.

Notation is important here. Let $\mathbf{a} = a_1 a_2 \cdots a_n$ and $\mathbf{b} = b_1 b_2 \cdots b_m$ be two sequences of length n and m. One way to think of alignment is that an alignment is produced when null elements, $-$, are inserted into the sequences; the new sequences must both be of the same length L. Then the two sequences are written, one over the other. $\mathbf{a} = a_1 a_2 \cdots a_n$ becomes, with insertion of $-$, $\mathbf{a}^* = a_1^* a_2^* \cdots a_L^*$ whereas $\mathbf{b} = b_1 b_2 \cdots b_m$ becomes $\mathbf{b}^* = b_1^* b_2^* \cdots b_L^*$. The subsequence of $\mathbf{a}^*$ or $\mathbf{b}^*$ whose elements are not equal to $-$ is the original sequence. The alignment is

$$a_1^* \; a_2^* \; \cdots \; a_L^*$$
$$b_1^* \; b_2^* \; \cdots \; b_L^*.$$

To see this process, let $\mathbf{a} = $ ATAAGC and $\mathbf{b} = $ AAAAACG. To obtain an alignment, one of many possibilities is to set $\mathbf{a}^* = $ $-$ATAAGC$-$ and $\mathbf{b}^* = $ AAAAA$-$ CG. For example, $b_1^* = $ A, $b_6^* = -$, and $b_7^* = $ C whereas $b_1 = $ A, $b_6 = $ C and $b_7 = $ G. The alignment is written

$$\mathbf{a}^* = -\text{ATAAGC}-,$$
$$\mathbf{b}^* = \text{AAAAA}-\text{CG}.$$

Here, $b_1^* = $ A is said to be inserted into the first sequence or deleted from the second, depending on the point of view. $a_2^* = $ A matches $b_2^* = $ A to make an identity, whereas $a_3^* = $ T matches $b_3^* = $ A to make a mismatch.

The problem of this section is to find how many ways a can be aligned with b. No alignment terms ($\bar{\ }$) are allowed, as there is no point in matching two deletions. This makes it clear that $\max[n,m] \le L \le n+m$. The case $L = n+m$ comes by first deleting all a_i and then deleting all b_j:

$$a_1\ a_2\ \cdots\ a_n\ -\ -\ \cdots\ -$$
$$-\ -\ \cdots\ -\ b_1\ b_2\ \cdots\ b_m.$$

Combinatorial insight comes by recognizing that alignments of two sequences can end in exactly one of three ways

$$\begin{array}{ccc} \cdots a_n & \cdots a_n & \cdots - \\ \cdots - & \cdots b_m & \cdots b_m, \end{array}$$

where $\binom{a_n}{-}$ corresponds to an insertion/deletion of a_n, $\binom{a}{b}$ corresponds to an indentity or substitution, and $\binom{-}{b_m}$ corresponds to an insertion/deletion of b_m. Note that the fate of the unseen bases (those not displayed) is not specified. Define
$$f(i,j) = \text{number of alignments of one sequence of } i$$
$$\text{letters with another of } j \text{ letters.}$$

Theorem 9.1 *Let $f(n,m)$ be defined as above. Then*

(i) $f(n,m) = f(n-1,m) + f(n-1,m-1) + f(n,m-1)$

and

(ii) $f(n,n) \sim (1+\sqrt{2})^{2n+1} n^{-1/2}$,

as $n \to \infty$, where $c(n) \sim d(n)$ means $\lim_{n \to \infty} c(n)/d(n) = 1$.

Proof. To obtain (i), review the three situations above. The idea is to focus on the end of the alignment. If a_n is deleted, then there exist $f(n-1,m)$ alignments of the earlier part of the sequence. If a_n and b_m are aligned, $f(n-1,m-1)$ alignments result. If b_m is deleted, then $f(n,m-1)$ alignments result. Therefore,

$$f(n,m) = f(n-1,m) + f(n-1,m-1) + f(n,m-1).$$

The proof of (ii) is lengthy and is omitted here. ∎

Two sequences of length 1000, for example, have

$$f(1000,1000) \approx (1+\sqrt{2})^{2001} \sqrt{1000} = 10^{767.4\ldots}$$

alignments! There are approximately 10^{80} elementary particles in the universe; Avogadro's number is on the order of 10^{23}. Obviously, we cannot examine all alignments even if every molecule in the universe is operating in parallel at the speed of the fastest possible computer.

If it is agreed not to count

$$\begin{matrix} C & - \\ - & G \end{matrix} \quad \text{and} \quad \begin{matrix} - & C \\ G & - \end{matrix}$$

as distinct, the situation improves (slightly). Let $g(n,m)$ denote this smaller number of alignments. $\binom{-}{b_m}$ has three possibilities

$$\begin{matrix} \cdots & a_n & - \\ \cdots & b_{m-1} & b_m \end{matrix} \quad \begin{matrix} \cdots & - & - \\ \cdots & b_{m-1} & b_m \end{matrix} \quad \begin{matrix} \cdots & a_n & - \\ \cdots & - & b_m, \end{matrix}$$

whereas $\binom{a_n}{-}$ has

$$\begin{matrix} \cdots & a_{n-1} & a_n \\ \cdots & b_m & - \end{matrix} \quad \begin{matrix} \cdots & a_{n-1} & a_n \\ \cdots & - & - \end{matrix} \quad \begin{matrix} \cdots & - & a_n \\ \cdots & b_m & -. \end{matrix}$$

The new version of the recursion equation is

$$\begin{aligned} g(n,m) &= g(n-1,m) + g(n,m-1) + g(n-1,m-1) - g(n-1,m-1) \\ &= g(n-1,m) + g(n,m-1), \end{aligned}$$

subtracting the double count. The result is given in the next theorem, where the asymptotics are derived from Stirling's formula.

Stirling's formula

$$n! \sim \sqrt{2\pi n}\, n^{n+1/2} e^{-n}.$$

Theorem 9.1 gave results for counting alignments whereas Theorem 9.2 gives results for counting traces.

Theorem 9.2 *If $g(n,m)$ is defined as above, $g(0,0) = g(0,1) = g(1,0) = 1$, and $g(n,m) = \binom{n+m}{n}$. If $n = m$,*

$$g(n,n) = \binom{2n}{n} \sim 2^{2n}(\sqrt{n\pi})^{-1}, \quad \text{as } n \to \infty.$$

Proof. The recursion

$$g(n,m) = g(n-1,m) + g(n,m-1)$$

has solution $\binom{n+m}{n} = g(n,m)$. The result can also be derived as follows: The new way of counting alignments (actually traces) is to identify aligned pairs $\genfrac{}{}{0pt}{}{a_i}{b_j}$ and to ignore permutations of $\genfrac{}{}{0pt}{}{a_l\ a_{l+1}}{-\ -}\ \genfrac{}{}{0pt}{}{-}{b_k}\cdots$. The key is to realize that there must be k aligned pairs, $0 \leq k \leq \min\{n,m\}$. There are $\binom{n}{k}$ ways to chose a's and

Dynamic Programming Alignment of Two Sequences 189

$\binom{m}{k}$ ways of choosing b's, so there are $\binom{n}{k}\binom{m}{k}$ alignments with k aligned pairs. Therefore,

$$g(n,m) = \sum_{k \geq 0} \binom{n}{k}\binom{m}{k} = \binom{n+m}{k},$$

where the last equality is an exercise. ∎

Two sequences $n = m = 1000$ have $g(1000, 1000) \sim 10^{600}$ traces so that direct search is still impossible.

It is possible to further reduce the number of alignments by requiring matches (identities and mismatches) to occur in blocks of length at least b without interruptions by deletions. The motivation for this is that biologists sometimes reject alignments with small groups of matches. The counting scheme of Theorem 9.1 can be used with this new requirement.

Let $g(b, n)$ denote the number of alignments of two sequences of length n in which all matching blocks have size at least b. Equivalently, $g(b, n)$ is the number of $(0, 1)$ matrices with two rows and an unspecified number of columns such that both rows contain precisely n 1's, each column contains at least one 1, and columns with two 1's occur in adjacent blocks of size b or more. We are interested in the asymptotic behavior of $g(b, n)$ for fixed b as $n \to \infty$, as a function of b.

Observe that alignments where no column sum equals 2 are simply permutations of n columns with a single 1 in row 1 and n columns with a single 1 in row 2. These alignments with no matchings satisfy the criteria for any b. Thus, for all b and n,

$$g(b, n) \geq \binom{2n}{n}.$$

Applying Stirling's formula as above (with b fixed),

$$g(b, n) \geq ((\pi n)^{-1/2})(4^n + o(1)) \text{ as } n \to \infty.$$

The next theorem is given without proof.

Theorem 9.3 *Let $g(b, n)$ be the number of alignments of two sequences of length n where matches must occur in blocks of length at least $b \geq 1$. Define $\phi(x) = (1-x)^2 - 4x(x^b - x + 1)^2$, and let ρ be the smallest positive real root of $\phi(x) = 0$. Then*

$$g(b, n) \sim (\gamma_b n^{-1/2}) D_b^n, \text{ as } n \to \infty,$$

where $D_b = \rho^{-1}$ and

$$\gamma_b = (\rho^b - \rho + 1)(-\pi \rho \phi'(\rho))^{-1/2}.$$

To see the relationship to Theorem 9.1, note that $g(1, n) = f(n, n)$. For $b = 1$, then,

$$\phi(x) = (1 - x)^2 - 4x = 1 - 6x + x^2,$$

b	D_b	γ_b
1	5.8284	0.57268
2	4.5189	0.53206
3	4.1489	0.54290
4	4.0400	0.55520
5	4.0103	0.56109
10	4.00001	0.564183

Table 9.1: *Number of alignments for block length b*

and $\phi(x) = 0$ when $x = 3 \pm 2\sqrt{2}$. Therefore, $\rho = 3 - 2\sqrt{2}$ and

$$\gamma_b = 0.5727$$

and

$$D_b = (3 - 2\sqrt{2})^{-1} = 5.828.$$

We have, then,

$$f(n,n) = g(1,n) \sim (0.5727)n^{-1/2}(5.828)^n$$

Note then $(1 + \sqrt{2})^{2n} = (5.828)^n$.

Table 9.1 shows the behavior for $b \geq 1$. When $b = 2$, for example,

$$g(2,n) \sim (0.53206)n^{-1/2}(4.5189)^n.$$

Corollary 9.1 *As $b \to \infty$, $D_b \to 4$ and $\gamma_b \to \pi^{-1/2}$.*

9.2 Shortest and Longest Paths in a Network

In this section, we present a general algorithm for finding the shortest or longest path in a network. Let the vertices of the graph be $v_0, v_1, v_2, \ldots, v_N$. We assume the arcs are directed, from v_i to v_j where $i < j$. v_0 is, of course, the origin. Define

$$W(i,j) = \text{weight of the edge } (v_i, v_j),$$

$$L(j) = \min\{W(0, i_1) + W(i_1, i_2) + \cdots + W(i_k, j) :$$

$$0 < i_1 < i_2 \cdots < i_k < j;\ \text{all } k\}.$$

Algorithm 9.1 (Shortest Path)
```
input:   W(i,j), 0 ≤ i < j ≤ N
output:  L(1) ··· L(N)
```

Dynamic Programming Alignment of Two Sequences

for $j = 1$ to N

$\quad L(j) \leftarrow W(0, j)$

$\quad$ for $i = 1$ to $j - 1$

$\quad\quad L(j) \leftarrow \min\{L(j), L(i) + W(i, j)\}$

Proof. Assume $L(0), L(1), \ldots, L(j-1)$ are all computed correctly. Then the last loop of the algorithm states that

$$L(j) = \min\{W(0, j), L(1) + W(1, j), \ldots, L(j-1) + W(j-1, j)\}.$$

This equation is correct, as the shortest path from 0 to j has a last edge, $W(i_k, j)$, $i_k < j$. Then,

$$L(j) = W(0, i_1) + \cdots + W(i_k, j)$$

$$= \bigl(W(0, i_1) + \cdots + W(i_{k-1}, i_k)\bigr) + W(i_k, j).$$

The quantity in parentheses must be the shortest path from 0 to i_k, or else $L(j)$ is not minimal. Thus,

$$L(j) = L(i_k) + W(i_k, j).$$

∎

The running time of the algorithm is proportional to

$$\sum_{j=1}^{N}\sum_{i=1}^{j-1} 1 = \sum_{j=1}^{N}(j-1) = O(N^2).$$

Sequence alignments can be formulated as a path in a network. The nodes of the network are (i, j), where each (i, j) node denotes an aligned pair $\genfrac{}{}{0pt}{}{a_i}{b_j}$. There are nm nodes in the graph. There are edges between nodes (i, j) and (k, l) if $i \leq k$ and $j \leq l$, so the algorithm requires a slight generalization. The weight of the edges will be discussed in the next section, but the complexity of finding a shortest path by the algorithm shortest path is $O(n^2m^2)$ by our running time analysis. It turns out to be easy to reduce this to $O(nm)$ for many cases, by exploiting the special properties of our edge weights.

To show more explicitly the connection between alignment and the network graph, it is necessary to add a source and sink to the network. In Figure 9.1, we show an example network for $\mathbf{a} =$ TAGGCA and $\mathbf{b} =$ ATGGAA. The source in the upper left-hand corner is denoted by "o" and the sink in the lower right-hand corner is denoted by "*." The solid arcs correspond to alignments with aligned letters $\binom{A}{A}$, $\binom{G}{G}$, and $\binom{A}{A}$. This does not determine the precise alignment however. Only the matches are specific edges in our example:

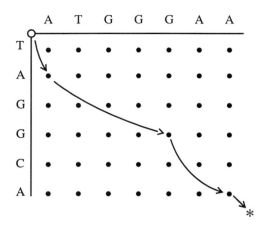

Figure 9.1: *Alignment as a directed network*

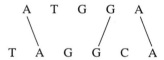

What remains for defining alignment from trace is to specify the order of the deletions. Weight of the arcs must depend on the inserted/deleted letters as well as the visited nodes. Deletion of both sequences corresponds to a single arc from source to sink.

9.3 Global Distance Alignment

The sequences $\mathbf{a} = a_1 a_2 \cdots a_n$ and $\mathbf{b} = b_1 b_2 \cdots b_m$ are written over the finite alphabet $\mathcal{A}$. In particular, the 20-letter, amino acid alphabet of proteins or the purine/pyrimidine alphabet for DNA can be used. Let $d(a, b)$ be a distance on the alphabet and let $g(a)$ be the positive cost of inserting or deleting of the letter a. The distance $d(a, b)$ represents the cost of a mutation of a into b. If $d(a, b)$ is extended so that $d(a, -) = d(-, a) = g(a)$, then define

$$D(\mathbf{a}, \mathbf{b}) = \min \sum_{i=1}^{L} d(a_i^*, b_i^*),$$

where the minimum is extended over all alignments of $\mathbf{a}$ with $\mathbf{b}$. See Figure 9.2 for a schematic of *global* alignment, global denoting the feature that all letters of $\mathbf{a}$ and $\mathbf{b}$ must be accounted for in the alignment.

Dynamic Programming Alignment of Two Sequences

Figure 9.2: *A schematic global alignment*

Theorem 9.4 *If* $\mathbf{a} = a_1 a_2 \cdots a_n$ *and* $\mathbf{b} = b_1 b_2 \cdots b_m$, *define*

$$D_{i,j} = D(a_1 a_2 \cdots a_i, b_1 b_2 \cdots b_j).$$

Also set

$$D_{00} = 0, D_{0,j} = \sum_{k=1}^{j} d(-, b_k), \text{ and } D_{i,0} = \sum_{k=1}^{i} d(a_k, -).$$

Then

$$D_{i,j} = \min\{D_{i-1,j} + d(a_i, -), D_{i-1,j-1} + d(a_i, b_j), D_{i,j-1} + d(-, b_j)\}. \quad (9.2)$$

If $d(\cdot, \cdot)$ *is a metric on the alphabet, then* $D(\cdot, \cdot)$ *is a metric on the set of finite sequences.*

Proof. We verify Equation (9.2) with reasoning similar to that for verifying the recursion equation for $f(n, m)$ in Section 10.1. The alignment of $a_1 \cdots a_i$ and $b_1 \cdots b_j$ can end in one of three ways

$$\begin{array}{ccc} \cdots a_i & \cdots a_i & \cdots - \\ \cdots - & \cdots b_j & \cdots b_j. \end{array}$$

If the optimal alignment ends in $\binom{a_i}{-}$, the cost must be $D_{i-1,j} + d(a_i, -)$ because the initial part of the alignment must itself be optimal and align $a_1 \cdots a_{i-1}$ with $b_1 \cdots b_j$.

If the optimal alignment ends in $\binom{a_i}{b_j}$, the cost must be $D_{i-1,j-1} + d(a_i, b_j)$ because $a_1 \cdots a_{i-1}$ and $b_1 \cdots b_{j-1}$ must be optimally aligned.

The case $\binom{-}{b_j}$ is identical in reasoning with the case $\binom{a_i}{-}$.

The optimal alignment has least cost of these three possibilities and (1) is proven. ∎

Another statement of Equation (9.1) with $\delta = d(a, -) = d(-, a)$ is

Algorithm 9.2 (Global Distance)
```
input:   a, b, d(a, b)
output:  D_{i,j}, 0 ≤ i ≤ n,  0 ≤ j ≤ m
```

$D_{i,j} = 0$ for $i = 0, j = 0, \ldots, m$ and $j = 1, i = 0, 1, \ldots, n$.
for $i = 1$ to n
    for $j = 1$ to m
        $D_{i,j} \leftarrow \min\{D_{i-1,j} + \delta, D_{i-1,j-1} + d(a_i, b_j), D_{i,j-1} + \delta\}$.

Corollary 9.2 *The running time of Algorithm 9.2 is $O(nm)$.*

To obtain all optimal alignments, there are two techniques. The first involves saving pointers at each (i, j). The pointers show which of $(i-1, j), (i-1, j-1)$, or $(i, j-1)$ were involved with the optimal $D_{i,j}$. Then, when $D_{n,m}$ is found, the pointers are followed to produce alignments. The second technique, called a traceback, is to ask, at each (i, j) on an optimal path, which of $(i-1, j)$, $(i-1, j-1)$, or $(i, j-1)$ are optimal by recomputing the three terms. The alignments in these cases are produced by depth first search with stacking, where the stacks are managed by last in–first out. An example of distance alignment is shown in Table 9.2 with $\mu = -1$ and $\delta = 1$. The identities in the optimal alignment are shown by bars where the optimal alignment is

```
- G C T G A T A T A G C T
  | | | | |   | | | | |
G G G T G A T - T A G C T.
```

9.3.1 Indel Functions

Frequently in sequence evolution, deletion (or insertion) of several adjacent letters are not the sum of single deletions (or insertions) but the result of one event. Thus,

	–	G	G	G	T	G	A	T	T	A	G	C	T
–	0	1	2	3	4	5	6	7	8	9	10	11	12
G	1	0	1	2	3	4	5	6	7	8	9	10	11
C	2	1	1	2	3	4	5	6	7	8	9	9	10
T	3	2	2	2	2	3	4	5	6	7	8	9	9
G	4	3	2	2	3	2	3	4	5	6	7	8	9
A	5	4	3	3	3	3	2	3	4	5	6	7	8
T	6	5	4	4	3	4	3	2	3	4	5	6	7
A	7	6	5	5	4	4	4	3	3	3	4	5	6
T	8	7	6	6	5	5	5	4	3	4	4	5	5
A	9	8	7	7	6	6	5	5	4	3	4	5	6
G	10	9	8	7	7	6	6	6	5	4	3	4	5
C	11	10	9	8	8	7	7	7	6	5	4	3	4
T	12	11	10	9	8	8	8	7	7	6	5	4	3

Table 9.2: *Distance alignment*

Dynamic Programming Alignment of Two Sequences

it is sometimes required to weight these multiple indels differently from summing single letter indel weights. Let $g(k)$ be the indel weight for an indel of k bases. It is reasonable that $g(k) \leq kg(1)$ holds.

Theorem 9.5 Set $D_{i,j} = D(a_1 \cdots a_i, b_1 \cdots b_j)$, $D_{0,0} = 0$, $D_{0,j} = g(j)$, and $D_{i,0} = g(i)$. Then

$$D_{i,j} = \min \left\{ \begin{array}{l} D_{i-1,j-1} + d(a_i, b_j), \\ \min_{1 \leq k \leq j}\{D_{i,j-k} + g(k)\}, \\ \min_{1 \leq l \leq i}\{D_{i-l,j} + g(l)\} \end{array} \right\}.$$

Proof. The proof proceeds as in Theorem 9.4, except here, for example, the alignments can end in deletions of up to i letters of the sequence $a_1 a_2 \cdots a_i$. Therefore, to replace $D_{i-1,j} + g(1)$, we have

$$\min_{1 \leq l \leq i}\{D_{i-l,j} + g(l)\}.$$

∎

The computation time of this algorithm is $\sum_{i,j}(i+j) = O(n^2 m + nm^2)$, or $O(n^3)$ if $n = m$. For sequences of length 1000 or more, it is important to reduce this running time. This is easily accomplished for this case of $g(k)$ linear.

Theorem 9.6 Let $g(k) = \alpha + \beta(k-1)$ for constants α and β. Set $E_{0,0} = F_{0,0} = D_{0,0} = 0$, $E_{i,0} = D_{i,0} = g(i)$, and $F_{0,j} = D_{0,j} = g(j)$. If $E_{i,j}$ and $F_{i,j}$ satisfy

$$E_{i,j} = \min\{D_{i,j-1} + \alpha, E_{i,j-1} + \beta\}$$

and

$$F_{i,j} = \min\{D_{i-1,j} + \alpha, F_{i-1,j} + \beta\},$$

then

$$D_{i,j} = \min\{D_{i-1,j-1} + d(a_i, b_j); E_{i,j}; F_{i,j}\}.$$

Proof. The identities that we need to establish are

$$E_{i,j} = \min_{1 \leq k \leq j}\{D_{i,j-k} + g(k)\}$$

and

$$F_{i,j} = \min_{1 \leq l \leq i}\{D_{i-l,j} + g(l)\}.$$

We will prove the identity for $E_{i,j}$. ($F_{i,j}$ follows in a similar manner.)
Assume

$$E_{i,j^*} = \min_{1 \leq k \leq j^*}\{D_{i,j^*} + g(k)\} \text{ for } 0 \leq j^* < j.$$

Then

$$\min_{1\leq k\leq j}\{D_{i,j-k} + g(k)\}$$

$$= \min\left\{\min_{2\leq k\leq j}\{D_{i,j-k} + g(k)\}, D_{i,j-1} + g(1)\right\}$$

$$= \min\left\{\min_{1\leq k-1\leq j-1}\{D_{i,(j-1)-(k-1)} + g(k-1)\} + \beta, D_{i,j-1} + \alpha\right\}$$

$$= \min\{E_{i,j-1} + \beta, D_{i,j-1} + \alpha\}.$$

∎

We say $g(k)$ is *concave* if $g(k+m+l) - g(k+m) \leq g(k+l) - g(k)$ for all $k, l, m \geq 0$. If $g(k)$ is convex, an $O(nm)$ algorithm is easier to obtain than when $g(k)$ is concave. However, a convex indel function seems an unlikely situation in biology. The next theorem is given without proof.

Theorem 9.7 *If $g(k)$ is concave, such as $g(k) = \alpha + \beta \log(k)$, then $O(n^2 \log(n))$ running times can be obtained with a somewhat more complex algorithm.*

A key to thinking about these algorithms is to identify steps in the matrix (or the alignment network) with steps in the algorithm. First of all, recall the general shortest path algorithm took time $O(n^2 m^2) = O(n^4)$ when $n = m$, whereas our algorithms take time $O(n^2)$ and $O(n^3)$. The reason is in the portion of the matrix that must be searched when finding $D_{i,j}$. In the shortest path algorithm 9.1 we must search over the entire rectangle of area $(i+1)(j+1) - 1$ (excluding (i,j)) in Figure 9.3. In Algorithm 9.2 and Theorem 9.6 we just search over the three cells $(i-1, j-1)$ (match) and $(i, j-1)\&(i-1, j)$ (indels). In Theorem 9.5 for general indel weights, we search over the i-th row and j-th column and $(i-1, j-1)$ for a total of $(1 + i + j)$ cells. What has happened in all these more efficient algorithms is that long edges in Algorithm 9.1 have been represented as the sum of short edges, and we have carefully taken only those last short edges entering (i,j).

Observe that any indel function $0 \leq g$ can be composed to a minimum path $\hat{g}$:

$$\hat{g}(k) = \min\{g(l_1) + \cdots + g(l_k) : \Sigma l_i = k, 0 \leq l_i \leq k\}.$$

Then the correct weight for length k indels is $\hat{g}(k)$.

Proposition 9.1 *$\hat{g}$ is subadditive.*

Proof. $\hat{g}(k+l) \leq \hat{g}(k) + \hat{g}(l)$ by definition. ∎

An easy application of Algorithm 9.1 gives

Algorithm 9.3 ($\hat{g}$)
```
input:    g
```

Dynamic Programming Alignment of Two Sequences

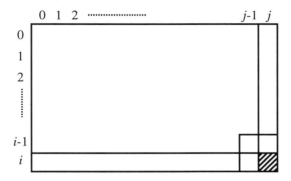

Figure 9.3: *Computation of all* (i, j)

```
for  k = 0, . . . , max{n, m}
    ĝ(k) ← g(k)
    for l = 1 to k − 1
        ĝ(k) ← min{ĝ(l) + g(k − l), ĝ(k)}
```

If, for example, $g(k) = \alpha + \beta(k-1)$ with $0 \le \alpha \le \beta$, then $\hat{g}(k) = \alpha k$, whereas if $\beta < \alpha$, then $\hat{g}(k) = g(k)$. It cannot be assumed that $\hat{g}$ is monotone or concave: For $g(k) = k$, $k \ne 3$, and $g(3) = 0$, we obtain $\hat{g}(k) = k$ modulo 3. It does follow in general from subadditivity that $\lim_{k \to \infty} \frac{\hat{g}(k)}{k} = \gamma \ge 0$.

9.3.2 Position-Dependent Weights

In some applications, the weights given to substitutions are dependent on the positions in the sequences. It is routine to generalize our algorithms to fit this general setting. Define $s_{i,j}(a, b)$ to be the substitution function for aligning a and b in the i-th and j-th positions of $\mathbf{a} = a_1 \cdots a_n$ and $\mathbf{b} = b_1 \cdots b_m$. Indels are slightly more subtle to set up. At the i-th position of the a sequence, the weight is α_i if deleting a_i begins the indel, or β_i if deleting a_i extends the indel. Similarly, the weight is γ_j if deleting b_j begins the indel, and δ_j if deleting b_j extends the indel. The extension of the recursion of Theorem 9.6 is

$$E_{i,j} = \min\{D_{i,j-1} + \gamma_j, E_{i,j-1} + \delta_j\},$$
$$F_{i,j} = \min\{D_{i-1,j} + \alpha_j, F_{i,j-1} + \beta_j\},$$

and

$$D_{i,j} = \min\{D_{i-1,j-1} + s_{i,j}(a_i, b_j), E_{i,j}, F_{i,j}\}.$$

We remark that if $\alpha_i \neq \beta_i$ or $\gamma_j \neq \delta_j$, it is not guaranteed that

$$D(a_1 \cdots a_n, b_1 \cdots b_m) = D(a_n \cdots a_1, b_m \cdots b_1),$$

as these indel functions are dependent on which end "starts" the indel. No such problem arises if $\alpha_i = \beta_i$ and $\gamma_j = \delta_j$, all i and j, or if $\alpha_i \equiv \alpha, \beta_i \equiv \beta, \gamma_j \equiv \gamma$, and $\delta_j \equiv \delta$.

9.4 Global Similarity Alignment

Now take $s(a, b)$ to be a similarity measure on the alphabet, that is, we have $s(a, a) > 0$ for all a; for some (a, b) pairs, it is necessary that $s(a, b) < 0$. The idea is that similarity (e.g., matching a with a) is rewarded by a positive score, whereas aligning dissimilar letters is penalized by a negative score. Let $-\hat{g}(a)$ be the indel penalty associated with a. ($\hat{g}$ is used to distinguish this function from the indel weight of distance alignment.) Set $s(a, -) = s(-, a) = -\hat{g}(a)$. Then define the similarity of **a** and **b** by

$$S(\mathbf{a}, \mathbf{b}) = \max \sum_{i=1}^{L} s(a_i^*, b_i^*),$$

where the maximum is over all alignments. The result analogous to Theorem 10.6 for similarity alignments is

Theorem 9.8 *If* $\mathbf{a} = a_1 a_2 \cdots a_n$ *and* $\mathbf{b} = b_1 b_2 \cdots b_m$, *define*

$$S_{i,j} = S(a_1 a_2 \cdots a_i, b_1 b_2 \cdots b_j).$$

Also, set

$$S_{0,0} = 0, S_{0,j} = \sum_{k=1}^{j} s(-, b_k), \text{ and } S_{i,0} = \sum_{i=1}^{i} s(a_i, -).$$

Then,

$$S_{i,j} = \max\{S_{i-1,j} + s(a_i, -), S_{i-1,j-1} + s(a_i, b_j), S_{i,j-1} + s(-, b_j)\}.$$

If $s(a, -) = s(-, a) = -\hat{\delta}$, *for all* a, *then*

$$S_{i,j} = \max\{S_{i-1,j} - \hat{\delta}, S_{i-1,j-1} + s(a_i, b_j), S_{i,j-1} - \hat{\delta}\}.$$

Proof. This theorem is proved exactly as was Theorem 10.6. ∎

To illustrate the similarity algorithm, we align the same sequences as above, *E. coli* threonine tRNA and *E. coli* valine tRNA. We use a single letter indel algorithm and choose the parameters $s(a, a) = +1$; $s(a, b) = -1$ if $a \neq b$; and $\hat{g} = 2$. Table 9.3 shows the matrix (S_{ij}) for the 5' ends of the sequences. The optimal alignment is

Dynamic Programming Alignment of Two Sequences

	–	G	G	G	T	G	A	T	T	A	G	C	T
–	0	–2	–4	–6	–8	–10	–12	–14	–16	–18	–20	–22	–24
G	–2	1	–1	–3	–5	–7	–9	–11	–13	–15	–17	–19	–21
C	–4	–1	0	–2	–4	–6	–8	–10	–12	–14	–16	–16	–18
T	–6	–3	–2	–1	–1	–3	–5	–7	–9	–11	–13	–15	–15
G	–8	–5	–2	–1	–2	0	–2	–4	–6	–8	–10	–12	–14
A	–10	–7	–4	–3	–2	–2	1	–1	–3	–5	–7	–9	–11
T	–12	–9	–6	–5	–2	–3	–1	2	0	–2	–4	–6	–8
A	–14	–11	–8	–7	–4	–3	–2	0	1	1	–1	–3	–5
T	–16	–13	–10	–9	–6	–5	–4	–1	1	0	0	–2	–2
A	–18	–15	–12	–11	–8	–7	–4	–3	–1	2	0	–1	–3
G	–20	–17	–14	–11	–10	–7	–6	–5	–3	0	3	1	–1
C	–22	–19	–16	–13	–12	–9	–8	–7	–5	–2	1	4	2
T	–24	–21	–18	–15	–12	–11	–10	–7	–6	–4	–1	2	5

Table 9.3: *Similarity alignment of threonine tRNA (**a**) with valine tRNA (**b**)*

```
- G C T G A T A T A G C T
|   | | | |   | | | | |
G G G T G A T - T A G C T.
```

The multiple letter indel case is covered by the next theorem. Indel penalty $\hat{g}(k)$ is again assumed to be a function of indel length k.

Theorem 9.9 Set $S_{i,j} = S(a_1 a_2 \cdots a_i, b_1 b_2 \cdots b_j)$, $S_{0,0} = 0$, $S_{0,j} = -\hat{g}(j)$, and $S_{i,0} = -\hat{g}(i)$. Then

$$S_{i,j} = \max \begin{cases} S_{i-1,j-1} + s(a_i, b_j), \\ \max_{1 \leq k \leq j}\{S_{i,j-k} - \hat{g}(k)\}, \\ \max_{1 \leq l \leq i}\{S_{i-l,j} - \hat{g}(l)\} \end{cases}.$$

The following result obtains the $O(nm)$ running time for the case of linear indel weights.

Theorem 9.10 Let $\hat{g}(k) = \alpha + \beta(k-1)$ for constants α and β. Set $E_{0,0} = F_{0,0} = S_{0,0} = 0$, $E_{i,0} = S_{i,0} = -\hat{g}(i)$, and $F_{0,j} = S_{0,j} = -\hat{g}(j)$. If

$$E_{i,j} = \max\{S_{i,j-1} - \alpha, E_{i,j-1} - \beta\},$$

and

$$F_{i,j} = \max\{S_{i-1,j} - \alpha, F_{i-1,j} - \beta\},$$

then

$$S_{i,j} = \max\{S_{i-1,j-1} + s(a_i, b_j), E_{i,j}, F_{i,j}\}.$$

The parallels between the formulas for distance and similarity alignment lead to a natural question. When are similarity and distance algorithms equivalent? When full sequences are aligned by distance (similarity) in global alignments, there is a similarity (distance) algorithm that gives the same set of optimal alignments; that is, finding similarity and distance alignments are dual problems.

Theorem 9.11 *Let a similarity measure be given with $s(a, b)$ and indel penalties $\hat{g}(k)$ and a distance measure be given with $d(a, b)$ and indel weight $g(k)$. Assume there is a constant c, such that $s(a, b) = c - d(a, b)$ and $\hat{g}(k) = g(k) - (kc)/2$. Then an alignment is similarity optimal if and only if it is distance optimal.*

Proof. Now, by elementary counting,

$$n + m = 2\#\text{matches} + \sum_k k\Delta_k,$$

where matches means aligned letters and Δ_k is the number of indels of length k. Using this simple equation,

$$D(\mathbf{a}, \mathbf{b}) = \min\left\{\sum_{\text{matches}} d(a, b) + \sum_k g(k)\Delta_k\right\}$$

$$= \min\left\{\sum_{\text{matches}} c + \sum_k k\Delta_k c/2 - \sum_{\text{matches}} s(a, b) + \sum_k \hat{g}(k)\Delta_k\right\}$$

$$= \min\left\{c(n+m)/2 - \sum_{\text{matches}} s(a, b) + \sum_k \hat{g}(k)\Delta_k\right\}$$

$$= c(n+m)/2 - \max\left\{\sum_{\text{matches}} s(a, b) - \sum_k \hat{g}(k)\Delta_k\right\}$$

$$= c(n+m)/2 - S(\mathbf{a}, \mathbf{b}).$$

∎

Usually, $0 \le c \le \max_{a'b'} d(a', b')$ so that we have both positive and negative similarity values. Note that

$$D(\mathbf{a}, \mathbf{b}) + S(\mathbf{a}, \mathbf{b}) = c(n + m)/2,$$

so "large distance" is "small similarity." After seeing this equivalence, it is surprising that there are problems with a simple similarity algorithm for which no equivalent distance algorithm exists. This situation arises in Section 9.6.

9.5 Fitting One Sequence into Another

Next, the algorithms are modified to solve a new problem: Find the best fit of a "short" sequence into a "larger" sequence. An example of when this might be of interest is in locating a regulatory pattern in a nucleotide sequence, such as TATAAT in a bacterial promoter. The algorithm finds where the short pattern approximately appears in the longer sequence.

First consider the problem of fitting $\mathbf{a} = a_1 a_2 \cdots a_n$ into $\mathbf{b} = b_1 b_2 \cdots b_m$. For the purpose of visualizing the problem think of n as much smaller than m. (The relative sizes of n and m are irrelevant to the mathematics.) The problem is to find
$$T(\mathbf{a}, \mathbf{b}) = \max \{ S(\mathbf{a}, b_k b_{k+1} \cdots b_{l-1} b_l) : 1 \leq k \leq l \leq m \}.$$
See Figure 9.4 for a schematic for this problem. The solution to this problem as indicated in the definition will take time proportional to
$$\sum_{k=1}^{m} \sum_{k=l}^{m} n(l - k) = O(nm^3).$$

Figure 9.4: *A schematic fitting alignment*

Instead, we take another approach. Note that deletions of the beginning and end of **b** are without penalty. Certainly, deletions of the beginning of **b** are encoded in the boundary values of our usual matrix setup. Define
$$T_{i,j} = \max \{ S(a_1 a_2 \cdots a_i, b_k b_{k+1} \cdots b_j) : 1 \leq k \leq j \}.$$

Theorem 9.12 *Define $T_{0,j} = 0$, $0 \leq j \leq m$, and $T_{i,0} = -i\delta$. Then*
$$T_{i,j} = \max \{ T_{i-1,j-1} + s(a_i, b_j), T_{i,j-1} - \delta, T_{i-1,j} - \delta \}.$$

Proof. The main idea is in the boundary conditions. Initial deletions of **b**, $b_1 b_2 \cdots b_j$, are set at $T_{0,j} = 0$. Each letter of **a** must be accounted for; hence, $T_{i,0} = -i\delta$. After the initial match, deletions in **b** as well as **a** must be accounted for. The recursion for $T_{i,j}$ does just that. ∎

Corollary 9.3
$$T(\mathbf{a}, \mathbf{b}) = \max\{T(n, j), j : 1 \leq j \leq m\}.$$

Proof. To obtain unweighted ends of the alignment of **a** with $b_k \cdots b_j$, we just choose the best score $T_{n,j}$. ∎

To illustrate this algorithm, we take as sequence **b** the *E. coli* promoter sequence of *lacI*. In *E. coli* promoter sequences, the -10 signal TATAAT is well-known to have functional significance. We take **a** = TATAAT. As above $s(a, a) = 1$, $s(a, b) = -1$ if $a \neq b$, and $\delta = 2$. The matrix $(S(i, j))$ is shown in Table 9.4. Searching the last row of the matrix gives two solutions of $\max_{1 \leq j \leq 58} S(6, j) = 2$, at (6,13) and (6,43). The pattern at (6,43) has alignment

$$\text{TATAAT}$$
$$\text{CATGAT}$$

and is CATGAT in the promoter sequence, the canonical -10 pattern. The pattern at (6,13) has alignment

$$\text{TATAAT}$$
$$\text{TCGAAT}$$

with TCGAAT in the promoter sequence, an equally good fit.

This illustrates the utility of the algorithm, in that it locates the putative -10 signal CATGAT in *lacI*. It also emphasizes the difficulty of promoter signal analysis by finding an equally good pattern TCGAAT 30 bases $5'$ of the -10 pattern.

```
    G  A  C  A  C  C  A  T  C  G  A  A  T  G  G  C  G  C  A  A  A  A  C  C  T  T
T  -1 -1 -1 -1 -1 -1 -1  1 -1 -1 -1 -1  1 -1 -1 -1 -1 -1 -1 -1 -1 -1 -1 -1  1  1
A  -3  0 -2  0 -2 -2  0 -1  0 -2  0  0 -1  0 -2 -2 -2 -2  0  0  0  0 -2 -2 -1  0
T  -5 -2 -1 -2 -1 -3 -2  1 -1 -1 -2 -1  1 -1 -1 -3 -3 -3 -2 -1 -1 -1 -3 -1  0
A  -7 -4 -3  0 -2 -2 -2 -1  0 -2  0 -1 -1  0 -2 -2 -4 -4 -2 -1  0  0 -2 -2 -3 -2
A  -9 -6 -5 -2 -1 -3 -1 -3 -2 -1 -1  1 -1 -2 -1 -3 -3 -5 -3 -1  0  1 -1 -3 -3 -4
T -11 -8 -7 -4 -3 -2 -3  0 -2 -3 -2 -1  2  0 -2 -2 -4 -4 -5 -3 -2 -1  0 -2 -2 -2
```

```
    T  C  G  C  G  G  T  A  T  G  G  C  A  T  G  A  T  A  G  C  G  C  C  C  G  G  A  A  G  A  G  A  G  T
   -1 -1 -1 -1 -1 -1  1 -1  1 -1 -1 -1  1 -1  1 -1 -1 -1 -1 -1 -1 -1 -1 -1 -1 -1 -1 -1 -1 -1 -1 -1  1
    0  0 -2 -2 -2 -2 -1  2  0  0 -2 -2  0 -1  0  0 -1  2  0 -2 -2 -2 -2 -2 -2  0  0 -2  0 -2  0 -2 -1
   -1 -1 -3 -3 -3 -1  0  3  1 -1 -3 -2  1 -1 -1  1  0  1 -1 -3 -3 -3 -3 -3 -2 -1 -1 -2 -1 -2 -1 -1
   -1  0 -2 -2 -4 -4 -3  0  1  2  0 -2 -2 -1  0  0 -1  2  0  0 -2 -4 -4 -4 -4 -2 -1 -2  0 -2  0 -2 -2
   -3 -2 -1 -3 -3 -5 -5 -2 -1  0  1 -1 -1 -3 -2  1 -1  0  1 -1 -1 -3 -5 -5 -5 -3 -1 -2 -1 -1 -1 -1 -3
   -3 -4 -3 -2 -4 -4 -4 -4 -1 -2 -1  0 -2  0 -2 -1  2  0 -1  0 -2 -2 -4 -6 -6 -6 -5 -3 -2 -3 -2 -2 -2  0
```

Table 9.4: *Matrix for best fit of* TATAAT *into the* E. coli *promoter of* lacI

9.6 Local Alignment and Clumps

Surprising relationships have been discovered between sequences that overall have little similarity. These are several dramatic cases when unexpectedly long

Dynamic Programming Alignment of Two Sequences

matching segments have been located between viral and host DNA. The subject of this subsection is a dynamic programming algorithm to find these similar segments. This is probably the most useful dynamic programming algorithm for current problems in molecular biology. These alignments are called local alignments. See Figure 9.5 for a schematic. For a mathematical statement of the problem, it is necessary to assume a similarity function $s(a,b)$. The object is to find

$$H(\mathbf{a}, \mathbf{b}) = \max\{S(a_i a_{i+1} \cdots a_{j-1} a_j, b_k b_{k+1} \cdots b_{l-1} b_l) : 1 \leq i \leq j \leq n, 1 \leq k \leq l \leq m\},$$

where $H(\emptyset, \emptyset) = 0$. This amounts to $\binom{n}{2}\binom{m}{2}$ sequence alignment problems, and a new algorithm must be devised. Computing H using the method of Theorem 9.8 to find S takes time $O(n^6)$, when $n = m$.

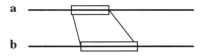

Figure 9.5: *A schematic local alignment*

When work first began on problems of this type, the problem formulations were based on distance functions and the algorithms involved forward and backward recursions, each recursion requiring a matrix. Those algorithms are quite complex. The similarity formulation given above is solved in a straightforward way. Define $H_{i,j}$ to be the maximum similarity of two segments ending at a_i and b_j:

$$H_{i,j} = \max\{0; S(a_x a_{x+1} \cdots a_i, b_y b_{y+1} \cdots b_j) : 1 \leq x \leq i, 1 \leq y \leq j\}$$

A recursion similar to those given for the similarity problems discussed above is obtained for H.

Theorem 9.13 *Set $H_{i,0} = H_{0,j} = 0$ for $1 \leq i \leq n$ and $1 \leq j \leq m$. Then*

$$H_{i,j} = \max\{0, H_{i-1,j-1} + s(a_i, b_j), \max_{1 \leq k \leq i}\{H_{i-k,j} - g(k)\},$$
$$\max_{1 \leq l \leq j}\{H_{i,j-l} - g(l)\}\}.$$

Proof. Just as before, think of possible segments ending at (i, j). If there is one positive scoring segment ending at (i, j), then they must satisfy the usual recursion. If the usual three-term recursion ends with all negative values, then there is no positive scoring segment ending at (i, j) and the value in $H_{i,j} = 0$. ∎

Of course, single or linear indels can be treated as discussed above.

Corollary 9.4

$$H(\mathbf{a}, \mathbf{b}) = \max\{H_{k,l} : 1 \leq k \leq n, 1 \leq l \leq m\}.$$

Corollary 9.5 *If* $g(k) = \alpha + \beta(k-1)$, $E_{i,j} = F_{i,j} = H_{i,j} = 0$ *when* $i \cdot j = 0$, *with*

$$E_{i,j} = \max\{H_{i,j-1} - \alpha, E_{i,j-1} - \beta\},$$

and

$$F_{i,j} = \max\{H_{i-1,j} - \alpha, F_{i-1,j} - \beta\},$$

it follows that

$$H_{i,j} = \max\{0, H_{i-1,j-1} + s(a_i, b_j), E_{i,j}, F_{i,j}\}.$$

This method will be referred to as the local or the maximum segments algorithm. What about other highly similar segments?

The following procedure finds one alignment from all alignments with the highest score and then continues to find the next best alignment with no matches or mismatches of its alignment in common with those already output. The algorithm corresponding to Theorem 9.13 stops after the best local alignment is output. The algorithm given here extends the straightforward application of Theorem 9.13. As we will see in Section 9.8, there are many alignments virtually identical to the optimal local alignment but that differ in minor detail. It is usually not feasible nor even interesting to write out all of these alignments. Instead, we ask if there are other "interesting" alignments. This is accomplished by our definition of a *clump* of alignments as a set of alignments that share one or more match (identity or mismatch) with a given alignment. We then take one alignment from each clump for the output.

When calculating the matrix H, stack all (i, j, Y) with $Y = H_{i,j}$ and $H_{i,j} \geq C =$ cutoff value. The stack is ordered by $\succ$ where $(i, j, H_{i,j}) \succ (k, l, H_{k,l})$ if

(i) $H_{i,j} > H_{k,l}$,
(ii) $H_{i,j} = H_{k,l}$ and $i + j < k + l$,

or

(iii) $H_{i,j} = H_{k,l}$, $i + j = k + l$, and $i < k$.

During tracebacks for some stack entry, we only output one alignment. For this, one additional concept is needed, that of minimum length alignment. Define the length of an alignment beginning at (p, q) and ending at (i, j) to be $|i + j - (p + q)|$. We only output minimum length alignments, although this is entirely a matter of choice.

The algorithm begins with the top (i, j, Y), i.e., the largest under "$\succ$," and that alignment is output. Next, we must find the next largest scoring alignment that

Dynamic Programming Alignment of Two Sequences

has no matches or mismatches in common with those alignments already output. The simple concept of recomputing the matrix, not allowing matches (identities or mismatches) already used, is employed. This simple device *declumps*, that is, removes the effects of all alignments in the clump. This does involve more calculation. As the elements below and to the right of an alignment's end [i.e. (i, j)] must be recomputed, each succeeding alignment takes $nm/4$ matrix entry recomputations on the average. Although this might be worth the cost because of simplicity, much more efficient algorithms can be given in the cases of single and linear indels.

For simplicity, take the single letter indel case, and let (k, l) be the upper leftmost position of a match in the alignment. (The alignment must end in a match or it is not optimal.) The new matrix H^* satisfies

$$H^*_{i,j} = H_{i,j}, \ i < k \text{ or } j < l.$$

Define $H^*_{k,l}$ by

$$H^*_{k,l} = \max\{0, H^*_{k-1,l} - g(1), H^*_{k,l-1} - g(1)\}.$$

Note that the match ending the alignment is not allowed. Consider the row $(k, j), l < j$. Recompute each entry $j = l+1, l+2, \ldots$, until $H_{k,j} = H^*_{k,j}$. Then it is clear that $H_{k,l} = H^*_{k,l}$ for the rest of the row. Similar considerations hold for the remainder of the alignment. Note that on each row and column it is necessary to go at least to the position that was necessary for the preceding row or column. By this device a much more efficient algorithm is obtained. If an alignment has length L, the recomputation required is approximately L^2, and if several alignment clumps are output, the recomputation is proportional to $\sum_i L_i^2$.

To generalize the declumping algorithm to the case $g(k) = \alpha + \beta(k-1)$ is quite easy. Recall that three matrices are required in Corollary 9.5: H, E, and F. We calculate three new matrices: H^*, E^*, and F^*, proceeding along rows and columns as above until all three matrices agree; $H^* = H, E^* = E$, and $F^* = F$. To illustrate the algorithm, we set $s(a, a) = 2$, $s(a, b) = -1$ if $a \neq b$, and $g(k) = 2k$. Two sequences are compared and Table 9.5(a) gives the matrix $H(\times 10)$, where the best matching segments are

```
TGAG-ATA
|| | |||
TGCGAATA
```

with score 9. The matrix N is shown in Table 9.5(b) where the recomputed entries are shown with ∗ to the right. For this step, the best matching segments are

```
GATACT
| | ||
GAT-CT
```

with score 5.

	-	G	C	T	C	T	G	C	G	A	A	T	A
-	0	0	0	0	0	0	0	0	0	0	0	0	0
C	0	0	2	0	2	0	0	2	0	0	0	0	0
G	0	2	0	1	0	1	2	0	4	2	0	0	0
T	0	0	1	2	0	2	0	1	2	3	1	2	0
T	0	0	0	3	1	2	1	0	0	1	2	3	1
G	0	2	0	1	2	0	4	2	2	0	0	1	2
A	0	0	1	0	0	1	2	3	1	4	2	0	3
G	0	2	0	0	0	0	3	1	5	3	3	1	1
A	0	0	1	0	0	0	1	2	3	7	5	3	3
T	0	0	0	3	1	2	0	0	1	5	6	7	5
A	0	0	0	1	2	0	1	0	0	3	7	5	9
C	0	0	2	0	3	1	0	3	1	1	5	6	7
T	0	0	0	4	2	5	3	1	2	0	3	7	5

(a) First local comparison

	-	G	C	T	C	T	G	C	G	A	A	T	A
-	0	0	0	0	0	0	0	0	0	0	0	0	0
C	0	0	2	0	2	0	0	2	0	0	0	0	0
G	0	2	0	1	0	1	2	0	4	2	0	0	0
T	0	0	1	2	0	2	0	1	2	3	1	2	0
T	0	0	0	3	1	0	1	0	0	1	2	3	1
G	0	2	0	1	2	0	0	0	2	0	0	1	2
A	0	0	1	0	0	1	0	0	0	4	2	0	3
G	0	2	0	0	0	0	3	1	0	2	3	1	1
A	0	0	1	0	0	0	1	2	0	2	1	2	3
T	0	0	0	3	1	2	0	0	1	0	1	0	1
A	0	0	0	1	2	0	1	0	0	3	2	0	0
C	0	0	2	0	3	1	0	3	1	1	2	1	0
T	0	0	0	4	2	5	3	1	2	0	0	4	2

Table 9.5: *(b) Second local comparison with the first clump denoted by boxes.*

9.6.1 Self-Comparison

There is an easy application of the local algorithm to find repeats within a sequence. Since we will compare **a** with **a** but do not need to discover that $\mathbf{a} = \mathbf{a}$, it is necessary to set $H_{i,i} = 0$ and to compute $H_{i,j}$ only for $i < j$.

Dynamic Programming Alignment of Two Sequences

Algorithm 9.4 (Repeat)
input: **a, b**

1. $H_{i,i} = 0$, $H_{0,i} = 0$, all $i = 1$ to n
2. for $i = 1$ to n
     for $j = i+1$ to n
       $H_{i,j} \leftarrow \max\{H_{i-1,j} - \delta, H_{i-1,j-1} + s(a_i, a_j), H_{i,j-1} - \delta, 0\}$.

This simple algorithm can find overlapping repeats. If ATATATATATAT is a substring, then, depending on the scoring parameters,

ATATATATAT
ATATATATAT

might be the best repeat, the substring matched to a shift by two letters. The second best repeat would then be

ATATATA T
ATATATAT,

the substring shifted by 4. Clearly, for tandem or adjacent repeated units, this method can find the repeat and its "periodicity" even when the pattern is not always perfectly periodic.

Suppose we wish to find nonoverlapping repeats. They are, of course, also found by algorithm repeats, but now we restrict our search to only nonoverlapping repeats. At the expense of increased computing time there is an easy algorithm for this task too.

Algorithm 9.5 (Nonoverlapping Repeat)
input: **a, b**

   $M \leftarrow 0$
   for $i = 1$ to $n-1$
     $M \leftarrow \max\{M, H(a_1 \cdots a_i, a_{i+1} \cdots a_n)\}$

Clearly this algorithm takes $O(n^3)$ time.

9.6.2 Tandem Repeats

In the last section (9.6.1), we studied finding repeats within a sequence $\mathbf{a} = a_1 a_2 \cdots a_n$. Here, we will be given the pattern $\mathbf{b} = b_1 b_2 \cdots b_k$ and wish to find the maximum repeats of **b** within **a**, where the repeats of **b** must be adjacent or tandem. Define $\mathbf{b}^l = \mathbf{bb} \cdots \mathbf{b}$ to have **b** repeated or concatenated l times. Let us take the similarity function

$$R(\mathbf{a}, \mathbf{b}) = \max\{S(a_i a_{i+1} \cdots a_j, b_x b_{x+1} \cdots b_k \mathbf{b}^l b_1 b_2 \cdots b_y :$$
$$1 \leq i \leq j \leq n, 1 \leq x \leq k+1, 0 \leq y \leq k, l \geq 0\},$$

where $b_1 b_2 \cdots b_y = \emptyset$ when $y = 0$, and $b_x b_{x+1} \cdots b_k = \emptyset$ when $x = k+1$. Solving this problem in the obvious way is very time-consuming, as $\mathbf{b}^l$ might be very long. Each repeat of $\mathbf{b}$ must have at least one match with $\mathbf{a}$ or the score could be improved by omitting that $\mathbf{b}$ repeat. Therefore, $l \leq kn$ and we see that $R(\mathbf{a}, \mathbf{b}) = H(\mathbf{a}, \mathbf{b}^{kn})$ which takes time $O(kn^2)$. Instead, there is a clever algorithm that runs in time $O(kn)$. This algorithm is based on two observations.

Consider the previous discussion, where the pattern is duplicated at least kn times. Let the pattern $\mathbf{b}$ have length k. The first observation is that the value in any cell is independent of the value in the cell k columns to its left on the same row. Let the cell under consideration be C_j, and the cell k columns to the left on the same row be C_{j-k}. Note that there is an alignment ending at C_{j-k} that produces that cell's score. But, *there is an identical alignment shifted k columns right which gives the same score in C_j* (because of the repetitive nature of the pattern sequence). Now, any alignment that passes through C_{j-k} on the way to C_j involves k deletions. Because deletions receive a negative score, the score of this alignment must be less than the score at C_{j-k}. But we know that the score at C_j is at least as large as the score at C_{j-k}, a contradiction.

In the style of previous dynamic programming algorithms, we need to define an appropriate $R_{i,j}$. Assume that we have completed the best score of an alignment ending at a_i and b_j taken from $a_1 a_2 \cdots a_i$ and $b_1 b_2 \cdots b_j$, respectively. For b_j, this assumes that our ability to align repeats of $\mathbf{b}$ with $a_1 a_2 \cdots a_i$ is not constrained by lack of $\mathbf{b}$ repeats. Formally,

$$R_{i,j} = \text{best score of an alignment that ends at } a_i \text{ and } b_j.$$

Note that now $1 \leq j \leq k$. For the induction step, assume that $R_{i-1,j}, 1 \leq j \leq k$, is known. We compute $R^*_{i,j}$ by the recursion

$$R^*_{i,j} = \max\{0, R_{i,j-1} - \delta, R_{i-1,j} - \delta, R_{i-1,j-1} + s(a_i, b_j)\},$$

beginning with $j = 1$. Now, for $j = 1$, $R_{i,j-1} = R_{i,0}$ and $R_{i-1,j-1} = R_{i-1,0}$ are not given. By periodicity, $R_{i,0} = R_{i,k}$ and $R_{i-1,0} = R_{i-1,k}$. Because initially we do not know $R_{i,0} = R_{i,k}$, we set $R_{i,0} = 0$ and compute $R^*_{i,1}$. Therefore, $R^*_{i,1} = R_{i,1}$ unless the correct alignment comes from extending an alignment ending at a_i and b_k. As the recursion proceeds to $R^*_{i,k}$, we have $R_{i,k} = R^*_{i,k}$ unless the alignment comes from $R_{i,0}$ by deleting across $b_1 \cdots b_k$. The previous paragraph excludes that possibility and, therefore,

$$R_{i,k} = R^*_{i,k}.$$

The second observation is that if we recompute the row this time using the correct value of $R_{i,0} = R_{i,k}$, we get the correct values of $R_{i,j}, 1 \leq j \leq k$. By letting the array "wrap" around in this fashion we compute $R(\mathbf{a}, \mathbf{b})$ in time $O(kn)$.

Dynamic Programming Alignment of Two Sequences

Algorithm 9.6 (wrap)
input: **a, b**

1. $R_{0,j} = 0$ for $0 \le j \le k$
   $R_{i,0} = 0$ for $0 \le i \le n$
2. for $i = 1$ to n
    for $j = 1$ to k
        $R_{i,j} = \max\{0, R_{i,j-1} - \delta, R_{i-1,j-1} + s(a_i, b_j), R_{i-1,j} - \delta\}$
        $R_{i,0} = R_{i,k}$
    for $j = 1$ to k
        $R_{i,j} = \max\{0, R_{i,j-1} - \delta, R_{i-1,j-1} + s(a_i, b_j), R_{i-1,j} - \delta\}$
3. $R(\mathbf{a}, \mathbf{b}) = \max\{R_{i,j} : 1 \le i \le n, 1 \le j \le k\}$.

9.7 Linear Space Algorithms

Most of the algorithms presented so far find alignments in quadratic time and quadratic space. It is often the case that the space requirements are so large that the arrays cannot be stored in RAM, and the algorithm spends a good deal of time swapping data between disk and RAM. In this section, we present two techniques to reduce space requirements. The first gives a reduction to $O(n)$ and the second a reduction to $O(D_{n,m} \min\{n, m\})$.

In our first method, the cost of the reduction to $O(n)$ storage is approximately a doubling of computing time. For simplicity, this section will use the algorithm for similarity S with $g(k) = k\delta$.

Algorithm 9.7 (S)
input: $n, m, \mathbf{a}, \mathbf{b}, s(\cdot, \cdot), \delta$
output: S.

1. $S_{i,0} \leftarrow -i\delta$, $i = 0, 1, \ldots, n$
   $S_{0,j} \leftarrow -j\delta$, $j = 0, 1, \ldots, m$
2. for $i = 1$ to n
    for $j = 1$ to m
        $S_{i,j} = \max\{S_{i-1,j-1} + s(a_i, b_j), S_{i-1,j} - \delta, S_{i,j-1} - \delta\}$
3. $S \leftarrow S_{n,m}$.

To compute $S = S_{n,m}$ without quadratic space requires a simple modification of S.

Algorithm 9.8 (S*)
input: $n, m, \mathbf{a}, \mathbf{b}, s(\cdot, \cdot), \delta$
output: $S_{n,1}, S_{n,2}, \ldots, S_{n,m} = \mathbf{S}$

1. $T_{1,j} \leftarrow -j\delta$ for $j = 0$ to m
    for $i = 1$ to n
        for $j = 0$ to m
            $T_{0,j} \leftarrow T_{1,j}$
        $T_{1,0} \leftarrow -i\delta$
        for $j = 1$ to m
            $T_{1,j} = \max\{T_{0,j-1} + s(a_i, b_j), T_{0,j} - \delta, \ T_{1,j-1} - \delta\}$
2. $S_{n,j} \leftarrow T_{1,j}, \ j = 0, \ldots, m$.

The shortcoming of algorithm S^* is that it does not allow us to determine the optimal alignments that have score $S_{n,m}$. The next algorithm overcomes that difficulty by determining midpoints of alignments. Define

$$\mathbf{a}_{1,i} = a_1 a_2 \cdots a_i,$$
$$\mathbf{b}_{1,j} = b_1 b_2 \cdots b_j,$$

and

$$\hat{\mathbf{a}}_{n,i+1} = a_n a_{n-1} \cdots a_{i+1},$$
$$\hat{\mathbf{b}}_{m,j+1} = b_m \cdots b_{j+1}.$$

Finally, define $\mathbf{e} \| \mathbf{f}$ to be the continuation of the finite sequences $\mathbf{e}$ and $\mathbf{f}$. The output $\mathbf{c}$ is one optimal alignment, which we take to be equivalent to the set of aligned pairs (i, j).

In the Algorithm SL, in step SL2, the output $\mathbf{S}1$ from $S^*(i, m, \mathbf{a}_{1,i}, \mathbf{b}_{1,m}, \mathbf{S}1)$ is $S(\mathbf{a}_{1,i}, \emptyset), S(\mathbf{a}_{1,i}, b_1), \ldots, S(\mathbf{a}_{1,i}, b_1 \cdots b_m)$.

Algorithm 9.9 (SL)
input: $n, m, \mathbf{a}, \mathbf{b}, s(\cdot, \cdot), \delta$
output: $\mathbf{c}$.

1. if $m = 0$

        $\mathbf{c} \leftarrow -$

    if $n = 1$

            if j satisfies $s(a_1, b_j) = \max_k s(a_1, b_k) > -2\delta$
                then $\mathbf{c} \leftarrow (1, j)$
                else $\mathbf{c} \leftarrow -$

2. $i \leftarrow \lfloor n/2 \rfloor$
    $S^*(i, m, \mathbf{a}_{1,i}, \mathbf{b}_{1,m}, \mathbf{S}1)$
    $S^*(n - i, m, \hat{\mathbf{a}}_{m,i+1}, \hat{\mathbf{b}}_{m,1}, \mathbf{S}2)$

3. $M \leftarrow \max_{0 \leq j \leq m} S1(j) + S2(m - j)$
    $k \leftarrow \min\{j : S1(j) + S2(m - j) = M\}$

Dynamic Programming Alignment of Two Sequences

    4. algorithm $SL(i, k, \mathbf{a}_{1,i}, \mathbf{b}_{1k}, \mathbf{c}_1)$
       algorithm $SL(m - i, n - k, \mathbf{a}_{i+1,n} b_{k+1,n}, \mathbf{c}2)$
    5. $\mathbf{c} \leftarrow \mathbf{c}1 \| \mathbf{c}2$.

Proof of Algorithm SL. If $m = 0$, $\mathbf{c} = -$. If $n = 1$ and $+\max_k s(a_1, b_k) = s(a_1, b_j) > -2\delta$, then $S(a_1, b_1 \cdots b_m) = -(j-1)\delta + s(a_1, b_j) - (m-j)\delta$ and $\mathbf{c} = (1, j)$. Otherwise, $\mathbf{c} = -$. Because the score of the alignment is additive when $g(k) = k\delta$, $M = S_{n,m}$. ■

The time complexity of Algorithm SL is easy to calculate. The first pass is done in time proportional to

$$2\left(\frac{n}{2} \cdot m\right) = nm.$$

Then the problem divides into problems of size $n/2$, k, and $n/2, m - k$. They take time proportional to

$$2\left(\frac{n}{4} \cdot k\right) + 2\left(\frac{n}{4} \cdot (m - k)\right) = \frac{n}{2}m.$$

Thus, the algorithm converges in time proportional to

$$O\left(\sum_{k \geq 0} nm2^{-k}\right) = O(2nm) = O(nm).$$

This analysis shows that the running time of algorithm SL is double that of algorithm S.

The case of $g(k) = \alpha + \beta(k - 1)$ is an exercise. For local algorithms, the following device gives us an optimal local alignment in linear space. In the first pass, an algorithm corresponding to S^* called H^* can find the "first" occurrence of $H(\mathbf{a}, \mathbf{b}) = H_{i,j}$ (according to the ordering $\succ$ given after Corrollary 9.4). Then run $H^*(a_i a_{i-1} \cdots a_1, b_j b_{j-1} \cdots b_1)$ to find the first (according to $\succ$) of score $H_{i,j}$ at $H^*(a_i \cdots a_k, b_j \cdots b_k)$. Now, the best local alignment is of the substrings $a_k a_{k+1} \cdots a_i$ and $b_l b_{l+1} \cdots b_j$. Algorithm $SL(a_k \cdots a_i, b_l \cdots b_j)$ gives the alignment in linear space. At worst, the algorithm takes four times as long as that of Algorithm H. Declumping to get the k best local alignments in linear space is a much more difficult job.

Now we turn to our second $O(n)$ space method. It is natural to observe that only a portion of the matrix is used for calculation of the final comparison value. We return to distance for an elaboration of this observation.

$$D_{ij} = \min\left\{D_{i-1,j-1} + d(a_i, b_j), \min_{k \geq 1}\{D_{i,j-k} + g(k)\}, \min_{k \geq 1}\{D_{i-k,j} + g(k)\}\right\}$$

and assume $d(a, b) = 1$ unless $a = b$.

Lemma 9.1 *For all* (i,j), $D_{i,j} - 1 \leq D_{i-1,j-1} \leq D_{i,j}$.

Proof. The proof is by induction on $i+j$. The left-hand side is immediate from the recursion. If $D_{i,j} = D_{i-1,j-1} + d(a_i, b_j)$, then $D_{i,j} \geq D_{i-1,j-1}$ follows. Otherwise, without loss of generality, assume $D_{i,j} = D_{i-k,j} + g(k)$. The induction hypothesis implies $D_{i-k,j} \geq D_{i-(k+1),j-1}$ so that $D_{i,j} \geq D_{i-1-k,j-1} + g(k)$. The recursion equation now implies $D_{i,j} \geq D_{i-1-k,j-1} + g(k) \geq D_{i-1,j-1}$. ∎

Lemma 9.1, which is elementary, is the key to an elegant method. The lemma states that $D_{i,i+c}$ is a nondecreasing function of i. This implies a structure for the matrix $D_{i,j}$: It is shaped like a valley with increasing elevations along lines of constant $j - i$. The lowest elevation is $D_{00} = 0$. The focus below is on the boundaries of elevation changes when $D_{i,i+c} = k$ changes to $D_{i+1,i+1+c} = k+1$.

Suppose all indels have cost 1, i.e., $g(k) = k$. The basic idea of the algorithm is to start at $D_{00} = 0$ and extend along $j - i = 0$ until $D_{i,i} = 1$. In general, there will be $2k + 1$ boundaries of the region $D_{i,j} \leq k$. Each boundary $j - i = c$ is extended until $D_{i,j} = k+1$ (for $j-i = c$). The extension of the boundary to $k+1$ can be determined from the boundaries for $k, k - 1, \ldots$ and checking $a_i = b_j$. This procedure is followed until $D_{n,m}$ is reached. If $D_{n,m} = s$, it is clear that no more than $(2s + 1) \min\{n, m\}$ entries have been computed. It is sufficient to only store these boundaries so that required storage is $O(s^2)$. Therefore, we can compute $D_{n,m}$ in linear time equal to $O\left(D_{n,m} \times \min\{n, m\}\right)$.

9.8 Tracebacks

So far, not much attention has been given to actually producing the alignments. There are two methods to produce an alignment: saving pointers and recomputation. We will treat the pointer method first.

In the case of single indels, $g(k) = k\delta$, pointers can be easily handled. Recall that there are generally four possibilities for alignment, including option (3) for local alignment:

(0) $H_{i-1,j-1} + s(a_i, b_j)$,

(1) $H_{i,j-1} - \delta$,

(2) $H_{1-i,j} - \delta$,

(3) 0.

Option (3), 0, means that the alignment ends. Therefore, any subset of $\{0, 1, 2\}$ is possible, where $\emptyset$ = option (3). It is convenient to use integers $t \in [0, 7]$ to correspond to the eight subsets of $\{0, 1, 2\}$.

Computed tracebacks simply repeat the recursion at each entry, beginning with the optimal score at the lower right end of the alignment and checking to see which options resulted in the given score. Just as above, multiple options arise.

Dynamic Programming Alignment of Two Sequences

It is often the case that many alignments result from one score. At each (i,j) in the traceback where multiple options exist, the unexplored options with the location (i,j) are placed into a stack. The stack is managed in a *last in – first out* (LIFO) manner. We return to the stack when an alignment is finished. Of course, LIFO allows us to utilize the alignment already obtained up to the (i,j) found in the stack. When the stack is empty, all optimal alignments have been output.

The time efficiency of producing alignments is hard to estimate precisely, because of the efficiency of the LIFO stack. One, unique, alignment can be found in time $O(L)$, where L = length of the alignment, for the single indel case. Computed tracebacks in the multiple indel case take time $O(\max\{n,m\}L)$. The next theorem shows how to reduce this when $g(k) = \alpha + \beta(k-1)$.

Theorem 9.14 *If $g(k) = \alpha + \beta(k-1)$ and $S_{i,j} > S_{i,j-k^*} - \beta k^*$, then*

$$S_{i,j} > S_{i,j-k} - g(k), \text{ all } k \geq k^* + 1.$$

Proof.

$$S_{i,j} > S_{i,j-k^*} - \beta k^*$$
$$\geq \max_{1 \leq q \leq j-k^*} \{S_{i,j-k^*-q} - \alpha - \beta(q-1)\} - \beta k^*$$
$$= \max_{1+k^* \leq q+k^* \leq j} \{S_{i,j-(k^*+q)} - \alpha - \beta((k^*+q)-1)\}$$
$$= \max_{1+k^* \leq k \leq j} \{S_{i,j-k} - g(k)\}.$$

∎

Near Optimal Alignments (1)

The optimal alignments depend on the input sequences and the algorithm parameters. The weights assigned to mismatches and indels are determined by experience and an effort is made to use biological data to infer meaningful values. Of course, in addition to assigning weights, there are sometimes unknown constraints on the sequences that cause the correct alignment to differ from the optimal alignment given by an algorithm. Hence, it is of some interest to produce all alignments with score (distance or similarity) within a specified distance of the optimum score. Two algorithms are given here for the similarity algorithm.

To be explicit, let $S = (S_{i,j})$ be the single indel similarity matrix with

$$S_{i,j} = \max\{S_{i-1,j-1} + s(a_i, b_j), S_{i-1,j} - \delta, S_{i,j-1} - \delta\}.$$

The task is to find all alignments with score within $e > 0$ of the optimum value $S_{n,m}$. All optimum alignments are included.

At position (i, j) assume a traceback from (n, m) to $(0, 0)$ is being performed that can result in an alignment with score greater than or equal to $S_{n,m} - e$. The score of the current alignment from (n, m) to but not including (i, j) is $T_{i,j}$. $T_{i,j}$ is the sum of the possibly nonoptimal alignment weights to reach (i, j). From (i, j), as usual, three steps are possible: $(i - 1, j)$, $(i - 1, j - 1)$, and $(i, j - 1)$. Each step is in a desired alignment if and only if

$$T_{i,j} + S_{i-1,j} - \delta \geq S_{n,m} - e,$$
$$T_{i,j} + S_{i-1,j-1} + s(a_i, b_j) \geq S_{n,m} - e,$$
$$T_{i,j} + S_{i,j-1} - \delta \geq S_{n,m} - e,$$

respectively.

- If $T_{i,j} + S_{i-1,j} - \delta \geq S_{n,m} - e$,
  move to $(i - 1, j)$ with $T_{i-1,j} = T_{i,j} - \delta$.

- If $T_{i,j} + S_{i-1,j-1} + s(a_i, b_j) \geq S_{n,m} - e$,
  move to $(i - 1, j - 1)$ with $T_{i-1,j-1} = T_{i,j} + s(a_i, b_j)$.

- If $T_{i,j} + S_{i,j-1} - \delta \geq S_{n,m} - e$,
  move to $(i, j - 1)$ with $T_{i,j-1} = T_{i,j} - \delta$.

Multiple near-optimal alignments can be produced by stacking unexplored directions. Of course, multiple insertions and deletions can be included.

The sequences displayed below are chicken hemoglobin mRNA sequences, nucleotides 115–171 from the chain (upper sequence) and 118–156 from the chain (lower sequence).

```
UUUGCGUCCUUUGGGAACCUCUCCAGCCCCACUGCCAUCCUUGUCACACGGCAACCCCAUGGUC
UUUCCCCACUUCG    AUCUUUGUCACAC                         GGCUCCGCUCAAAUC
```

This alignment is presumed correct from the analysis of the many known amino acid sequences for which such RNA sequences code.

Using global distance, with a mismatch weight of 1 and $g(k) = 2.5 + k$, where k is the length of the insertion or deletion, the biologically correct alignment is found among the 14 optimal alignments. To indicate the size of neighborhoods in this example, there are 14 alignments within 0% of the optimum, 14 within 1%, 35 within 2%, 157 within 3%, 579 within 4%, and 1317 within 5%. For mismatch weight of 1 and a multiple insertion or deletion function $2.5 + 0.5k$ the correct alignment is not in the list of the two optimal alignments. This example illustrates the sensitivity of alignment to weighting functions.

Near-Optimal Alignments (2)

The method described above requires all near-optimal alignments to be explored. There are often an exponential number of them. A less precise but frequently more

Dynamic Programming Alignment of Two Sequences

practical approach is to ask if (i, j) is a match on *any* near-optimal alignment. The best score of all alignments with (i, j) matched is

$$S(a_1 \cdots a_{i-1}, b_1 \cdots b_{j-1}) + s(a_i, b_j) + S(a_{i+1} \cdots a_n, b_{j+1} \cdots b_n).$$

The position (i, j) can, for example, be highlighted if this quantity is $\geq S_{n,m} - e$. This analysis can be done by computing the matrices for $S(a_1 \cdots a_n, b_1 \cdots b_m)$ and $S(a_n a_{n-1} \cdots a_1, b_m b_{m-1} \cdots b_1)$.

9.9 Inversions

Our goal here is to describe an algorithm for optimal alignment of two DNA sequences that allows inversions. An inversion of a DNA sequence is defined to be the reverse complement of the sequence. Although the number of inversions is not restricted, the inversions will not be allowed to intersect one another. Later we will discuss the case of intersecting inversions. Although we could describe other versions of our algorithm, including full or global sequence alignment, here we present the local alignment algorithm with a linear indel weighting function.

When we allow inversions, the inverted regions will not exactly match and must themselves be aligned. In addition, one of the inverted regions must be complemented to preserve the polarity of the DNA sequence. Let us define

$$Z(g, h; i, j) = S_1(a_g a_{g+1} \cdots a_i, \bar{b}_j \bar{b}_{j-1} \cdots \bar{b}_h),$$

where $\bar{A} = T$, $\bar{C} = G$, $\bar{G} = C$, and $\bar{T} = A$. In the original sequences the segments $a_g a_{g+1} \cdots a_i$ and $b_h b_{h+1} \cdots b_j$ are matched after an inversion. This means that $a_g a_{g+1} \cdots a_i$ and $\bar{b}_j \bar{b}_{j-1} \cdots \bar{b}_h$ are aligned. $Z(g, h; i, j)$ is indexed by the beginning (g, h) and ending (i, j) coordinates of the sequences in their original order. The function S_1 is the alignment score defined in Theorem 9.10 using the matching function $s_1(a, b)$ and indel function $g_1(k) = \alpha_1 + \beta_1(k - 1)$. Each inversion is charged an additional cost γ. The noninverted alignment uses matching function $s_2(a, b)$ and indel function $g_2(k) = \alpha_2 + \beta_2(k - 1)$.

The recursion for the best score W, with inversions, is given in

Algorithm 9.10 (All Inversions)
```
input:  a, b.
    set  U(i, j) = V(i, j) = W(i, j) = 0  if  i = 0 or j = 0.
    for j = 1 to m
      for i = 1 to n
        U(i, j) = max{U(i − 1, j) + β₂, W(i − 1, j) + α₂}
        V(i, j) = max{V(i, j − 1) + β₂, W(i, j − 1) + α₂}
        for g = 1 to i
          for h = 1 to j
```

compute $Z(g,h;i,j)$

$$W(i,j) = \max\{ \max_{\substack{1\leq g\leq i \\ 1\leq h\leq j}} \{W(g-1,h-1) + Z(g,h;i,j)\} + \gamma, \quad (9.3)$$

$$W(i-1,j-1) + s_2(a_i,b_j), U(i,j), V(i,j), 0\}$$

best inversion score = $\max\{W(i,j) : 1\leq i\leq n, 1\leq j\leq m\}$.

The proof that the recursion gives the optimal score for nonintersecting inversions follows the usual proof for dynamic programming alignment algorithms.

The system of recurrences in (9.3) is expensive in computation time. If $Z(g,h;i,j)$ is computed for each (g,h) where $1 \leq g \leq i$ and $1 \leq h \leq j$, this takes time $O(i^2 j^2)$, and the full Algorithm 9.10 takes time $O(n^6)$, when $n = m$. If general $g_i(k)$ is used, the corresponding version of Algorithm 9.10 takes time $O(n^7)$, when $n = m$.

Clearly Algorithm 9.10 is too costly in time for any problem of interest. Essentially the reason for this is that many very poor quality inversions are calculated and rejected. Biologists are only interested in longer, high quality inversions. Fortunately, there is a computationally efficient way to choose these inversions and dramatically speed up the alignment algorithm.

We first apply the local algorithm with $s_1(a,b)$ and $g_1(k) = \alpha_1 + \beta_1(k-1)$ to the sequences $\mathbf{a} = a_1 \cdots a_n$ and the inverted sequence $\mathbf{b}^{(\text{inv})} = \bar{b}_m \bar{b}_{m-1} \cdots \bar{b}_1$. The local algorithm gives the best K (inversion) local alignments with the property that no match (identity or mismatch) is used more than once in the alignments. Each time a best alignment is located the matrix must be recalculated to remove the effect of the alignment. If alignment i has length L_i, the time required to produce the list $\mathcal{L}$ of the best K inversion alignments is $O\left(nm + \sum_{i=1}^{K} L_i^2\right)$. To reduce the time requirement we choose an appropriate value of K. To further reduce running time we could impose a score threshold C_1 chosen so that the probability that two random sequences have a K-th best alignment score $\geq C_1$ is small. See the discussion of statistical significance in Chapter 11.

Algorithm 9.11 (Best Inversions)
input: **a, b**

1. apply local algorithm to **a** and $\mathbf{b}^{(\text{inv})}$ to get K alignments

$$\mathcal{L} = \{(Z(g,h;i,j),(g,h),(i,j)) : \alpha \in [1,K]\}$$

2. set $U(i,j) = V(i,j) = W(i,j) = 0$ if $i = 0$ or $j = 0$

Dynamic Programming Alignment of Two Sequences

for $j = 1$ to m
    for $i = 1$ to n

$$U(i,j) = \max\{U(i-1,j) + \beta_2, W(i-1,j) + \alpha_2\}$$

$$V(i,j) = \max\{V(i,j-1) + \beta_2, W(i,j-1) + \alpha_2\}$$

$$W(i,j) = \max\{\max_{\mathcal{L}}\{W(g-1,h-1) + Z(g,h;i,j)\} + \gamma,$$

$$W(i-1,j-1) + s_2(a_i,b_j), U(i,j), V(i,j), 0\}$$

best inversion score $= \max\{W(i,j) : 1 \leq i \leq n, 1 \leq j \leq m\}$.

We have greatly reduced computation time. Part 1 of the algorithm can be done in time $O(nm)$. Part 2 requires time proportional to nm times a constant plus the average number of elements in $\mathcal{L}$ that "end" at (i, j). It might seem that only one best inversion has this property, but recall that we align $a_g a_{g+1} \cdots a_i$ with $\bar{b}_j \bar{b}_{j-1} \cdots \bar{b}_h$. This allows the possibility of several elements with "end" (i, j). Still the list $\mathcal{L}$ is restricted to K elements. In the illustrative example discussed next, $|\mathcal{L}| = 2$. Clearly, part 2 of the Algorithm Best Inversions runs in time $O(nm|\mathcal{L}| + O(\sum_{i=1}^{|\mathcal{L}|} L_i^2)$.

For maximum flexibility we have allowed $s_1(a, b)$ and $s_2(a, b)$ as well as $w_1(k)$ and $w_2(k)$ to have different values. This might be advisable if the inversion segments are thought to have evolved differently from the rest of the alignment, but usually $s_1 = s_2$ and $g_1 = g_2$.

To illustrate our algorithm we use the sequences **a** = CCAATCTACTACT GCTTGCA and **b** = GCCACTCTCGCTGTACTGTG. The matching functions are

$$s_1(a,b) = s_2(a,b) = \begin{cases} 10 & \text{when } a = b, \\ -11 & \text{when } a \neq b, \end{cases}$$

while

$$w_1(k) = w_2(k) = -15 - 5k.$$

The inversion penalty is $\gamma = -2$ whereas the list $\mathcal{L}$ is defined by $K = 2$. The two alignments in $\mathcal{L}$ are shown in Table 9.6, where the matched pairs are boxed. Table 9.6 shows the matrix H for **a** and $\mathbf{b}^{(\text{inv})}$ and the two alignments in $\mathcal{L}$. The best local alignment with inversion, along with the matrix W, is shown in Table 9.7.

We now turn to modifying our algorithm for inversions to yield the J best alignments. Our object is to produce the J best alignments that do not share a match, mismatch, or inversion. Therefore, when an inversion from $\mathcal{L}$ is used in an alignment, it cannot be used in a succeeding alignment. This is accomplished

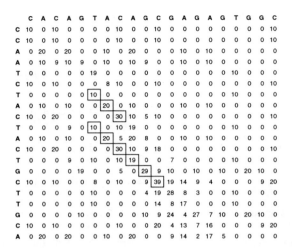

	C	A	C	A	G	T	A	C	A	G	C	G	A	G	A	G	T	G	G	C
C	10	0	10	0	0	0	0	10	0	0	10	0	0	0	0	0	0	0	0	10
C	10	0	10	0	0	0	0	10	0	0	10	0	0	0	0	0	0	0	0	10
A	0	20	0	20	0	0	10	0	20	0	0	0	10	0	10	0	0	0	0	0
A	0	10	9	10	9	0	10	0	10	9	0	10	0	10	0	10	0	0	0	0
T	0	0	0	0	0	19	0	0	0	0	0	0	0	0	0	0	10	0	0	0
C	10	0	10	0	0	0	8	10	0	0	10	0	0	0	0	0	0	0	0	10
T	0	0	0	0	0	10	0	0	0	0	0	0	0	0	0	10	0	0	0
A	0	10	0	10	0	0	20	0	10	0	0	0	10	0	10	0	0	0	0	0
C	10	0	20	0	0	0	0	30	10	5	10	0	0	0	0	0	0	0	0	10
T	0	0	0	9	0	10	0	10	19	0	0	0	0	0	0	10	0	0	0
A	0	10	0	10	0	0	20	5	20	8	0	0	10	0	10	0	0	0	0	0
C	10	0	20	0	0	0	0	30	10	9	18	0	0	0	0	0	0	0	0	10
T	0	0	0	9	0	10	0	10	19	0	0	7	0	0	0	0	10	0	0	0
G	0	0	0	0	19	0	0	0	5	0	29	9	10	0	10	0	10	0	20	10	0
C	10	0	10	0	0	8	0	10	0	9	39	19	14	9	4	0	0	0	9	20
T	0	0	0	0	0	10	0	0	0	4	19	28	8	3	0	0	10	0	0	0
T	0	0	0	0	0	10	0	0	0	0	14	8	17	0	0	0	10	0	0	0
G	0	0	0	0	10	0	0	0	0	0	10	9	24	4	27	7	10	0	20	10	0
C	10	0	10	0	0	0	0	10	0	0	20	4	13	7	16	0	0	0	9	20
A	0	20	0	20	0	0	10	0	20	0	0	9	14	2	17	5	0	0	0	0

Table 9.6: *Best local alignment with inversions of* **a** = *CCAATCTACTACT-GCTTGCA and* **b** = *GCCACTCTCGCTGTACTGTG*

	G	C	C	A	C	T	C	T	C	G	C	T	G	T	A	C	T	G	T	G
C	0	10	10	0	10	0	10	0	10	0	10	0	0	0	0	10	0	0	0	0
C	0	10	20	0	10	0	10	0	10	0	10	0	0	0	0	0	0	0	0	0
A	0	0	0	30	10	5	0	0	0	0	0	0	0	10	0	0	0	0	0	0
A	0	0	0	10	19	0	0	0	0	0	0	0	0	10	0	0	0	0	0	0
T	0	0	0	5	0	29	9	10	0	0	0	10	0	10	0	0	10	0	10	0
C	0	10	10	0	15	9	39	19	20	9	10	0	0	0	0	10	0	10	0	0
T	0	0	0	0	0	25	19	49	29	24	19	20	9	10	0	0	20	0	10	0
A	0	0	0	10	0	5	14	29	38	18	13	8	9	0	20	0	0	9	0	0
C	0	10	10	0	20	0	15	24	39	27	28	9	4	0	28	30	10	5	0	0
T	0	0	0	0	0	30	10	25	19	28	16	38	18	14	8	17	40	20	15	10
A	0	0	0	10	0	10	19	14	14	8	17	18	27	7	24	5	20	29	9	4
C	0	10	10	0	20	5	20	9	24	4	18	13	7	16	4	34	15	9	18	0
T	0	0	0	0	0	30	10	30	10	13	0	28	8	17	5	14	44	24	19	14
G	10	0	0	0	0	10	19	10	19	20	2	8	38	18	13	9	24	54	34	29
C	0	20	10	0	10	5	20	8	20	8	30	10	18	27	76	56	51	46	43	36
T	0	0	9	0	0	20	0	30	10	9	10	40	20	28	56	65	66	46	56	36
T	0	0	0	0	0	10	9	10	19	0	5	20	29	30	51	45	75	55	56	45
G	10	0	0	0	0	0	0	5	0	29	9	15	30	18	46	40	55	85	65	66
C	0	20	10	0	10	0	10	0	15	9	39	19	14	19	41	56	50	65	74	54
A	0	0	9	20	0	0	0	0	0	4	19	28	8	3	36	36	45	60	54	63

Table 9.7: *The matrix W and alignment*

by appropriately changing $\mathcal{L}$ as the algorithm proceeds. As in the local algorithm, the computation proceeds along the p-th row from column q to column r until

$$U^*(p,r) = U(p,r),$$
$$V^*(p,r) = V(p,r),$$
(9.4)

and
$$W^*(p,r) = W(p,r).$$

Then the computations are performed on the q-th column. In this way, an isolated island of the matrix is recalculated.

A second, more complicated feature of the algorithm arises at this point. The validity of the recalculation procedure in (9.4) is justified by the basic property of the dynamic programming recursion equations. However, these recursions allow $W(g-1, h-1)$ to affect $W(i,j)$ if $(Z(g,h;i,j),(g,h),(i,j)) \in \mathcal{L}$. Therefore, after an island of recalculation is performed, we examine $\mathcal{L}$ for (i,j) such that the value $W^*(i,j)$ might be changed; that is, where $W^*(g,h) \neq W(g,h)$. Finding such an entry begins a new island of recalculation. Therefore, the procedure of recalculation can initiate a cascade of such islands. If $\mathcal{L}$ has a large number of entries, the likelihood of new islands is increased. It is, therefore, difficult to give a rigorous analysis of the running time of finding the J best alignments. If $\mathcal{L}$ is short, it remains $O(nm|\mathcal{L}| + \sum_{i=1}^{J} M_i^2)$, where M_i is the length of the i-th alignment. Including the running time for computing $\mathcal{L}$ adds $O(nm + \sum_{i=1}^{K} L_{J+i}^2)$.

9.10 Map Alignment

Before DNA is sequenced, it is often the case that the approximate location of certain features are mapped on the DNA. In this book we do not discuss genetic mapping where the approximate location of genes are determined. Sometimes the localization is very rough – locating a gene anywhere on a specific chromosome, for example. The locations are determined by recombination frequency. Another type of mapping, physical mapping, is also important to molecular biology and is discussed in Chapter 6. Here the distances between features of interest are estimated from measurements of DNA fragments themselves, hence the name physical map. Recall from Chapter 2 that restriction sites are small sequence-specific patterns in the DNA at which restriction enzymes cut the DNA into two molecules. Restriction sites are often the features mapped in a physical map. The first genome to be mapped with an extensive physical map was the 7000+ restriction site map of *E. coli*. There eight restriction enzyme sites were mapped. Although our analysis is not restricted to restriction maps, it will guide our problem formulation, as well as give several modifications for the basic algorithm.

First we will define a map. Each map site has two characteristics, site location and site feature name. The map $\mathbf{A} = A_1 A_2 \cdots A_n$ consists of a sequence of pairs $A_i = (a_i, r_i)$, where $a_i = $ the location of the i-th site in number of basepairs

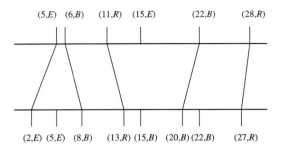

Figure 9.6: *Map alignment*

and r_i = the feature name at the i-th site. Similarly, $\mathbf{B} = B_1 B_2 \cdots B_m$ is a map where $B_j = (b_j, s_j)$.

Before setting up a similarity measure between maps, some observations about the character of these maps are appropriate. The transformations that evolve macromolecular sequences have their analogues here. A site can appear or disappear, corresponding to indels or substitutions. Although one site can be substituted for another, this is usually thought to be an unlikely event, and sites substitutions will not usually be included. The beginning and ends of a map might not correspond to a restriction site even though the map is a restriction map. In those cases we create special site names to denote beginning and end.

As emphasized above, locations are only approximate. From that reason, it is valuable to have a similarity function that rewards for identical site alignment and for intersite distances that are similar. If, as is natural, the similarity function were to measure the identity of sites with the distances from the beginning of the maps, an error in distance at one region of the map could propagate down the entire map.

Suppose the map alignment is

$$\mathbf{M} = \begin{matrix} r_{i_1} r_{i_2} \cdots r_{i_d}, \\ s_{j_1} s_{j_2} \cdots s_{j_d}, \end{matrix}$$

where $r_{i_k} = s_{j_k}$. This alignment score $S(\mathbf{M})$ is defined by

$$\begin{aligned} S(\mathbf{M}) = {}& \nu \times d - \mu |a_{i_1} - b_{j_1}| \\ & - \mu \times \sum_{t=2}^{d} |(a_{i_t} - a_{i_{t-1}}) - (b_{j_t} - b_{j_{t-1}})| \\ & - \mu |(a_n - a_{i_d}) - (b_m - b_{j_d})| - \lambda \times (n + m - 2d), \end{aligned}$$

where ν, λ, and μ are nonnegative. In this scoring definition, each matched (identical) pair is rewarded by ν. The difference between distances between adjacent aligned sites (i_{t-1}, j_{t-1}) and (i_t, j_t) is penalized by a factor of μ, as are the discrepancies of the leftmost and rightmost aligned pairs. Also each unaligned site is penalized by λ. See Figure 9.6 for an example. To preserve the form of our sequence alignment algorithms, set

$$s(A_{i_j}, B_i) = \begin{cases} \nu & \text{if } r_i = s_j, \\ -\infty & \text{if } r_i \neq s_j. \end{cases}$$

The algorithm we present is designed to find $S(\mathbf{A}, \mathbf{B})$, the maximum score of all alignments between two maps.

When $r_i = s_j$, let $X(i, j)$ be the score of the best scoring global alignment with rightmost pair (r_i, s_j) not including the term $\mu|(a_m - a_i) - (b_m - b_j)|$. Then

$$Y(i, j) = \max\{X(g, h) - \mu|(a_i - a_g) - (b_j - b_h)| : g < i, h < j \text{ and } r_g = s_h\}$$
$$X(i, j) = \max\{\nu - \lambda(n + m - 2) - \mu|a_i - b_j|, Y(i, j) + \nu + 2\lambda\}.$$

In the recursions for $X(i, j)$, the first term $\nu - \lambda(n + m - 2) - \mu|a_i - b_j|$ is the score of an alignment with only (i, j) matched. The second term, $y + \nu + 2\lambda$, is the score when there is at least one matched pair to the left of (i, j). The last step is to add back the excluded term:

$$S(\mathbf{A}, \mathbf{B}) = \max\{X(i, j) - \mu|(a_n - a_i) - (b_m - b_j)| : 1 \leq i \leq n, 1 \leq j \leq m\}.$$

Algorithm 9.12 (Map)
```
input:   n, m, A, B, μ, λ, ν
output:  i, j, S(A, B)
    S ← -μ(aₙ + bₘ) - λ(m + n)
    for i ← 1 to n
        for j ← 1 to m
            if rᵢ = sⱼ
                y ← -μ(aₙ + bₘ) - λ(m + n)
                for g ← 1 to i - 1
                    for h ← 1 to j - 1
                        if r_g = s_h
                            y ← max{y, X(g, h) - μ|(aᵢ - a_g) - (bⱼ - b_h)|}
                X(i, j) ← max{ν - λ(n + m - 2) - μ|aᵢ - bⱼ|, y + ν + 2λ}
                S ← max{S, X(i, j) - μ|(aₙ - aᵢ) - (bₘ - bⱼ)|} .
```

The computational complexity of this algorithm is $O(n^2 m^2)$. By limiting the number of sites between matches to α, the computation can be sped up by changing the nested loss for g and h, so

Algorithm 9.13 (Map*)
```
input:   n, m, A, B, μ, λ, ν
output:  i, j, S(A, B).
    S ← -μ(aₙ + bₘ) - λ(m + n)
    for i ← 1 to n
```

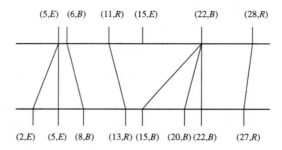

Figure 9.7: *Multiple matching*

```
for j ← 1 to m
    if r_i = s_j
        y ← -μ(a_n + b_m) - λ(m + n)
        for g ← max{1, i - α} to i - 1
            for h ← max{1, j - α} to j - 1
                if r_g = s_h
                    y ← max{y, X(g, h) - μ|(a_i - a_g) - (b_j - b_h)|}
        X(i, j) ← max{ν - λ(n + m - 2) - μ|a_i - b_j|, y + ν + 2λ}
        S ← max{S, X(i, j) - μ|(a_n - a_i) - (b_m - b_j)|}.
```

Evidently, this algorithm could be modified to produce local map alignments, or any of the other sequence alignment algorithms presented earlier. Instead we turn to an aspect of alignment particular to map alignment. It arises from one method of constructing restriction maps. The DNA is end labeled, and with partial digestion of an enzyme, the distances to the cut sites measured on a gel. Recall that the migration of DNA on a gel under an electric field is not too precise. Sometimes closely spaced sites will have associated DNAs that appear as one band in the resulting gel. Therefore, the map pair $A_i = (a_i, r_i)$ might actually represent several sites (a^α, r^α) such that the differences between α's are small. This introduces the notation of multiple matching of sites which is easily added to the algorithm. See Figure 9.7 for alignment with multiple matching.

Algorithm 9.14 (Multiple Match)
```
input:   n, m, A, B, μ, λ, ν
output:  i, j, S(A, B)
    S ← -μ(a_n + b_m) - λ(m + n)
    for i ← 1 to n
        for j ← 1 to m
            if r_i = s_j
```

```
        y ← −μ(aₙ + bₘ) − λ(m + n)
        for g ← max{1, i − α} to i − 1
            for h ← max{1, j − α} to j − 1
                if rg = sh
                    y ← max{y, X(g, h) − μ|(aᵢ − a_g) − (b_j − b_h)|}
        X(i, j) ← max{ν − λ(n + m − 2) − μ|aᵢ − b_j|, y + ν + 2λ}
```

$$X(i,j) \leftarrow \max\left\{X(i,j), \lambda + \max\left\{\begin{array}{l} X(i,j-1) \\ -\mu(s_j - s_{j-1}), \\ X(i-1,j) \\ -\mu(r_i - r_{i-1}) \end{array}\right\}\right\}$$

$$S \leftarrow \max\{S, X(i,j) - \mu|(a_n - a_i) - (b_m - b_j)|\}$$

9.11 Parametric Sequence Comparisons

One of the greatest difficulties in applying dynamic programming sequence comparison algorithms to biology is the choice of algorithm penalty parameters. In some cases, small changes in amino acid weights or in the indel function create large changes in the resulting alignments. In other cases, the alignments are very robust to changes in the algorithm parameters. There is no one set of "correct" parameters: Parameters that will find significant matches of one statistical quality for one pair of sequences are not useful for another type of matching. It is of interest, therefore, to consider sequence comparison for a large set of parameter values. Ideally, we would like to compute optimal alignments for *all* possible parameter values. At first glance, this would seem to require an infinite number of sequence comparisons, and, therefore, appears to be a completely unrealistic goal. In this section, we describe an algorithm to do this.

The general setting of these problems is a linear score function for a d-dimensional parameter space

$$S(\mathbb{A}) = k_0 + \sum_{i=1}^{d} k_i \lambda_i,$$

where $(k_0, k_1, \ldots, k_d)$ is a function of the alignment $\mathbb{A}(\mathbf{a}, \mathbf{b})$ of two sequences $\mathbf{a}$ and $\mathbf{b}$ and $\lambda \in ([0, \infty])^d$.

Proposition 9.2 *The score* $S(\lambda) = \max\{S(\mathbb{A}) : \mathbb{A} \text{ is an alignment of } \mathbf{a}, \mathbf{b}\}$ *is a continuous, concave, piecewise linear function of* λ.

Proof. The function

$$S(\lambda) = \max\left\{k_0 + \sum_{i=1}^{d} k_i \lambda_i : k_1, \ldots, k_d \text{ determined from } \mathbb{A}(\mathbf{a}, \mathbf{b})\right\}$$

has these properties because there are a finite number of alignments. ∎

We will restrict ourselves to one- and two-dimensional parameter systems. Our approach to parametric alignment is to discover the piecewise linear "pieces" or regions of the function $S(\lambda)$, and therefore we will be able to find the function in finite time. Before we present an algorithm, it is interesting to study which regions there are. For notational convenience, we use the parameters μ (for mismatch) and δ (for indel) as earlier in the chapter. There are two results, one for global alignment and one for local alignment.

Theorem 9.15 *For global alignment, lines forming a boundary between two regions have the form $\delta = c + (c + 1/2)\mu$ for some $c > -1/2$.*

Proof. Global alignments with w identities, x mismatches, and y indels (one letter each) have score $w - \mu x - \delta y$ and satisfy $2w + 2x + y = n + m$ (by Theorem 9.11). Rewriting the last equation,

$$w + x + \frac{1}{2}y = \frac{n+m}{2}$$

or

$$w - (-1)x - \frac{1}{2}y = \frac{n+m}{2}.$$

Therefore, every alignment plane $w - \mu x - \delta y$ meets at $(\mu, \delta) = (-1, 1/2)$. Clearly, every boundary or intersection of alignment planes also pass through $(-1, -1/2)$. Let $\delta = c + b\mu$ be such a boundary line. We have $-1/2 = c - 1 \times b$ or $b = c + 1/2$ and $\delta = c + (c+1/2)\mu$. Because we want (μ, δ) in $[0, \infty) \times [0, \infty)$, $c + 1/2 > 0$ is required. ∎

Proposition 9.3 *There are at most $n + 1$ regions in global alignment.*

Proof. Note that whenever $\mu > 2\delta$, there will be no mismatches in an optimal alignment and optimal alignments are fixed on the lines $\{(\mu, \delta) : \mu > 2\delta\}$. Therefore no optimal alignment boundary intersects the μ axis. Hence all boundaries intersect the positive δ axis. With $\mu = 0$, the equations of the lines are $w_i - y_i\delta$ as we move up the δ axis from $\delta = 0$. When $\delta_i < \delta_{i+1}$,

$$w_i - y_i\delta_i > w_{i+1} - y_{i+1}\delta_i$$

and

$$w_i - y_i\delta_{i+1} < w_{i+1} - y_{i+1}\delta_{i+1},$$

it is easy to show that $w_i > w_{i+1}$ and $y_i > y_{i+1}$.

Clearly, the w_i can only take on $\min\{n, m\} + 1$ distinct values, $0, 1, \ldots, \min\{n, m\}$. ∎

The number of regions can be shown to be $O(n^{2/3})$ with some more work. There is a corresponding result for local alignment.

Proposition 9.4 *There are at most $O(n^2)$ regions in local alignment.*

Dynamic Programming Alignment of Two Sequences

Proof. Let a local alignment have w identities, x mismatches, and y indels. Then $2w + 2x + y \leq n + m$. If two local alignments have (w, x, y_1) and (w, x, y_2), respectively, and $y_1 < y_2$, then alignment 1 dominates alignment 2 for all $(\mu, \delta) \in [0, \infty)^2$. Therefore, no two optimal regions have $(w_i, x_i) = (w_j, x_j)$ and there are $O(n^2)$ regions. ∎

9.11.1 One-Dimension Parameter Sets

First, we present an elementary case of the local alignment algorithm. Let $\mathbf{a} = a_1 a_2 \cdots a_n$ and $\mathbf{b} = b_1 b_2 \cdots b_m$ be the sequences we compare. The score of aligning letters a and b is $s(a, b) = 1$ if $a = b$ and $s(a, b) = -\mu$ if $a \neq b$. The penalty $w(k)$ for deletion or insertion of k letters is given by $w(k) = \delta k$. This is a special case of $w(k) = \alpha + \beta k$. The local algorithm proceeds recursively by finding the best score H_{ij} ending at a_i and b_j by the equation

$$H_{ij} = \max\{H_{i-1,j-1} + s(a_i, b_j); H_{i-1,j} - \delta; H_{i,j-1} - \delta; 0\}. \tag{9.5}$$

The algorithm is initialized by $H_{0,j} = H_{i,0}$ for $0 \leq i \leq n, 0 \leq j \leq m$. The score for $\mathbf{a}$ aligned with $\mathbf{b}$ is, of course,

$$H = H(\mathbf{a}, \mathbf{b}) = \max_{i,j} H_{i,j}. \tag{9.6}$$

To further simplify this algorithm, take $\mu = 2\delta$ so that the alignment score is a function of one parameter, $\lambda = \delta$. We emphasize the dependence on the parameter by writing $H = H(\lambda)$. Recalling Proposition 9.2 above:

Proposition 9.5 $H(\lambda)$ *is decreasing, piecewise linear and concave. The rightmost linear segment of $H(\lambda)$ is constant.*

Recall that Equation (9.5) allows us to find $H(\delta)$ for any fixed δ. A brief sketch of the algorithm is given next. It is easy to find $H(0)$ and $H(\infty)$. The line segment through $(\infty, H(\infty))$ is easy to find, because $H(\infty) =$ length of longest exact match between $\mathbf{a}$ and $\mathbf{b}$. As many alignment lines often satisfy $H(0) = H(0) - s \cdot 0$, it is necessary to choose the one with the minimum score s, because that determines the optimal alignment just to the right of 0. An algorithm to find the minimum s is given below. Thus, we find the leftmost and rightmost segments of $H(\delta), 0 \leq \delta \leq \infty$. If their intersection (x, y) satisfies $H(x) = y$, we know the entire function $H(\delta)$. Otherwise, $H(x) > y$. Computing $H(x)$ allows us to find all lines through $(x, H(x))$. Below, we show how to find the line that dominates, to the left, all these lines. It is part of the final solution $H(\delta)$. We continue intersections with our line containing $(0, H(0))$ until we have found the line segment L_1 of H that intersects at $(x_1, H(x_1))$ with that containing $(0, H(0))$. Then we take $(x_1, H(x_1))$ and L_2, the line segment of H found just before L_1, and repeat the procedure.

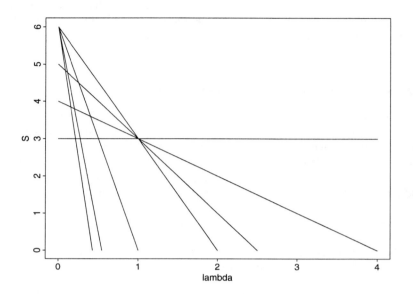

Figure 9.8: *Alignment lines*

The parametric algorithm then depends on our ability to find which alignment line through $(x, H(x))$ is optimal to the left (or right) of x. To illustrate the problem, Figure 9.8 shows all alignment lines optimal at one or more $\lambda \in [0, \infty)$. There are multiple lines with $H(0) = 6$ and the multiple lines with $H(1) = 3$. To chose the line dominant for $\lambda > 0$, we introduce the idea of infinitesimal ϵ. Here, think of $\epsilon > 0$ as a small number, so small that any finite multiple remains smaller than any number that occurs in the algorithms described earlier. Our new numbers will have the form $u + v\epsilon$, where $u, v \in R$. The idea, for example, is that we will run the algorithm for $\lambda = \epsilon$ and find the line maximizing all those through $(0, H(0))$.

Before explicitly describing the infinitesimal version of the local algorithm, it is necessary to define a lexographic linear order on the numbers $u + v\epsilon$. Let $x_1 = u_1 + v_1\epsilon$ and $y_2 = u_2 + v_2\epsilon$. If $u_1 > u_2$, then $x_1 > y_2$. If $u_1 = u_2$ and $v_1 > v_2$, then $x_1 > y_2$. Of course, if $u_1 = u_2$ and $v_1 = v_2$, then $x_1 = y_2$. Thus, the new numbers are linearly ordered. Also, addition is easily defined $x_1 + y_1 = (u_1 + u_2) + (v_1 + v_2)\epsilon$. Of course, $-x_1 = (-u_1) + (-v_1)\epsilon$.

For $\delta = u + v\epsilon$, the algorithm of Equations (9.5) and (9.6) can be used to compute $H = H(\delta)$. It is clear that the algorithm is well defined: Only addition, subtraction, and maximums are involved. This algorithm will be referred to as the

	C	T	G	T	C	G	C	T	G	C	A	C	G
T	0,0	1,0	0,0	1,0	0,0	0,0	0,0	1,0	0,0	0,0	0,0	0,0	0,0
G	0,0	0,0	2,0	0,2	0,1	1,0	0,0	0,0	2,0	0,2	0,0	0,0	1,0
C	1,0	0,0	0,2	1,1	1,2	0,0	2,0	0,2	0,2	3,0	1,2	1,0	0,0
C	1,0	0,1	0,0	0,0	2,1	0,3	1,0	1,1	0,0	1,2	2,1	2,2	0,4
G	0,0	0,1	1,1	0,0	0,3	3,1	1,3	0,1	2,1	0,3	0,3	1,2	3,2
T	0,0	1,0	0,0	2,1	0,3	1,3	2,2	2,3	0,5	1,2	0,0	0,0	1,4
G	0,0	0,0	2,0	0,3	1,2	1,3	0,4	1,3	3,3	1,5	0,3	0,0	1,0

Table 9.8: $H(1 - \epsilon)$

infinitesimal algorithm. Note that ϵ is never specified and that in this sense this is symbolic computation. As the order on infinitesimals is consistent with the order on reals, it is easy to show that if $H(u + v\epsilon) = a + b\epsilon$, then $H(u) = a$. In this way, the usual algorithm is a special case of the infinitesimal algorithm.

Figure 9.8 shows several alignment lines through $(0, H(0))$ for $\mathbf{a} = $ TGC-CGTG and $\mathbf{b} = $ CTGTCGCTGCACG. Notice that if we move just to the right of $\delta = 0$, to $\delta = \epsilon = 0 + 1\epsilon$, we can find the optimal line. The idea then is to run the algorithm for H with the penalty set slightly larger than 0, that is at $\delta = \epsilon$. The values of $H_{i,j}(\delta) = u + v\epsilon$, and at $\delta = 0$, $H_{ij} = u$. Just as it is routine to run the new algorithm, $H(\delta) = H(\epsilon) = \max_{\substack{1 \leq i \leq n \\ 1 \leq j \leq m}} H_{ij}(\delta)$ is easily calculated. For the sequences in Figure 9.8, $H(\epsilon) = 6 - 3\epsilon$. The infinitesimal algorithm for H and the scalar algorithm for H are consistent as can be seen by calculating $H(0) = 6$ by the scalar algorithm.

For $(1, H(1))$ we see there are four competing lines. To choose the dominant line to the left, we calculate $H(1 - \epsilon)$. The matrix is shown as Table 9.8 where $H(1-\epsilon) = a+b\epsilon$ is represented by the pair (a, b). Thus, the four cells with $(3, b)$ correspond to the four lines. The maximal line is $(3, 3) = 3r + 3\epsilon$, dominating $(3, 1) = 3 + \epsilon$, $(3, 0) = 3$, and $(3, 2) = 3 + 2\epsilon$.

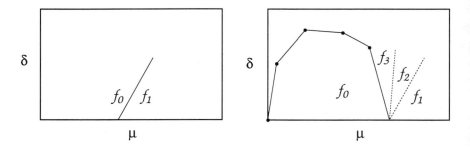

Figure 9.9: *Determining convex polygons:* $f = r - s\mu - t\delta$.

9.11.2 Into Two-Dimensions

Next we face the task of finding all alignment scores for the two-dimensional (μ, δ) parameter space. In the one-dimension parameter space, $(\mu, \delta, H(\delta))$ is a piecewise linear, convex function, whereas in the two-dimensional parameter space $H(\mu, \delta)$ is a convex surface in three-space. Recall that an alignment score satisfies

$$S(\mathbb{A}) = r - s\mu - t\delta, \tag{9.7}$$

where r = number of identities, s = number of mismatches, and t = number of indels. The function $f(\mu, \delta) = r - s\mu - t\delta$ is referred to as an alignment hyperplane. The simplicity of our one-dimensional algorithm does not carry over here due to the increase in dimension. It is necessary to introduce another order of infinitesimal and to impose a linear order on our new numbers. Then we derive a technique to find the unique optimal alignment hyperplane adjacent (to the left or right) of any infinitesimal vector from a given point (μ, δ). This algorithm is the basis of our method to find all convex polyhedron in (μ, δ)-space where the interior has a unique optimal alignment hyperplane.

First, we extend our numbers to include two orders of infinitesimals, ϵ_1 and ϵ_2. Let $x = u_1 + v_1\epsilon_1 + w_1\epsilon_2$ and $y = u_2 + v_2\epsilon_1 + w_2\epsilon_2$. If $u_1 > u_2$, then $x > y$. If $u_1 = u_2$ and $v_1 > v_2$, then $x > y$. Of course, if $u_1 = u_2$, $v_1 = v_2$, and $w_1 = w_2$, then $x = y$. As before, any finite multiple of ϵ_1 cannot exceed 1, and any finite multiple of ϵ_2 cannot exceed ϵ_1. Addition and subtraction are defined in the obvious way.

Our basic algorithm finds the unique optimal alignment hyperplane in the direction (a, b) from (μ, δ). Although the surface is in three-space, we are in the two-dimensional parameter space. We are allowing a point to represent the vector from $(0, 0)$ to that point. It is possible that the (a, b) direction coincides with an intersection of optimal alignment hyperplanes. To assure uniqueness, we must move a small distance, perpendicular to (a, b), that is in direction $(-b, a)$

Dynamic Programming Alignment of Two Sequences

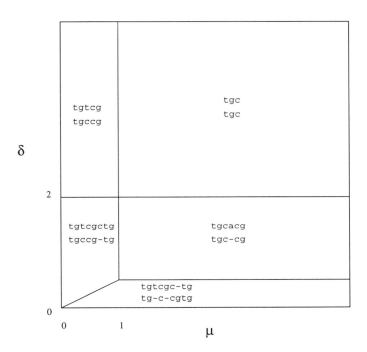

Figure 9.10: *Local alignment regions for* CTGTCGCTGCACG *vs.* TGCCGTG

or $(b, -a)$. The direction (a, b) is of length $\epsilon_1 \sqrt{a^2 + b^2}$, whereas the directions $(-b, a)$ or $(b, -a)$ are of length $\epsilon_2 \sqrt{a^2 + b^2}$. Therefore, the parameters are

$$(\mu^*, \delta^*) = (\mu, \delta) + \epsilon_1(a, b) + \epsilon_2(-b, a) \tag{9.8}$$

or

$$(\mu^*, \delta^*) = (\mu, \delta) + \epsilon_1(a, b) + \epsilon_2(b, -a).$$

See Figure 9.9 for a graphical representation of the parameters.

In order to find the convex polygons of constant alignment hyperplane in $[0, \infty] \times [0, \infty]$, think of the parameter space as a rectangle with four edges and four vertices. Begin at a vertex V_0, $(0, 0)$ or (∞, ∞), say. From V_0, use the basic two-dimensional algorithm along the line L in the counterclockwise direction (say). Initially L is the line from $(0, 0)$ to $(\infty, 0)$. The algorithm can find the alignment hyperplane $f_0(\mu, \delta) = r_0 - s_0\mu - t_0\delta$ immediately adjacent to V_0 in this direction. The goal is to trace out the convex polyhedron in (μ, δ) associated with this hyperplane, with vertex edge labels: $(V_0, e_0, V_1, e_1, \ldots, V_n = V_0)$. By a method similar to the one-dimensional algorithm, it is easy to find the vertex V_1 of the first corner point on line L. To find edge e_1, determine the alignment

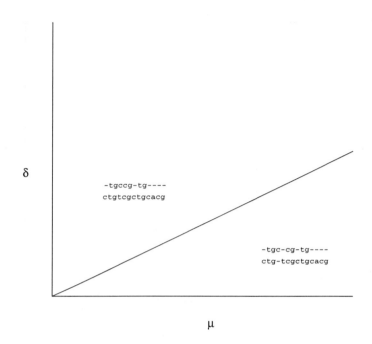

Figure 9.11: *Global alignment regions*

hyperplane f_1 adjacent to the line *beyond* V_1 on L. The intersection $l = f_0 \cap f_1$, which is a line, has optimal alignment hyperplane f_2 immediately adjacent and counterclockwise. If $f_2 = f_0$, then l is the equation of the line containing the edge e_1. Otherwise, intersect f_0 and f_2, repeating the process until the intersection contains e_1. The process is repeated along e_2 and continued until $V_n = V_0$. Figure 9.9 illustrates this process.

Having traced out the vertices and edges of one of the convex polygons of constant alignment hyperplane, it can be removed from $[0, \infty]^2$. The procedure is repeated at a vertex on the boundary of the remaining figure, until all convex polygons have been characterized.

To extend our methods to higher-dimensional parameter spaces is, of course, possible. For k-dimensional parameter spaces, we need $\epsilon = (\epsilon_1, \epsilon_2, \ldots, \epsilon_k)$, where $\epsilon_k < \epsilon_k < \cdots < \epsilon_1$. It is routine to describe the relevant vectors that generalize (9.8). It is necessary to have 1-,2-,..., $(k-1)$-dimensional algorithms at hand to obtain the k-dimension algorithm.

In Figure 9.10, the local alignment regions are given for two sequences, and in Figure 9.11, the global alignment regions are given.

Problems

Problem 9.1 Let $\mathbf{a} = a_1 a_2 \cdots a_n$ and $\mathbf{b} = b_1 b_2 \cdots b_m$ with $1 < n \leq m$. Count the alignments of length m.

Problem 9.2 Let $\mathbf{a} = a_1 a_2 \cdots a_n$ and $\mathbf{b} = b_1 b_2 \cdots b_m$ with $1 < n \leq m$. Count the alignments of length $m + 2$.

Problem 9.3 Prove that

$$\sum_{k \geq 0} \binom{n}{k} \binom{m}{k} = \binom{n+m}{n}$$

Problem 9.4 Find the global similarity alignment of $\mathbf{a}$ = CAGTATCGCA and $\mathbf{b}$ = AAGTTAGCAG with $s(x, y) = +1$ if $x = y$, -1 if $x \neq y$, and $\delta = 1$.

Problem 9.5 In scoring global similarity alignments, reward runs of k matches by a function $h(k) = n + \xi k$ [in addition to the sum of the match weights $s(a, b)$]. Generalize the algorithm of Theorem 9.8.

Problem 9.6 Find all global alignments that score within 1 of the optimal for the sequences in the previous problem.

Problem 9.7 Let $\mathbf{p} = (p_1, p_2)$ be a point in the plane and $\mathcal{L}$ be a line in the plane. Define the distance between $\mathbf{p}$ and $\mathcal{L}$: $\mathbb{D}(\mathbf{p}, \mathcal{L}) = \min\{\sqrt{(p_1 - x)^2 + (p_2 - y)^2} : (x, y) \in \mathcal{L}\}$. Now let $\mathbb{A}$ be an alignment of $\mathbf{a} = a_1 \cdots a_n$ and $\mathbf{b} = b_1 \cdots b_n$, defined by the set of aligned pairs (i_k, j_k), $1 \leq k \leq K$. For the diagonal line $\mathcal{D} = \{(i, i) : 1 \leq i \leq n\}$, give an efficient algorithm for $d_\mathbb{A} = \max\{\mathbb{D}((i_k, j_k), \mathcal{D}) : 1 \leq k \leq K\}$.

Problem 9.8 Find the sum of the similarity scores of all alignments of $\mathbf{a} = a_1 a_2 \cdots a_n$ and $\mathbf{b} = b_1 b_2 \cdots b_m$, where $s(\mathbf{a}, \mathbf{b})$ is the similarity function and indels are weighted $g(k) = \delta k$.

Problem 9.9 Let $h(k) = \min\{\left(\sum_i (g(l_i))^p\right)^{1/p} : \sum_i l_i = k, 0 \leq l_i \leq k\}$, where $p \geq 1$ and $g \geq 0$. Show h is subadditive.

Problem 9.10 Define $f_k(l) = f(l + k) - f(k)$. Show that f_k is subadditive for all k if and only if f is concave.

Problem 9.11 When $\hat{g}$ is formed from g in Section 9.3.1, we can view this as a mapping $\psi(g) = \hat{g}$. Show $\psi^2(g) = \psi(g)$ [that is, $\widehat{(\hat{g})} = \hat{g}$].

Problem 9.12 Fit $\mathbf{a}$ =ATTGAC into $\mathbf{b}$ =CAGTATCGCA with $s(x, y) = +1$ if $x = y$, -1 if $x \neq y$, and $\delta = 1$.

Problem 9.13 Generalize the algorithm for fitting one sequence into another to the case $g(k) = \alpha + \beta(k-1)$.

Problem 9.14 Find the best local alignment of **a** = CAGTATCGCA and **b** = AAGTTAGCAG with $s(x,y) = +1$ if $x = y$, -1 if $x \neq y$ and $\delta = 1$. Declump and find the second best local alignment.

Problem 9.15 Generalize Algorithm 9.4 (Repeat) to the case $g(k) = \alpha + \beta(k-1)$.

Problem 9.16 Algorithm 9.4 (Repeat) finds repeats within a sequence. If tandem repeats exist, describe an algorithm that declumps all other matches of the "repeat unit" with itself.

Problem 9.17 Generalize the linear space Algorithm 9.9 (SL) to the indel function $g(k) = \alpha + \beta(k-1)$. Prove your algorithm is correct.

Problem 9.18 Generalize Algorithm 9.12 (Map) to local map alignment.

Problem 9.19 For global alignments with $s(x,x) = 1, s(x,y) = -\mu, (x \neq y)$ and $g(k) = \alpha + \beta k$, show that any line forming a boundary between three or more regions has the form $\mu = c + (c+1/2)\mu, \alpha = d + d\mu$. Hint: Consider $(\mu, \alpha, \beta) = (-1, 0, -1/2)$.

Problem 9.20 For global alignment with single letter indels penalized by δ, generalize the algorithm to allow free insertion of "T" into the sequence **a**. For example, if **a** =AGA and **b** =ATTGTA, the score $S(\mathbf{a}, \mathbf{b}) = 3$ [if $s(x,x) = 1$].

Problem 9.21 Find the best local alignment of DNA sequence $\mathbf{a} = a_1 a_2 \cdots a_n$ with a protein sequence $\mathbf{b} = b_1 b_2 \cdots b_m$, where δ_P is the cost of deleting a letter from **a**, δ_N is the cost of deleting a letter from **b**, and triplets xyz of **a** are aligned to single letters q of **b** by score $s(g(xyz), q)$, where g is the genetic code. Also, $s(g(a_{i_1} a_{i_2} a_{i_3}), q) = -\infty$ unless $|i_3 - i_1| \leq 3$.

Chapter 10

Multiple Sequence Alignment

Sequence relationships are usually not restricted to those between two sequences. Rather, they extend to relatedness among a family of sequences. This quite naturally leads us to the study of the alignment of r sequences, where $r > 2$. To motivate this class of problems, we discuss a recently cloned gene that is defective in the childhood disease cystic fibrosis. Then, we study in Section 10.2 natural extensions of our dynamic programming algorithms to r sequences. In Section 10.3, information from pairwise alignments is utilized to reduce the computation necessary in Section 10.2. In Section 10.3, we give a method of approaching multiple alignment by pairwise alignments. The widely used search method of profiles is closely related to these ideas. In Section 10.6, a method not based on dynamic programming is presented for multiple sequence analysis.

10.1 The Cystic Fibrosis Gene

Cystic fibrosis (CF) is one of the most common genetic diseases in Caucasian populations. The disease occurs in 1 of 2000 live births and it is estimated that about 1 out of 20 Caucasians carry a defective gene. The disease is recessive, which means both chromosomes must be defective for the defect to result in the disease. The major symptoms of cystic fibrosis include chronic pulmonary disease and an increase in sweat electrolytes. The disease affects the lung airways, pancreas, and sweat glands. There are no effective known treatments and most CF children die by their twenties.

Genetic linkage mapping assigned the cystic fibrosis gene locus to the long arm of chromosome 7 in the mid-1980s. Both physical and genetic mapping further isolated the location of the CF locus. Many person-years of laboratory work went into mapping, cloning, and sequencing before, in 1989, the cystic fibrosis defective gene sequence was announced. The scientists who search for disease genes have been called gene hunters. The 1989 discovery was made by a group led by one of the leaders of this field, Francis Collins. We should mention that CF is a complex disease and that this gene defect, although the major defective allele, is not the only cause of CF.

The situation then is that experimentalists produced a gene sequence of approximately 6500 bps in length, which corresponds to an amino acid sequence of 1480 residues. the defective gene has a deletion of 3 bps that results in the deletion of a Phe residue in the amino acid sequence. Using the sequence alone and no additional knowledge of the protein, the experimentalists were able to classify the protein and to make useful biological hypotheses that led to important experiments. Here, we apply several of the tools we developed in the last chapter to the cystic fibrosis sequence which will motivate multiple sequence alignment.

For reasons we will soon discover, the cystic fibrosis sequence is known as CFTR. The unusual length of CFTR leads us to look for long repeats because this is a frequent mode of protein evolution. Because we are studying protein sequences, the simple $s(x,y)$ we have used in Chapter 8 are no longer adequate. A popular scoring scheme $s(x,y)$ for protein comparisons is the famous Dayhoff PAM matrix. We will use the PAM 120 matrix and $g(k) = 13k$. Using these parameters, we apply algorithm repeats to CFTR and find two long repeated regions, R_N for the one beginning at position nearer the N terminus and R_C for the one beginning at position nearer the C terminus. This repeat will be very useful to us as we examine the results of a database search.

CYSTIC FIBROSIS TRANSMEMBRANE(3)
MULTIDRUG RESISTANCE PROTEIN
HETEROCYST DIFFERENTIATION PROTEIN
MULTIDRUG RESISTANCE PROTEIN
MATING FACTOR A SECRETION PROTEIN
CYAB PROTEIN
PROBABLE ATP-BINDING TRANSPORT
MULTIDGRUG RESISTANCE PROTEIN (5)
HAEMOLYSIN SECRETION PROTEIN (2)
MULTIDRUG RESISTANCE PROTEIN (2)
LEUKOTOXIN SECRETION PROTEIN (2)
HAEMOLYSIN SECRETION PROTEIN (2)
PROTEASES SECRETION PROTEIN PROTEIN
BETA-(1 ← 2) GLUCAN EXPORT PROTEIN
COLICIN V SECRETION PROTEIN

Table 10.1: *Top 25 scoring sequences*

Next, with the above parameters, we search the protein database with the algorithm local. A number of very high scoring matching segments are found. In Table 10.1, we show the names of the top 25 scoring sequences. Duplicates are from different organisms. The three CFTR sequences were, of course, not

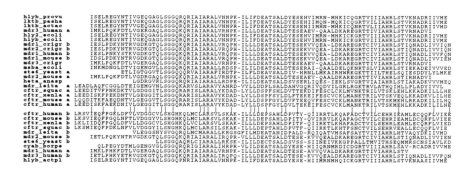

Figure 10.1: *Local alignments to R_N and R_C*

available to the Collins group, but their search had similar results. The search has highlighted similarities to a family of related ATP binding proteins that were already discovered and studied. Recall that ATP provides energy to the cell for many reactions. The proteins of the family are associated with a variety of biological activities in both prokaryotics and eukaryotes. Most of them are involved in the transport of small hydrophilic molecules across cell membranes. The family is defined by a conserved domain of about 200 amino acids that includes an ATP binding site. In Figure 10.1, we show an alignment of the repeat and of selected sequences from other highly similar sequences. These other sequences align best to R_N or R_C. This tells us that R_N and R_C comprise two ATP binding sites in CFTR. It is now clear why CFTR has the name cystic fibrosis transmembrane conductance regulator.

We have learned a great deal of biology by starting with an unknown sequence and making a database search. If CFTR had had similarity with only one of the members of this family, or if the similarity had not been to the ATP binding sites, then these powerful conclusions could not have been so easily drawn. Still, some troubling analytical questions can be asked of our alignment. Essentially, we have aligned r sequences by taking one sequence and performing $r-1$ local alignments. What if these local alignments were scattered along the initial sequence? Even with the current alignment, if we started from another member of the family, would we obtain the same alignment? Would some changes result in a better "overall" alignment? These difficult questions motivate this chapter.

10.2 Dynamic Programming in r-Dimensions

Suppose we have r sequences

$$\mathbf{a}_1 = a_{11}a_{12}\cdots a_{1n_1},$$
$$\mathbf{a}_2 = a_{21}a_{22}\cdots a_{2n_2},$$
$$\vdots$$
$$\mathbf{a}_r = a_{r1}a_{r2}\cdots a_{rn_r}.$$

In the next three sections we will study distance alignment. It is necessary to have a function

$$\rho : \{\mathcal{A} \cup \{-\}\}^r \to R.$$

A simple extension of our earlier dynamic programming algorithms suffices to find the minimum distance between the r sequences. An alignment $\mathbb{A}$ is obtained by inserting $-$'s into the sequence and obtaining a configuration

$$a_{11}^* a_{12}^* \cdots a_{1L}^*,$$
$$\vdots$$
$$a_{r1}^* a_{r2}^* \cdots a_{rL}^*,$$

and

$$D(\mathbf{a}_1, \mathbf{a}_2, \ldots, \mathbf{a}_r) = \min_{\mathbb{A}} \sum_{i=1}^{L} \rho(a_{1i}^*, a_{2i}^*, \ldots, a_{ri}^*).$$

Define

$$D_{i,j,\ldots,l} = D(a_{11}\ldots a_{1i}, a_{21}\ldots a_{2j}, \ldots, a_{r1}\ldots a_{rl}).$$

Then, exactly the same logic as in the two sequence case gives

$$D_{i,j,\ldots,l} = \min_{\epsilon \neq 0} \{ D_{i-\epsilon_1, j-\epsilon_2, \ldots, l-\epsilon_r} + \rho(\epsilon_1 \cdot a_{1i}, \epsilon_2 \cdot a_{2j}, \ldots, \epsilon_r \cdot a_{rl}) \},$$

where $\epsilon_i \in \{0, 1\}$ and

$$\epsilon \cdot a = \begin{cases} a & \text{if } \epsilon = 1, \\ - & \text{if } \epsilon = 0. \end{cases}$$

Note that the algorithm requires prohibitive time $-O\left(2^r \prod_{i=1}^{r} n_i\right) = O(n^r 2^r)$ – and space – $O\left(\prod_{i=1}^{r} n_i\right) = O(n^r)$. These pratical considerations of solving real problems in reasonable time and space motivate this chapter.

There are several methods to construct ρ. Assume a pairwise distance d on $\mathcal{A}$ is given. One appealing idea is to chose the "closest letter" to the existing one:

$$\rho(a_1, a_2, \ldots, a_r) = \min_{\alpha \in \mathcal{A}} \sum_{i=1}^{r} d(a_i, \alpha).$$

The minimizing α is analogous to a "center of gravity" letter.

10.2.1 Reducing the Volume

The objective function presented in Section 10.2 was the minimum sum of $\rho(a_1, a_2, \ldots, a_r)$ for aligned letters. It was suggested that the function ρ could be constructed from $d(\cdot, \cdot)$ by

$$\rho(a_1, a_2, \ldots, a_r) = \min_{\alpha \in \mathcal{A}} \sum_{i=1}^{r} d(a_i, \alpha).$$

This essentially chooses a minimum of the sum of mutation distances of each letter from a "common ancestor." This is far from the only way to evaluate optimal alignments, and we explore another scoring scheme next.

Define, for an alignment $\mathbb{A}$,

$$\mathbf{a}_1^* = a_{1,1}^* a_{1,2}^* \cdots a_{1,L}^*,$$
$$\mathbf{a}_2^* = a_{2,1}^* a_{2,2}^* \cdots a_{2,L}^*,$$
$$\vdots$$
$$\mathbf{a}_r^* = a_{r,1}^* a_{r,2}^* \cdots a_{r,L}^*.$$

The *sum of pairs* (SP) score for alignment $\mathbb{A}$ is

$$C(\mathbb{A}) = \sum_{i<j} \left(\sum_{l=1}^{L} d(a_{i,l}^*, a_{j,l}^*) \right)$$

$$= \sum_{i<j} C(\mathbb{A}_{ij}).$$

Whereas $C(\mathbb{A})$ is defined as the sum of the implied pairwise alignment scores of $\mathbb{A}$, $C(\mathbb{A}_{ij})$ is the score of the individual i-j alignment. Assume $\mathbb{B}$ is an optimal SP-alignment with $C(\mathbb{B}) = \min_A C(\mathbb{A})$. We require a number C' satisfying

$$C' \geq C(\mathbb{B}).$$

Then

$$C' \geq C(\mathbb{B}) = \sum_{i<j} C(\mathbb{B}_{i,j})$$
$$= C(\mathbb{B}_{x,y}) + \sum_{\substack{i<j \\ ij \neq xy}} C(\mathbb{B}_{i,j})$$
$$\geq \sum_{\substack{i<j \\ ij \neq xy}} D(\mathbf{a}_i, \mathbf{a}_j) + C(\mathbb{B}_{x,y})$$
$$= \sum_{i<j} D(\mathbf{a}_i, \mathbf{a}_j) + (C(\mathbb{B}_{x,y}) - D(\mathbf{a}_x, \mathbf{a}_y)).$$

Therefore,

$$\left(C' - \sum_{i<j} D(\mathbf{a}_i, \mathbf{a}_j) \right) \geq C(\mathbb{B}_{x,y}) - D(\mathbf{a}_x, \mathbf{a}_y). \tag{10.1}$$

This equation provides a bound on the pairwise score $C(\mathbb{B}_{xy})$.

The constant C' can be found by any greedy method to construct multiple alignments from pairwise alignments. The most simple such method fixes the alignment of the pair of sequences i-j with minimum distance. Of the remaining pairs (not i-j), the minimum distance pairwise alignment is fixed. If each member of the pair is already in a fixed alignment, then the new fixed alignment joins those two aligned groups. This seldom produces an optimal alignment but the resulting multiple alignment score C' is an upper bound.

This puts an upper bound on the score of the path of the SP-alignment projection onto the x-y sequences. How might we utilize the bound of Equation (10.1)? Our methods can easily compute the best score of any alignment passing through $(a_{x,i}, a_{y,j})$:

$$\text{Best}(x, i; y, j) = D(a_{x,1} \cdots a_{x,i-1}, a_{y,1} \cdots a_{y,j-1}) + d(a_{x,i}, a_{y,j})$$

$$+ D(a_{x,n_x} \cdots a_{x,i+1}, a_{y,n_y} \cdots a_{y,j+1}).$$

For each of the $\binom{r}{2}$ faces of the matrix, say the x-y face, find all (i, j) [that is, (x, i) and (y, j)] such that

$$C' - \sum_{i<j} D(\mathbf{a}_i, \mathbf{a}_j) \geq \text{Best}(x, i; y, j) - D(a_{x,1} \cdots a_{x,i}, a_{y,1} \cdots a_{y,j}).$$

This will give a band of (x, i), (y, j) in that face. When all these $\binom{r}{2}$ face restrictions are considered, the path through the r-dimensional matrix must lie in the intersection. This gives a substantial reduction of the r-dimensional volume to be computed.

10.3 Weighted-Average Sequences

We now look at the geometry of multiple sequence comparisons. These geometries are referred to as line geometries because any two points (sequences) can be joined by a straight line in the metric space. This geometry has some highly non-Euclidean properties and is not currently well understood. In the geometry of geodesics, spaces such as we study are referred to as "straight." We discuss the problems of aligning several sequences with these techniques. A useful application is a method for aligning two sets of sequences, each set of which has already been aligned. Although there does not seem to be much hope for optimal alignment of r sequences of unknown relationship, if the r sequences are related by a binary tree,

Multiple Sequence Alignment

they can be aligned in $O(rn^2)$ steps by a heuristic method naturally suggested by the geometry.

For our purposes, a new but simple concept of sequence is required along with a specific family of metrics on the letters of the sequence. First, if the original sequences are finite words over an alphabet $\mathcal{A}$, define a *weighted-average sequence* to be a finite sequence $\mathbf{a} = a_1 a_2 \cdots a_n$, where each a_i has the form $a_i = (p_0, p_1, p_2, \ldots)$ where $p_i \geq 0$ and

$$\sum_{i \geq 0} p_i = 1.$$

If p_i corresponds to the proportion of the ith element of A and p_0 to the proportion of $-$, it is then easy to convert a usual sequence into a weighted-average sequence by taking a statistical summary of the letters aligned at a given position. The letter $-$ is thought of as a space, indicating a deletion in the sequence in which it appears or an insertion in the opposite sequence. It is much more difficult to handle multiple gaps and, here, we just look position by position.

There are many possible methods to compare two letters $a = (p_0, p_1, \ldots)$ and $b = (q_0, q_1, \ldots)$. Here, we simply compute

$$d(a,b) = \left(\sum_{i \geq 0} w_i |p_i - q_i|^\alpha \right)^{1/\alpha},$$

where w_i are weighting factors and $\alpha \geq 1$ is a constant. It is well known that d is a metric on our set of letters.

To compute the global distance $D(\mathbf{a}, \mathbf{b})$ between two weighted-average sequences, the usual dynamic programming algorithm is employed. Here $\mathbf{a} = a_1 a_2 \cdots a_n$ and $\mathbf{b} = b_1 b_2 \cdots b_m$. If

$$D_{ij} = D(a_1 \cdots a_i, b_1 \cdots b_j),$$
$$D_{0j} = D(-, b_1 \cdots b_j),$$
$$D_{i0} = D(a_1 \cdots a_i, -),$$
$$D_{00} = 0,$$

then

$$D_{i,j} = \min\{D_{i-1,j} + d(a_i, -), D_{i-1,j-1} + d(a_i, b_j), D_{i,j-1} + d(-, b_j)\}.$$

Throughout, $- = (1, 0, \cdots)$ when used as a letter and $- = - - \cdots$ when used as a sequence. Of course, $D_{n,m} = D(\mathbf{a}, \mathbf{b})$.

For an optimal alignment of $\mathbf{a}$ and $\mathbf{b}$ define $\mathbf{c}(\lambda) = \lambda \mathbf{a} \oplus (1 - \lambda) \mathbf{b}$, where $c_i(\lambda) = \lambda a_i^* + (1 - \lambda) b_i^*$ and the last "+" sign is a simple vector addition. In case $\lambda = \frac{1}{2}$, $\mathbf{c}\left(\frac{1}{2}\right)$ is an equal weighting of a_i^* and b_i^* from an optimal alignment of $\mathbf{a}$ and $\mathbf{b}$, and more can be shown in this direction. Theorem 10.1 states that the resulting metric space is a line geometry.

Theorem 10.1 *Let*
$$\mathbf{c}(\lambda) = \lambda \mathbf{a} \oplus (1-\lambda)\mathbf{b}.$$
Then
$$D(\mathbf{a},\mathbf{b}) = D[\mathbf{a},\mathbf{c}(\lambda)] + D[\mathbf{b},\mathbf{c}(\lambda)]$$
and
$$D(\mathbf{a},\mathbf{c}(\lambda)) = ((1-\lambda)D(\mathbf{a},\mathbf{b})).$$

Proof. Recall that a_i^* and b_i^* are aligned in the optimal alignment of $\mathbf{a}$ and $\mathbf{b}$.

$$D[\mathbf{a},\mathbf{c}(\lambda)] \leq \sum_{i=1}^{L} d[a_i^*, c_i(\lambda)]$$

$$= \sum_{i=1}^{L} [\Sigma_j w_j |p_j - [\lambda p_j + (1-\lambda)q_j]|^\alpha]^{1/\alpha}$$

$$= (1-\lambda)\sum_{i=1}^{L} d(a_i^*, b_i^*) = (1-\lambda)D(\mathbf{a},\mathbf{b}).$$

In the same manner, $D[\mathbf{c}(\lambda),\mathbf{b}] \leq \lambda D(\mathbf{a},\mathbf{b})$ and $D[\mathbf{a},\mathbf{c}(\lambda)] + D[\mathbf{c}(\lambda),\mathbf{b}] \leq D(\mathbf{a},\mathbf{b})$. The triangle inequality implies each of the inequalities are equalities. ∎

Corollary 10.1 *Let $\mathbf{c}(\lambda)$ be defined by the same optimal alignment of $\mathbf{a}$ and $\mathbf{b}$ for all λ. Then*
$$D[\mathbf{c}(\lambda_1),\mathbf{c}(\lambda_2)] = |\lambda_1 - \lambda_2|D(\mathbf{a},\mathbf{b}).$$

Proof. First note that if $\lambda_1 \geq \lambda_2$,

$$D(\mathbf{c}(\lambda_1),\mathbf{b}) = \lambda_1 D(\mathbf{a},\mathbf{b}),$$
$$D(\mathbf{c}(\lambda_2),\mathbf{b}) = \lambda_2 D(\mathbf{a},\mathbf{b}),$$

and
$$D(\mathbf{c}(\lambda_1),\mathbf{c}(\lambda_2)) + D(c(\lambda_2),\mathbf{b}) \geq D(\mathbf{c}(\lambda_1),\mathbf{b}).$$

Therefore,
$$D(\mathbf{c}(\lambda_1),\mathbf{c}(\lambda_2)) \geq D(\mathbf{c}(\lambda_1),\mathbf{b}) - D(\mathbf{c}(\lambda_2),\mathbf{b}) = (\lambda_1 - \lambda_2)D(\mathbf{a},\mathbf{b}).$$

Also, because $\mathbf{c}(\lambda_1)$ and $\mathbf{c}(\lambda_2)$ are defined by the same optimal $\mathbf{a}$-$\mathbf{b}$ alignment,

$$D(\mathbf{c}(\lambda_1),\mathbf{c}(\lambda_2)) \leq \sum_{i=1}^{L} d(c_i(\lambda_1), c_i(\lambda_2)) = (\lambda_1 - \lambda_2)D(\mathbf{a},\mathbf{b}). \quad \blacksquare$$

Multiple Sequence Alignment

The theorem implies that a weighted-average sequence can be found to represent any point on the line between two sequences. Although the converse of the theorem is not true, it has a coordinate by coordinate version.

Theorem 10.2 *If* **c** *satisfies* $D(\mathbf{a}, \mathbf{c}) + D(\mathbf{c}, \mathbf{b}) = D(\mathbf{a}, \mathbf{b})$, *then each* $c_i = \lambda_i a_i^* + (1 - \lambda_i) b_i^*$ *for some optimal alignment of* **a** *and* **b**.

Proof. By inserting $\genfrac{}{}{0pt}{}{\phi}{\phi}$ into optimal **a**, **c** and **c**, **b** alignments, the alignments can be assumed to be of equal length:

$$a_1^* a_2^* \cdots a_L^*,$$
$$c_1^* c_2^* \cdots c_L^*,$$
$$c_1^* c_2^* \cdots c_L^*,$$
$$b_1^* b_2^* \cdots b_L^*.$$

Because $D(\mathbf{a}, \mathbf{b}) = D(\mathbf{a}, \mathbf{c}) + D(\mathbf{c}, \mathbf{b})$, the implied **a**, **b** alignment is optimal. Moreover, $d(a_i^*, b_i^*) = d(a_i^*, c_i^*) + d(c_i^*, b_i^*)$ and the result follows. ∎

At this point it might be conjectured that the geometry for more than two sequences immediately follows. Unfortunately, the geometrical properties of even three sequences is far from simple. Let $\mathbf{a}_1, \mathbf{a}_2$ and $\mathbf{a}_3$, be given sequences and define $\mathbf{b}(\lambda) = \lambda \mathbf{a}_1 \oplus (1 - \lambda) \mathbf{a}_2$ and $\mathbf{c}(\lambda) = \lambda \mathbf{a}_1 \oplus (1 - \lambda) \mathbf{a}_3$ for $\lambda \in [0, 1]$. Now

$$D(\mathbf{b}(1), \mathbf{c}(1)) = 0$$

and

$$D(\mathbf{b}(0), \mathbf{c}(0)) = D(\mathbf{a}_2, \mathbf{a}_3),$$

and if $\mathbf{a}_1, \mathbf{a}_2$, and $\mathbf{a}_3$ formed a triangle on the plane, $D(\mathbf{b}(\lambda), \mathbf{c}(\lambda)) = (1 - \lambda) D(\mathbf{a}_2, \mathbf{a}_3)$ would hold. This equation may only hold at $\lambda = 0, 1$. See Figure 10.2.

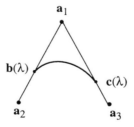

Figure 10.2: $D(\mathbf{b}(\lambda), \mathbf{c}(\lambda)) > (1 - \lambda) D(\mathbf{a}_2, \mathbf{a}_3)$

If all sequences are of equal length and the deletion weight is large enough, then the i-th column in any alignment is composed of the i-th members of the original sequences. In this extreme case, the resulting line geometry is Euclidean.

We now turn to consideration of algorithms for r sequences where $r \geq 3$. These ideas do not seem to suggest practical methods for aligning r sequences of unknown relationship. However, the problem of aligning r sequences, when a binary tree relating the sequences is assumed, does have a practical heuristic solution. We turn first to a simple but important problem.

10.3.1 Aligning Alignments

Suppose two sets of sequences $\mathbf{a}_1, \mathbf{a}_2, \ldots, \mathbf{a}_k$ and $\mathbf{b}_1, \mathbf{b}_2, \ldots, \mathbf{b}_l$ have been aligned by some method. Each such alignment can be easily made into a weighted-average sequence $\mathbf{a}_*$ and $\mathbf{b}_*$. The metric, $D(\cdot, \cdot)$, can be applied to align these alignments. Note that $\lambda \mathbf{a}_* \oplus (1 - \lambda)\mathbf{b}_*$ can be formed from any alignment which gives $D(\mathbf{a}_*, \mathbf{b}_*)$ but that the number of sequences involved, k and l, do not contribute to the complexity of computing $D(\mathbf{a}_*, \mathbf{b}_*)$.

10.3.2 Center of Gravity Sequences

Consider three sequences $\mathbf{a}_1, \mathbf{a}_2$, and $\mathbf{a}_3$. Let them be related by a tree in Figure 10.3(a) where $\mathbf{a}_1$ and $\mathbf{a}_2$ are nearest neighbors. Thus, $\mathbf{e}_2 = \frac{1}{2}\mathbf{a}_1 \oplus \frac{1}{2}\mathbf{a}_2$ occupies the midpoint of a line between $\mathbf{a}_1$ and $\mathbf{a}_2$. If all distances had the properties of Euclidean geometry, the center of gravity is a point on a line from the midpoint $\mathbf{e}_2$ to $\mathbf{a}_3$, two-thirds of the length from $\mathbf{a}_3$, and one-third from $\mathbf{e}_2$. Therefore, the desired sequence is $\mathbf{e}_3 = \frac{1}{3}\mathbf{a}_3 \oplus \frac{2}{3}[\mathbf{e}_2]$. This algorithm generalizes to r sequences. Other weightings can be used.

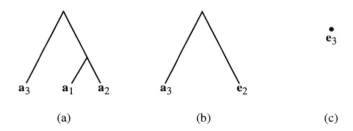

Figure 10.3: *(a) Three sequences (b) $\mathbf{a}_1$ and $\mathbf{a}_2$ replaced by $\mathbf{e}_2 = \frac{1}{2}\mathbf{a}_1 \oplus \frac{1}{2}\mathbf{a}_2$ and (c) $\mathbf{e}_3 = \frac{1}{3}\mathbf{a}_3 \oplus \frac{2}{3}\mathbf{e}_2$*

10.4 Profile Analysis

Section (10.3) contained three very practical ideas. First of all, in the setup was the idea of taking all the data in an multiple alignment position and summarizing that information in a sequence of probability vectors. To be explicit, take the multiple alignment $\mathbb{A}$

$$\mathbf{a}_1 = a_{11}^* a_{12}^* \cdots a_{1L}^*,$$
$$\mathbf{a}_2 = a_{21}^* a_{22}^* \cdots a_{2L}^*,$$
$$\vdots$$
$$\mathbf{a}_r = a_{r1}^* a_{r2}^* \cdots a_{rL}^*,$$

and set

$$p_{jl} = \sum_{i=1}^{r} \mathbb{I}\{a_{ij}^* = l\}/r$$
$$= n_{jl}/r,$$

for $l \in \mathcal{A}$ and $1 \leq j \leq L$. It is, of course, possible to weight individual sequences in this definition just as in Section 10.3.2. If $\mathbf{p}_j = (p_{j0}, p_{j1}, \ldots)$, then the alignment $\mathbb{A}$ produces the weighted-average sequence $\mathbf{p}_1 \mathbf{p}_2 \cdots \mathbf{p}_L$. Here, we have slightly abused notation and taken $\mathcal{A} = \{0, 1, \ldots\}$ where $0 \equiv$ "–." The second idea of the previous section was that if there is a distance or similarity function on $a = (p_0, p_1, \ldots)$ and $b = (q_0, q_1, \ldots)$, $d(a, b)$ or $s(a, b)$, then the various dynamic programming algorithms for two sequences can be applied to multiple alignment. Finally, if the sequences are related by a tree, pairwise alignment algorithms can be applied to produce a multiple alignment.

In this section, we describe one of the most fruitful implementation of these ideas that has become known as profile analysis. In profile analysis, the goal is to use a multiple alignment of a family of related sequences to search a database for more examples of the family. To be successful, the method should find more examples of the family with fewer false positives than a search with any individual member of the alignment. Although these simple methods have the obvious drawbacks of loss of some of the information that is in the alignment, they have proved quite useful.

Next, we describe how to align the profile $P = \mathbf{p}_1 \mathbf{p}_2 \cdots \mathbf{p}_L$ and the sequence $\mathbf{b} = b_1 \cdots b_m$, where $\mathbf{p}_j = (p_{0j}, p_{1j}, \ldots)$. We assume a similarity s defined on $\mathcal{A}^2$. The similarity $\hat{s}$ of letter b at position j can be defined by

$$\hat{s}(\mathbf{p}_j, b) = \sum_{l \in \mathcal{A}} s(l, b) p_{l,j}$$

which is just the expectation of $s(A, b)$ under $\mathbf{p}_j$. Log-likelihood weighting is also popular where

$$\hat{s}(\mathbf{p}_j, b) = -\sum_{l \in \mathcal{A}} s(l, b) \log(\max\{p_{l,j}, \Delta\}),$$

where $\Delta > 0$ might be chosen as $1/r$ to prevent $\log(0)$. Obviously, nothing is to prevent an even larger range of variations on this simple theme.

The discussion of $\hat{s}(\mathbf{p}, b)$ does not include deletions and insertions. Certain positions of the profile might be much more essential than others so that we naturally consider position-dependent indel weighting, at least for the profile sequence. Define the sequence indel functions by

$$g_{seq}(k) = \alpha + \beta(k-1)$$

and the profile indel functions by penalty of $-\gamma_i$ for initiating an indel at $\mathbf{p}_i$ and a penalty of $-\delta_i$ for extending an indel through $\mathbf{p}_i$. See Section 9.3.2.

To be specific, we chose the dynamic programming algorithm that finds the best (similarity) fit of the profile P into the sequence $\mathbf{b} = b_1 b_2 \cdots b_m$. We describe the adaptation of the algorithm in Section 9.5 to solve this problem. Recall that we initialize by $T_{0,j} = 0$, $0 \leq j \leq m$ and $T_{i,0} = g_{pro}(i) = \gamma_1 + \sum_{k=2}^{i} \delta_k$. Then

$$E_{i,j} = \max\{T_{i,j-1} - \alpha,\ E_{i,j-1} - \beta\},$$
$$F_{i,j} = \max\{T_{i-1,j} - \gamma_i,\ F_{i-1,j} - \delta_i\},$$

and

$$T_{i,j} = \max\{T_{i-1,j-1} + \hat{s}(\mathbf{p}_i, b_j), E_{i,j}, F_{i,j}\}.$$

This gives the best alignment with score

$$T(P, \mathbf{b}) = \max\{T_{L,j} : 1 \leq j \leq m\}$$

for position-dependent gaps in both the profile and sequence. If local or global alignment is required, the appropriate changes from Chapter 9 give the desired algorithm.

10.4.1 Statistical Significance

Profiles are used to search databases so that issues of statistical significance arise very naturally. When a local alignment is made using a sequence and a profile, then the results from Chapter 11 apply. Poisson approximations can be employed to give excellent estimates of statistical significance. The other case of most interest is when the profile is fit into the sequence. This was computed by the algorithm for $T(P, \mathbf{b})$ in Section 10.4.

For mathematical reasons as well as for ease of exposition, we do not allow indels. In the case of indels, simulations must be used to estimate the limiting distribution. In this case, the score $T(P, \mathbf{b})$ is easy to compute as the maximum of overlapping sums. Taking $\mathbf{b} = b_1 b_2 \cdots b_{L+n-1}$,

$$T(P, \mathbf{b}) = \max \left\{ \sum_{k=1}^{L} \hat{s}(\mathbf{p}_k, b_{j+k-1}) : 1 \leq j \leq n \right\}.$$

For random sequences $\mathbf{B} = B_1 B_2 \cdots B_{L+n-1}$, with iid letters, we are interested in the distribution of the scores $T(P, \mathbf{B})$. There is an easy heuristic. Each score

$$X_j = \sum_{k=1}^{L} \hat{s}(\mathbf{p}_k, B_{j+k+1})$$

is the sum of L independent random variables. If the $\mathbf{p}_k$ are well behaved, such as identical, each X_k will have an approximate normal (μ, σ^2) distribution by the Central Limit Theorem. This requires L to be moderately large. In addition, if $|j - k| \geq L$, X_j and X_k are independent. This is known as L-dependence. The maximum of n L-dependent normals, properly normalized, has an asymptotic extreme value distribution with distribution function $e^{-e^{-x}}$. Here is the procedure:

$$Y_i = (X_i - \mu)/\sigma, \quad i = 1, 2, \ldots, n.$$

Then

$$M_n = \max\{Y_i : 1 \leq i \leq n\}.$$

Set

$$a_n = (2 \log n)^{1/2}$$

and

$$b_n = (2 \log n)^{1/2} - \frac{1}{2}(2 \log n)^{-1/2}(\log \log n + \log 4\pi).$$

The following is a standard theorem.

Theorem 10.3 (Extreme value) *Let $X_1, X_2, \ldots, X_n$ be iid normal (μ, σ^2) and define $Y_1, \ldots, Y_n$, M_n, a_n, and b_n as above. Then*

$$\lim_{n \to \infty} \mathbb{P}(a_n(M_n - b_n)) = e^{-e^{-x}}.$$

This gives a rapid and practical way of assigning statistical significance to values of $T(P, \mathbf{B})$.

10.5 Alignment by Hidden Markov Models

Weighted-average sequences and, in particular, profile analyses proceed from a given multiple alignment to produce a sequence capturing the statistical details of the multiple alignment. The weighted-average sequence can then be used to discover more sequences that belong to the multiple alignment. Instead of using weighted-average sequences and profiles to discover new sequences, in this section we turn these ideas around and ask if a weighted-average sequence can be used to improve the multiple alignment from which it was derived. A naive version goes as follows. Take $a_1, \ldots, a_r$ aligned as $\mathbb{A}_1$. Then produce the generalized sequence $p_1 p_2 \cdots p_L$ from $\mathbb{A}_1$. If each sequence $a_1 a_2 \cdots a_r$ is aligned with $p_1 p_2 \cdots p_L$, a new multiple alignment $\mathbb{A}_2$ is obtained. Then a new weighted-average sequence can be derived. This process can be continued until the alignment stabilizes.

A more sophisticated setup proves to be valuable. Each sequence $a_1, a_2, \ldots, a_r$ is assumed to be produced by a statistical model, called a hidden Markov model (HMM). This simply means that there is a "hidden" Markov process producing the sequences. The outline of the alignment method is given as an algorithm:

```
1. Choose an initial model (i.e., HMM).

2. Align each sequence aᵢ, 1 ≤ i ≤ r, to the model.

3. Reestimate the parameters of the model.

4. Repeat steps 2 and 3.
```

It should be emphasized that Markov models seldom are correct models for sequences produced by evolution where a tree structure is involved. See Chapter 14 for a discussion of these topics. Still, HMM alignment is often useful for a multiple alignment heuristic.

We now describe the HMM in detail. The HMM is a finite Markov chain with a starting state 0 (BEGIN) and a stopping state (END). There are L states called *match states* m_k, $k = 1$ to L, correspond to the positions of a multiple alignment. The BEGIN state is m_0 and the END state is m_{L+1}. At a match state m_k, a letter $a \in \mathcal{A}$ is generated with probability $\mathbb{P}(a|m_k)$. There are also *insertion states* i_k which insert additional letters with probability $\mathbb{P}(a|i_k)$. Finally, there are *deletion states* d_k which imply that the amino acid at position k is deleted. In this case, the HMM does not produce a letter at position k. Of course, in the observed sequence the knowledge of which position produced a given letter is hidden. Transitions between states are summarized as follows:

$$m_k \to d_{k+1}, i_k, m_{k+1},$$
$$d_k \to d_{k+1}, i_k, m_{k+1},$$
$$i_k \to d_{k+1}, i_k, m_{k+1},$$

for $k = 0, 1, \ldots$. These transitions are summarized by transition probabilities such as $\mathbb{P}(i_k|m_k)$, for example. A special situation occurs at the end state when the sequence of states are $y_0, y_1, \ldots, y_N, y_{N+1}$ with $y_0 = m_0 =$ BEGIN and $y_{N+1} = m_{L+1} =$ END, because state $m_{L+1} =$ END is absorbing. See Figure 10.4 for a schematic view.

Multiple Sequence Alignment

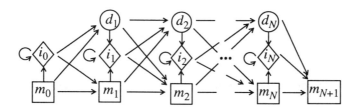

Figure 10.4: *The hidden Markov model*

The sequence of states $m_0 = y_0, y_1, \ldots, y_N, y_{N+1} = m_{L+1}$ produces a sequence of letters $x_1 x_2 \cdots x_M$, where, of course, $M \leq N$. When y_i is a match or insert state and has an associated letter, set $l(i)$ to be the subscript in $x_1 \cdots x_M$ of the letter produced by state y_i. The probability of the sequence of states $\mathbf{y} = y_0, y_1, \ldots, y_{N+1}$ and the sequence of letters $\mathbf{x} = x_1 \cdots x_M$ is

$$\mathbb{P}(\mathbf{y}, \mathbf{x}) = \mathbb{P}(m_{L+1}|y_N) \prod_{k=0}^{N-1} \mathbb{P}(y_{k+1}|y_k) \mathbb{P}(x_{l(k)}|y_k), \qquad (10.2)$$

where $\mathbb{P}(x_{l(k)}|y_k = d_k) \equiv 1$. The probability of the sequence $\mathbf{x} = x_1 x_2 \cdots x_M$ is the sum over all sequences of states that could produce the sequence:

$$\mathbb{P}(\mathbf{x}) = \sum_{\mathbf{y}} \mathbb{P}(\mathbf{y}, \mathbf{x}). \qquad (10.3)$$

In the model we have described, there are approximately $(9 + 2|\mathcal{A}|)(L+1)$ parameters. For amino acid sequences with $|\mathcal{A}| = 20$ and $L = 50$, there are about 2500 parameters. As our goal is to estimate the alignment by estimating the HMM, this is a daunting number of parameters. Still, we proceed to search for the model that maximizes $\mathbb{P}(\text{sequences}|\text{model})$. If we put a prior on models, we can use Bayes' theorem to obtain

$$\mathbb{P}(\text{model}|\text{sequences}) = \frac{\mathbb{P}(\text{sequences}|\text{model})\mathbb{P}(\text{model})}{\mathbb{P}(\text{sequences})},$$

but we do not pursue that here. Instead, we look for the maximum likelihood model.

Recall that Equation (10.3) gave the probability of producing sequence $\mathbf{x}$ from a given model,

$$\mathbb{P}(\mathbf{x}) = \sum_{\mathbf{y}} \mathbb{P}(\mathbf{y}, \mathbf{x})$$

or

$$\mathbb{P}(\mathbf{x}|\text{model}) = \sum_{\mathbf{y}} \mathbb{P}(\mathbf{y}, \mathbf{x}|\text{model}).$$

Assuming $\mathbf{a}_1, \mathbf{a}_2, \ldots, \mathbf{a}_r$ are all independent, then

$$\mathbb{P}(\mathbf{a}_1, \mathbf{a}_2, \cdots, \mathbf{a}_r | \text{model}) = \prod_{i=1}^{r} \mathbb{P}(\mathbf{a}_i | \text{model}).$$

Because $\mathbb{P}(\mathbf{a}|\text{model})$ is the sum over all paths, we just pick the path of highest probability to approximate it:

$$\mathbb{P}(\mathbf{a}|\text{model}) \approx \max\{\mathbb{P}(\mathbf{a}, \mathbf{y}|\text{model}) : \text{all paths } \mathbf{y}\}.$$

Solving this new problem is essentially an alignment problem, which is solved by

$$\max_{\mathbf{y}} \log \mathbb{P}(\mathbf{a}|\text{model}) = \max_{\mathbf{y}} \sum_{k=0}^{N} \left\{ \log \mathbb{P}(y_{k+1}|y_k) + \log \mathbb{P}(x_{l(k)}|y_k) \right\}.$$

This is just a position-dependent alignment problem, easily solved by dynamic programming, as in Section 9.3.2.

We now elaborate the alignment method we have outlined:

Algorithm 10.1 (HMM Alignment)

1. Choose an initial model. If no prior information is available, make all transitions equally likely.
2. Use dynamic programming to find the maximum likelihood path for each sequence $\mathbf{a}_1, \mathbf{a}_2, \ldots, \mathbf{a}_r$.
2'. Collect the count statistics:

    $n(y)$ = #paths through state y,
    $n(y'|y)$ = #paths that have $y \to y'$,
    $m(a|y)$ = #times letter a was produced at state y.

3. Reestimate the parameters of the model

    $$\hat{\mathbb{P}}(y'|y) = \frac{n(y'|y)}{n(y)},$$
    $$\hat{\mathbb{P}}(a|y) = \frac{m(a|y)}{n(y)}.$$

4. Repeat steps (3) and (4) until parameter estimates converge.

10.6 Consensus Word Analysis

Dynamic programming analyses of sequences are examples of what we call *analysis by position*. In analysis by position, we are concerned with whether or not

Multiple Sequence Alignment

we match positions $i_j, j = 1$ to r. Clearly, there are severe limitations on this approach for large r. For example, if $r = 100$ and we know the sequences are each in the correct position, or that the correct position is obtained by shifting one to the right, then there are $2^{100} \cong 10^{30}$ possible configurations. It is possible to approach multiple sequence analysis by *analysis by pattern*, asking, for example, if a certain k-tuple or its close neighbors appears in all the sequences. This section develops some ideas of analysis by pattern.

10.6.1 Analysis by Words

The basic idea is that of a pattern or word w and the neighborhood $N(w)$ of a word. In this section, w is taken to be a k-tuple although that is not a necessary limitation. The neighborhood is defined by distance (N_1), similarity (N_2), or combinatorics (N_3):

$$N_1(w) = \{w' : D(w, w') \leq e\},$$
$$N_2(w) = \{w' : S(w, w') \geq E\},$$
$$N_3(w) = \{w' : w' \text{ and } w \text{ differ by at most } d \text{ differences}\}.$$

In the definition N_3, "differences" can be defined as mismatches, so we have

$$|N_3(w)| = \sum_{l=0}^{d} \binom{k}{l} 3^l,$$

where the alphabet is {A,C,G,T}. If differences include indels, the combinatorics are a little more complex. For the DNA alphabet, $|N(w)| \leq 4^k$ always holds.

The idea is that whereas w may not occur too often with no errors, it might have close neighbors that do. To define an objective function, set

$$S(w) = \sum_{i=1}^{r} \max\{S(w, w') : w' \subset \mathbf{a}_i\}.$$

Then the best word score is

$$S = \max\{S(w) : \text{all } w\}.$$

To find S might seem prohibitive. If $S(w, w')$ costs c time, then S can be found in $4^k \sum_{i=1}^{r}(n_i - k + 1)c = 4^k r(n - k + 1)c$. This is much less than any method that takes time proportional to $\prod_{i=1}^{r} n_i = n^r$, but it can be improved. Before looking at the sequences, $S(w, w')$ can be calculated, in time no more than $4^{2k}c$ (often in much less time). Then at each word along a sequence, the neighborhood N is generated, and we can keep track of the best scoring occurrence of each of the 4^k words as we move along a sequence. After a sequence is analyzed, the cumulative score $S(w)$ can be updated. This process takes time $|N| \sum_{i=1}^{r}(n_i - k + 1) + 4^{2k}c$.

```
AAACAATTTCAGAATAGACAAAAACTCTGAGTGTAATAATGTAGCCTCGTGTCTTGCG
ACCGGAAGAAAACCGTGACATTTTAACACGTTTGTTACAAGGTAAAGGCGACGCCGCCC
TTTGTTTTTCATTGTTGACACACCTCTGGTCATGATAGTATCAATATTCATGCAGTATT
CATCCTCGCACCAGTCGACGACGGTTTACGCTTTACGTATAGTGGCGACAATTTTTTTT
TCCAGTATAATTTGTTGGCATAATTAAGTACGACGAGTAAAATTACATACCTGCCCGC
TTTCTACAAAACACTTGATACTGTATGAGCATACAGTATAATTGCTTCAACAGAACAT
TGCTATCCTGACAGTTGTCACGCTGATTGGTGTCGTTACAATCTAACGCATCGCCAATG
CCATCAAAAAAATATTCTCAACATAAAAAACTTTGTGTAATACTTGTAACGCTACATGGA
```

Figure 10.5: *Eight* E. coli *promoter sequences*

For an explicit example, we present, in Figure 10.5, eight promoter sequences from *E. coli*. There are patterns at two positions in these sequences, the so-called -10 and -35 patterns. For Figure 10.6, $k = 6$ and score matching patterns by $1 - d/k$, where $d =$ # mismatches. In our analysis, we restrict $\mathcal{W}$, the number of contiguous columns to be searched at a time, which we call the window width. We place $\mathcal{W}$ in all possible positions. The complexity for this search is $(n - \mathcal{W} + 1) \times r(\mathcal{W} - k + 1)N$, where n is the common sequence length.

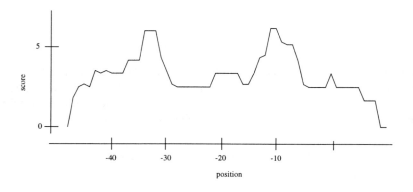

Figure 10.6: *Consensus analysis of* E. coli *promoter sequences*

10.6.2 Consensus Alignment

The idea of a consensus word can be utilized in alignment. Define a partial order on consensus words by their positions in the sequences as follows: $w^{(1)} \prec w^{(2)}$ if the occurrence of $w^{(1)}$ in sequence i is to the left of and not overlapping the occurrence of $w^{(2)}$ in sequence i for $i = 1$ to r. An optimal alignment $\mathbb{A}$ is defined

Multiple Sequence Alignment

to satisfy

$$S(\mathbb{A}) = \max\left\{\sum_i S(w_i) : w_1 \prec w_2 \prec w_3 \prec \cdots\right\}.$$

Only one version of this problem has a straightforward answer. Let $w_1|w_2$ mean that the consensus words are found in nonoverlapping windows:

$$R = \max\left\{\sum_i S(w_i) : w_1|w_2|w_3|\cdots\right\}.$$

This can be solved by a dynamic programming algorithm. Let W be the window width (which can be as large as $W = n$). Then, if R_i is the score for column 1 to i,

$$R_i = \max\{R_j + S(w_{j,i}) : i - W + 1 \leq j \leq i\},$$

where $w_{j,i}$ is the consensus word in columns j to i.

10.6.3 More Complex Scoring

Consensus analysis of protein sequences if of interest too. Let $s(i,j)$ be a scoring matrix for $\mathcal{A}$. Then the neighborhood

$$N(w) = \{w' : S(w, w') \geq E\}$$

with $w = w_1 w_2 \cdots w_k$ and $w' = w'_1 w'_2 \cdots w'_k$ is complex even with the simple scoring

$$S(w, w') = \sum_{i=1}^{k} s(w_i, w'_i).$$

Here, it is useful to compute each neighborhood of the sequences because 20^k is rapidly increasing and 20^{2k} will be too computational expensive.

For each letter $a \in \mathcal{A}$, we order the alphabet in decreasing score with a. Then in a search that has 20^k possibilities, we employ branch and bound to limit the words examined.

A variation on these ideas have turned out to be very useful in the popular search program BLAST. Theorem 11.30 tells us the minimum score T of statistically significant local matches (without indels) to a protein sequence. Then the sequence is used to produce **all** words W that score T or more, and the database is searched for all exact matches to members of W.

Problems

Problem 10.1 Count the number of multiple alignments that have only one letter in each column that is not equal to "$-$".

Problem 10.2 Define

$$N_3(w) = \{w' : w' \text{ differs from } w \text{ by at most } d \text{ one-letter indels}\}$$

Find $|N_3(w)|$, where w is a k-tuple over $\{A,C,G,T\}$.

Problem 10.3 Given a similarity function $s(\cdot, \cdot, \ldots, \cdot)$ on $(\mathcal{A} \cup \{-\})^r$, give the r-sequence generalization of the local dynamic programming algorithm. What is the time and space complexity?

Problem 10.4 Give a counterexample to Corollary 10.1 when $\mathbf{c}(\lambda_1)$ and $\mathbf{c}(\lambda_2)$ are defined by different optimal alignments.

Problem 10.5 For $1 > \lambda_1 \geq \lambda_2 > 0$, let $c(\lambda_i) = \lambda_i \mathbf{a} \oplus (1-\lambda_i)\mathbf{b}_k$ ($i = 1, 2$). Find λ such that for $d(\lambda) = \lambda \mathbf{c}(\lambda_1) \oplus (1 - \lambda)\mathbf{b}$, $D(c(\lambda_1), d(\lambda)) = (\lambda_1 - \lambda_2)D(\mathbf{a}, \mathbf{b})$.

Problem 10.6 In Section 10.4 on pofile analysis, we assumed that the best deletion of the first i letters of the profile has cost $g_{\text{pro}}(i) = \gamma_1 + \sum_{k=2}^{i} \delta_k$, which is a single deletion beginning with the first letter. Find a recursion for the optimal $g_{\text{pro}}(i)$.

Problem 10.7 (Motif 1) For the sequence $\mathbf{a} = a_1 \cdots a_n$ and the set of sequences $\{\mathbf{b}_1, \ldots, \mathbf{b}_r\}$, find an algorithm to compute $SM_1 = \max\{\sum_k S(a_{i_k} \ldots a_{j_k}, \mathbf{b}_{l_k}) : i_1 \leq j_1 < i_2 \leq j_2 < \cdots, l_k \in \{1, \ldots, r\}\}$. (Note that any $\mathbf{b}_i$ sequence can be used repeatedly.)

Problem 10.8 (Motif 2) We are going to splice sequences together. These are sets of sequences (indexed by $k = 1$ to m). Each set has r sequences (indexed by $i = 1$ to r). Each sequence $\mathbf{b}_{ik}$ has l letters (indexed by j). By choosing one sequence from each set, there are r^m sequences of length lm. Find a dynamic programming algorithm for

$$SM_2 = \max\{S(\mathbf{a}, \mathbf{b}) : \mathbf{b} \text{ is one of the } r^m \text{ sequences}\},$$

where $\mathbf{a} = a_1 a_2 \cdots a_n$. The time complexity should be $O(rnlm)$.

Chapter 11

Probability and Statistics for Sequence Alignment

In this chapter, we will study probabilistic aspects of sequences and of sequence comparison. A good deal has been learned about macromolecular sequence data since the first sequences were determined. Because evolution has preserved the essential features of these molecules, a significant sequence similarity between two macromolecules suggests a related function or origin. For this reason, the computer algorithms described in Chapter 9 and Chapter 10 have been devised to locate similar sequences or portions of sequences. The computer searches for similarity can be classified into two categories: search for a known pattern such as that of the hemoglobin family or search for unknown relationships with a newly determined sequence.

Statistical questions are natural in this setting. The scientist wants to find biologically significant relationships between sequences. Although statistical significance is neither necessary nor sufficient for biological significance, it is a good indicator. If there are 50,000 sequences in a database, we must have an automatic way to reject all but the most interesting results from a search of these sequences. Searching for a known pattern involves 50,000 comparisons. Searching all pairs of sequences involves $\binom{50,000}{2} \approx 1.3 \times 10^9$ comparisons. In neither case does a scientist want to look at all comparison results. Screening by statistical significance or p-value is a rational way to sort the results. There are many situations where naive calculations of statistical significance are misleading, so it is important to have "rigorous intuition." Along with learning the results contained in this chapter, it is hoped that the reader will develop this intuition.

The chapter is organized around the biological problems: global sequence comparisons, local sequence comparisons, and statistical behavior resulting from alignment penalty parameters. Several important results from probability are used in this chapter and we have highlighted them in separate sections. In this chapter, we adopt the following standard convention: Capital letters such as A_i will denote a random variable, such as a random letter from $\mathcal{A}$, whereas a_i will denote a value of the random variable, such as a specific letter $a_i \in \mathcal{A}$.

11.1 Global Alignment

In this section, we study probabilistic aspects of the global comparison of two sequences. For simplicity, the two sequences $\mathbf{A} = A_1 A_2 \cdots A_n$ and $\mathbf{B} = B_1 B_2 \cdots B_m$ will, unless noted otherwise, consist of letters drawn independently with identical distribution from a common alphabet (iid).

Recall that an alignment can be obtained by inserting gaps ("$-$") into the sequences, so that

$$A_1 A_2 \cdots A_n \to A_1^* A_2^* \cdots A_L^*$$

and

$$B_1 B_2 \cdots B_m \to B_1^* B_2^* \cdots B_L^*.$$

The subsequence of all $A_i^* \neq$ "$-$" is identical to $A_1 A_2 \cdots A_n$. Then, as the $*$-sequences have equal length, A_i^* is aligned with B_i^*. In Chapter 9, algorithms to achieve optimal alignments are discussed. Here, we are interested in the statistical distribution of these scores, not in how they are obtained. Global alignments refer to the situation where all the letters of each sequence must be accounted for in the alignments. There are two types of global alignments: where the alignment is given in advance and where it is determined by optimality.

11.1.1 Alignment Given

In this section, we assume the alignment is given with the sequences:

$$A_1 A_2 \cdots A_n,$$
$$B_1 B_2 \cdots B_n.$$

Of course, in this case the alignment is fixed in advance and there are no indels. Although it is routine and not of great biological interest, we give the statistical distribution of the alignment score for completeness. Let $s(A, B)$ be real-valued. Define the score S by

$$S = \sum_{i=1}^{n} s(A_i, B_i).$$

We assume the alphabet $\mathcal{A}$ is finite although that is not a necessary restriction.

Theorem 11.1 *Assume* $\mathbf{A} = A_1 A_2 \cdots A_n$ *and* $\mathbf{B} = B_1 B_2 \cdots B_n$ *with iid letters A_j and B_j. Define* $S = \sum_{i=1}^{n} s(A_i, B_i)$. *Then*

(i) $\mathbb{E}(S) = n\mathbb{E}(s(A, B)) = n\mu,$

(ii) $\text{Var}(S) = n\text{Var}(s(A,B)) = n\sigma^2$,

and

(iii) $\lim_{n\to\infty} \mathbb{P}(\frac{S-n\mu}{\sqrt{n}\sigma} \leq x) = \Phi(x) = \frac{1}{\sqrt{2\pi}} \int_{-\infty}^{x} e^{-t^2/2} dt$,

where $\Phi(x)$ is the cumulative distribution function (cdf) of a standard normal.

Proof. This is the usual result for sums of iid random variables. ∎

We remark that when $s(a,b) \in \{0,1\}$, then S is binomial $\mathcal{B}(n,p)$ with $p = \mathbb{P}(s(A,B) = 1)$. The iid assumptions can be relaxed in the theorem but we do not pursue that here. Instead, we turn to the much more interesting case of where the alignment is determined by optimality.

11.1.2 Alignment Unknown

The assumptions of the last section are carried over: $A_1 \cdots A_n$ and $B_1 \cdots B_m$ are composed of independent and identically distributed letters and $s(a,b)$ is a real-valued function on pairs of letters. We extend $s(\cdot,\cdot)$ to $s(a,-)$ and $s(-,b)$ so that deletions are included. Now an alignment score S is the maximum over all possible alignments,

$$S = \max\left\{\sum_{i=1}^{L} s(A_i^*, B_i^*) : \text{all alignments}\right\}.$$

The optimization over the large number of alignments destroys the classical normal distribution of alignment score, but the application of Kingman's subadditive ergodic theorem and the Azuma-Hoeffding lemma gives interesting results. Although we can give some analysis of the distribution of S, it is far from begin completely understood.

Kingman's Theorem

For easy reference, we give a statement of Kingman's theorem that appeared earlier as Theorem 3.1 in Chapter 3.

Theorem 11.2 (Kingman) *For s and t non-negative integers with $0 \leq s \leq t$, let $X_{s,t}$ be a collection of random variables which satisfy the following:*

(i) *Whenever $s < t < u$, $X_{s,u} \leq X_{s,t} + X_{t,u}$.*

(ii) *The joint distribution of $\{X_{s,t}\}$ is the same as that of $\{X_{s+1,t+1}\}$.*

(iii) *The expectation $h_t = \mathbb{E}[X_{0,t}]$ exists and satisfies $h_t \geq -Kt$ for some constant K and all $t > 1$.*

Then the finite $\lim_{t\to\infty} X_{0,t}/t = \rho$ exists with probability 1 and in the mean.

11.1.3 Linear Growth of Alignment Score

Returning to alignment score, recall again that an alignment score S is the maximum over all possible alignments

$$S = \max \left\{ \sum_{i=1}^{L} s(A_i^*, B_i^*) : \text{all alignments} \right\}$$

Here we handle the general case where deletions of length k are penalized by $-g(k)$, for a general non-negative subadditive g. Define $X_{s,t}$ by

$$-X_{s,t} = \text{score of } A_{s+1} \cdots A_t \text{ vs. } B_{s+1} \cdots B_t.$$

Then, evidently,

$$-X_{s,u} \geq (-X_{s,t}) + (-X_{t,u})$$

and

$$X_{s,u} \leq X_{s,t} + X_{t,u}.$$

We have $h_t = \mathbb{E}(X_{0,t})$ exists because the expectation of a single alignment exists and $-X_{0,t}$ is the maximum of a *finite* number of alignments scores. The final hypothesis to check is $h_t \geq -Kt$ for some constant K and all $t > 1$. Set $s^* = \max\{s(a,b) : a, b \in \mathcal{A}\}$. Clearly,

$$\mathbb{E}(-X_{0,t}) \leq \max\{ts^*, -2g(t)\} = t \max \left\{ s^*, \frac{-2g(t)}{t} \right\}.$$

If $s^* < -2g(t)/t$, note that $\lim(g(t)/t)$ exists by subadditivity. Therefore, there exists K satisfying

$$h_t \geq -Kt.$$

Our conclusion is that

$$\lim_{t \to \infty} (-X_{0,t}/t) = \rho$$

exists with probability 1 and in the mean. That ρ is constant follows from the fact that ρ is independent of the first k values of A_i and B_i. Therefore, optimal alignment score grows linearly with sequence length. Obviously, $\rho \geq \mathbb{E}(s(A,B))$. We have proved the next theorem.

Theorem 11.3 *Assume* $\mathbf{A} = A_1 A_2 \cdots A_n$ *and* $\mathbf{B} = B_1 B_2 \cdots B_n$ *with* A_i *and* B_j *iid. Define* $S_n = S(\mathbf{A}, \mathbf{B}) = \max\{\sum s(A_i^*, B_j^*) : \text{alignments}\}$. *Then, there is a constant* $\rho \geq \mathbb{E}(s(A,B))$ *such that*

$$\lim_{n \to \infty} \frac{S_n}{n} = \rho$$

with probability 1 and in the mean.

Since $\lim_{n \to \infty} \frac{S_n}{n} = \rho$ almost surely, $\frac{\mathbb{E}(S_n)}{n} \to \rho$ as well.

Probability and Statistics for Sequence Alignment

In the simplest case of interest, the alphabet has two uniformly distributed letters and $s(a,b) = 0$ if $a \neq b$, $s(a,a) = s(b,b) = 1$, and $\delta = 0$. An alignment with largest score is known as a *longest common subsequence*, and Chvátal and Sankoff wrote a seminal paper on this problem in 1975. In spite of much effort since then, ρ remains undetermined. Not too much is known about the variance of S_n either. (See Theorem 11.6 below.) There are bounds for ρ: $0.7615 \leq \rho \leq 0.8575$. Without alignment, the fraction of matching letters is $0.5 = \mathbb{E}(s(A,B))$ by the strong law of large numbers.

11.1.4 The Azuma-Hoeffding Lemma

This useful result, which is also known as the Azuma-Hoeffding inequality, gives a bound on "large deviations," or the probability that a random variable exceeds its mean by a specified amount. Its proof involves martingales but is elementary. Let $\mathcal{F}_0 \subseteq \mathcal{F}_1 \subseteq \cdots$ be an increasing family of sub-σ-algebras with X_n a random variable measurable with respect to $\mathcal{F}_n$. Then $\{X_n\}_{n\geq 0}$ is called a *martingale* (relative to $\{\mathcal{F}_n\}$) if (i) $\mathbb{E}|X_n| < \infty$ for all n and (ii) $\mathbb{E}(X_n|\mathcal{F}_{n-1}) = X_{n-1}$, a.s., $n \geq 1$.

As we have finite sets, these concepts are fairly easy. A sigma algebra is just a collection of sets containing the universal set that is closed under complements and unions. The conditional expectation $\mathbb{E}(f|\mathcal{F})$ is a function measurable on $\mathcal{F}$ (constant on the atoms of $\mathcal{F}$) so that $F \in \mathcal{F}$ implies $\mathbb{E}(\mathbb{I}_F f | \mathcal{F}) = \mathbb{I}_F(\mathbb{E}(f|\mathcal{F}))$. The next lemma requires a martingale with bounded increments.

Lemma 11.1 *Let $X_0 = 0, X_1, X_2, \ldots$ be a martingale relative to $\{\mathcal{F}_n\}$ so that $X_{n-1} = \mathbb{E}(Y|\mathcal{F}_{n-1}), n \geq 1$. If, for some sequence of positive constants c_n,*

$$|X_n - X_{n-1}| \leq c_n \quad \text{for } n \geq 1,$$

then

$$\mathbb{E}(e^{\beta X_n}) \leq e^{\beta^2/2 \sum_{k=1}^n c_k^2}.$$

Proof. We need two inequalities. The first inequality is

$$e^{\beta x} \leq \frac{c-x}{2c}e^{-\beta c} + \frac{c+x}{2c}e^{\beta c} \quad \text{for all } x \in [-c, c]. \tag{11.1}$$

Recall the definition of φ convex: $\varphi(\gamma x_1 + (1-\gamma)x_2) \leq \gamma\varphi(x_1) + (1-\gamma)\varphi(x_2)$, where $\gamma \in [0,1]$. Note that $\varphi(t) = e^{\beta t}$ is convex. Let $x_1 = -c, x_2 = c$,

$$\gamma = \frac{c-x}{2c},$$

and

$$1 - \gamma = \frac{c+x}{2c}.$$

Then Equation (11.1) follows from the definition of convexity.

The second inequality,
$$\frac{e^{-x} + e^{x}}{2} = \cosh x \leq e^{x^2/2} \quad \text{for all } x, \tag{11.2}$$
follows from comparing Taylor expansions of each side of the inequality.
Now the lemma easily follows.

$$\mathbb{E}\{e^{\beta X_n}\} = \mathbb{E}\{\mathbb{E}(e^{\beta X_n}|\mathcal{F}_{n-1})\}$$
$$= \mathbb{E}\{e^{\beta X_{n-1}}\mathbb{E}(e^{\beta(X_n - X_{n-1})}|\mathcal{F}_{n-1})\}$$

Because $|X_n - X_{n-1}| \leq c_n$, we apply Equation (11.1) to $e^{\beta(X_n - X_{n-1})}$, and using the martingale property obtain $\mathbb{E}(e^{\beta(X_n - X_{n-1})}|\mathcal{F}_{n-1}) \leq \cosh \beta c_n$. Therefore, applying Equations (11.1) and (11.2) repeatedly,

$$\mathbb{E}\{e^{\beta X_n}\} \leq \mathbb{E}\{e^{\beta X_{n-1}} \cosh(\beta c_n)\} \leq \cdots$$
$$\leq \left(\prod_{k=1}^{n} \cosh(\beta c_k)\right) \mathbb{E} e^{\beta X_0}$$
$$= \prod_{k=1}^{n} \cosh(\beta c_k)$$
$$\leq \prod_{k=1}^{n} e^{(\beta^2/2) c_k^2} = e^{(\beta^2/2) \sum_{k=1}^{n} c_k^2}.$$

∎

The next result follows from Lemma 11.1 and is applicable to global alignment.

Lemma 11.2 *Under the same assumptions as Lemma* 11.1, *for* $\lambda > 0$,
$$\mathbb{P}(X_n \geq \lambda) \leq e^{-\lambda^2/(2\sum_{k=1}^{n} c_k^2)}.$$

Proof. Markov's inequality states that for nondecreasing $g : R \to [0, \infty)$, $\mathbb{E}(g(Y)) \geq g(c)P(Y \geq c)$. Therefore, with $g(t) = e^{\beta t}$, $\beta > 0$,
$$\mathbb{P}(X_n \geq \lambda) \leq \frac{\mathbb{E}(e^{\beta X_n})}{e^{\beta \lambda}},$$
so that by Lemma 11.1,
$$\mathbb{P}(X_n \geq \lambda) \leq \exp\left\{\beta^2/2 \sum_{k=1}^{n} c_k^2 - \beta\lambda\right\}.$$
The value of β minimizing the exponent is $\beta = \lambda/\sum_{k=1}^{n} c_k^2$, so
$$\mathbb{P}(X_n \geq \lambda) \leq \exp\left\{-\lambda^2/(2\sum_{k=1}^{n} c_k^2)\right\}.$$

∎

11.1.5 Large Deviations from the Mean

Recall that $\mathbb{E}(S_n)/n \to \rho$, which even in the simplest nontrivial case is unknown. Nevertheless, the Azuma-Hoeffding lemma can be applied to show that large deviations from the mean and from λ have exponentially small probability. To bring Azuma-Hoeffding into play, we need to bound the martingale increments. This, in turn, requires a simple deterministic lemma about alignment scores. As usual, our alignment is scored by $s(a, b)$ and $g(k)$ which is subadditive ($g(k+l) \leq g(k) + g(l)$).

Lemma 11.3 *Let $S = S(a_1 a_2 \cdots a_k, b_1 b_2 \cdots b_k) = S(c_1 c_2 \cdots c_k)$ be the alignment score for k pairs of letters, $c_i = (a_i, b_i)$. Let $S' = S(c_1 \cdots c_{i-1}, c'_i, c_i \cdots c_k)$ be the score for k pairs of letters with only the i-th pair changed. Let $s^* = \max\{s(a,b) : a, b \in \mathcal{A}\}, s_* = \min\{s(a,b) : a, b \in \mathcal{A}\}$. Then,*

$$S - S' \leq \max\{\min\{2s^* + 4g(1), 2s^* - 2s_*\}, 0\} = c.$$

Proof. The i-th pair is $c_i(a_i, b_i)$ in S and $c'_i = (a'_i, b'_i)$ in S'. To bound the maximum difference $S - S'$, we consider several cases. Suppose a_i matches b_j, and b_i matches a_l, scoring at most $2s^*$. Changing a_i to a'_i and b_i to b'_i scores as little as $2s_*$ or all four letters could be deleted. By subadditivity, the additional deletion penalty, which could extend existing deletions, is at most $4g(1)$. The bound for this case is $S - S' \leq \min\{2s^* + 4g(1), 2s^* - 2s_*\}$. The cases of either a_i or b_i in a deletion is covered by the above bound. Finally, if a_i matches b_i, then a higher scoring match can be replaced by a lower scoring match or by deleting the two letters. This bound is $S - S' \leq \min\{s^* + 2g(1), s^* - s_*\}$.

There is one more detail. If $s^* + 2g(1) \leq 0$, then $S_k = 2g(k)$ independent of the sequences and $S - S' = 0$. Thus,

$$S - S' \leq \max\{\min\{2s^* + 4g(1), 2s^* - 2s_*\}, 0\} = c.$$

∎

Theorem 11.4 *Assume $\mathbf{A} = A_1 \cdots A_n$ and $\mathbf{B} = B_1 B_2 \cdots B_n$ have iid letters A_i and B_j. Define $S = S(\mathbf{A}, \mathbf{B})$ to be the global alignment score. Then, if c is the constant defined in Lemma 11.3,*

$$\mathbb{P}(S - \mathbb{E}S \geq \gamma n) \leq e^{-\gamma^2 n/2c^2}.$$

Proof. Our martingale is $X_i = \mathbb{E}(Y|\mathcal{F}_i)$, where $Y = S(C_1 C_2 \cdots C_n) - \mathbb{E}(S(C_1 C_2 \cdots C_n))$ and $\mathcal{F}_i = \sigma(C_1 C_2 \cdots C_i)$ is the σ-algebra generated by the first i pairs of random variables $C_i = (A_i, B_i)$. As Y is $\mathcal{F}_n$ measurable, $X_n = \mathbb{E}(Y|\mathcal{F}_n) = Y = S - \mathbb{E}(S)$. Because $\mathcal{F}_0$ is the trivial σ-algebra, $X_0 = \mathbb{E}(S) - \mathbb{E}(S) = 0$. The key part of the martingale is

$$\mathbb{E}(S|\mathcal{F}_i) = \sum_{c_{i+1},\ldots,c_n} S(C_1, \cdots, C_i, c_{i+1}, \ldots, c_n) \mathbb{P}(C_{i+1} = c_{i+1}, \ldots, C_n = c_n).$$

This holds because, to make $S(C_1 \cdots C_n)$ measurable on $\mathcal{F}_i = \sigma(C_1, \ldots, C_i)$, we need to average out the values of $C_{i+1} \cdots C_n$.

In fact,

$$X_i - X_{i-1} =$$
$$\sum_{c_{i+1}\cdots c_n} S(C_1, \ldots, C_i, c_{i+1}, \ldots, c_n) \times \mathbb{P}(C_{i+1} = c_{i+1}, \ldots, C_n = c_n)$$
$$- \sum_{c'_i c_{i+1}\cdots c_n} S(C_1, \ldots, C_{i-1}, c'_i, c_{i+1}, \ldots, c_n) \times \mathbb{P}(C_i = c'_i, C_{i+1} = c_{i+1}, \ldots, C_n = c_n)$$

Therefore,

$$|X_i - X_{i-1}| \leq$$
$$\sum_{c'_i c_{i+1}\cdots c_n} |S(C_1, \ldots, C_i, c_{i+1}, \ldots, c_n) - S(C_1, \ldots, C_{i-1}, c'_i, c_{i+1}, \cdots, c_n)|$$
$$\mathbb{P}(C_i = c'_i, C_{i+1} = c_{i+1}, \ldots, C_n = c_n).$$

From Lemma 11.3 then,

$$|X_i - X_{i-1}| \leq \max |S - S'| \leq c.$$

Finally, using Lemma 11.2,

$$\mathbb{P}(S - \mathbb{E}(S) \geq \gamma n) \leq e^{-(\gamma^2 n^2)/2nc^2} = e^{-\gamma^2 n / 2c^2}.$$

∎

Corollary 11.1 *Under the assumption of Theorem* 11.4,

$$\mathbb{P}(S_n/n - \rho \geq \gamma) \leq e^{-\gamma^2 n / 2c^2}.$$

Proof. The result from subadditivity is that $\rho = \lim_{n\to\infty} \frac{\mathbb{E}(S_n)}{n} = \sup_n \frac{\mathbb{E}(S_n)}{n}$. We have $\mathbb{E}(S_n) \leq n\rho$, so that

$$\mathbb{P}(S_n \geq (\gamma + \rho)n) \leq \mathbb{P}(S_n - \mathbb{E}(S_n) \geq \gamma n).$$

∎

The best known result on the variance of S follows from a general theorem of Steele.

Theorem 11.5 (Steele) *If $f(x_1, x_2, \ldots, x_n)$ is any function and $X_i, X'_i, 1 \leq i \leq n$ are $2n$ iid random variables, then*

$$\mathrm{Var}(f) \leq \frac{1}{2}\left\{\mathbb{E}\sum_{i=1}^{n}(f - f_{(i)})^2\right\},$$

where $f = f(X_1, X_2, \ldots, X_n)$ and $f_{(i)} = f(X_1, X_2, \ldots, X'_i, \ldots, X_n)$ is obtained by replacing X_i by X'_i.

Probability and Statistics for Sequence Alignment

This immediately applies to global alignment.

Theorem 11.6 *Assume* $\mathbf{A} = A_1 \cdots A_n$ *and* $\mathbf{B} = B_1 \cdots B_n$ *have iid letters* A_i *and* B_i. *Define* $S_n = S(\mathbf{A}, \mathbf{B})$ *to be the global alignment score. Then, if* $c^* = \max\{0, \min\{s^* + 2g(1), s^* - s_*\}\}$ *and* $p = \mathbb{P}(A_1 = B_1)$,

$$\mathrm{Var}(S_n) \leq n(1-p)c^*.$$

Proof. This proof differs slightly from one above, where we identify $C_i = (A_i, B_i)$. For each of the $2n$ random variables, we change one at a time. Therefore, the bound $|S - S'| \leq \max\{0, \min\{s^* + 2g(1), s^* - s_*\}\} = c^*$ holds here. ($2c^* = c$ of Lemma 11.3.) Note that with probability $\mathbb{P}(A_1 = B_1)$, the difference is 0. Therefore,

$$\mathbb{E}(S - S_{(i)}) \leq (1 - \mathbb{P}(A_1 = B_1))c^*.$$

Finally, there are $2n$ terms so $\frac{1}{2} \times 2n = n$. ∎

11.1.6 Large Deviations for Binomials

In this section, we study the simple case of

$$s(a, b) = \begin{cases} 1 & \text{if } a = b, \\ 0 & \text{if } a \neq b, \end{cases}$$

and $\delta = \infty$. The alignment score $S(\mathbf{A}, \mathbf{B})$, $\mathbf{A} = A_1 \cdots A_n$ and $\mathbf{B} = B_1 \cdots B_n$ with **fixed** alignment, is then a binomial random variable $\mathcal{B}(n, p)$, where $p = \mathbb{P}(A_i = B_j)$. There are many other cases in biology where binomial random variables arise. The property labeled as success or 1 might be hydrophobicity or positive charge of an amino acid in a protein or 'A', or purine in a DNA sequence. Success might be the presence of a consensus word or helix in a sequence. The scientist might, therefore, wish to estimate a p-value for observed values of the binomial random variable that is far from its mean. Large deviations give tractable estimates that are far more accurate than those given by the Central Limit Theorem. Let $Y_n \sim \mathcal{B}(n, p)$ and set a success fraction α for k of n successes,

$$\alpha = k/n,$$

where $p < \alpha < 1$. A fraction α of successes is larger than the expected fraction p.

Let $\mathcal{H}(a, p)$ be the *relative entropy*

$$\mathcal{H} \equiv \mathcal{H}(\alpha, p) \equiv (\alpha) \log\left(\frac{\alpha}{p}\right) + (1 - \alpha) \log\left(\frac{1 - \alpha}{1 - p}\right). \quad (11.3)$$

We observe that $\mathcal{H}(\alpha, p)$ increases from 0 to $\log(1/p)$ as α increases from p to 1. This value $\mathcal{H}$ is also called the Kullback-Liebler distance; it measures the distance from the $\mathcal{B}(n,p)$ distribution under which the data are generated to an alternative, the $\mathcal{B}(n,\alpha)$ distribution. The key concept and difficulty in understanding large deviations is dealing simultaneously with two probability measures on the same space of possible outcomes. The next theorem gives a useful upper bound valid for all n, p, and α.

Theorem 11.7 *For $p < \alpha < 1$, for $n = 1, 2, 3, \ldots$, with $\mathcal{H} = \mathcal{H}(\alpha, p)$ the relative entropy defined in Equation 11.3 above and $Y_n \sim \mathcal{B}(n, p)$, then*

$$\mathbb{P}(Y_n \geq \alpha n) \leq e^{-n\mathcal{H}}.$$

Proof. For all $\beta > 0$,

$$\mathbb{P}(Y_n \geq \alpha n) = \mathbb{P}(e^{\beta Y_n} \geq e^{\beta \alpha n})$$
$$\leq \mathbb{E}(e^{\beta Y_n})/e^{\beta \alpha n}$$
$$= (1 - p + pe^\beta)^n / e^{\beta \alpha n} = \{e^{-\alpha\beta}(1 - p + pe^\beta)\}^n,$$

where the inequality follows from Markov's inequality. Minimizing the quantity in the brackets gives the value $e^{-\mathcal{H}}$. ∎

We let

$$r \equiv \frac{p}{1-p} \bigg/ \frac{\alpha}{1-\alpha} = \frac{p}{\alpha}\frac{1-\alpha}{1-p}, \qquad (11.4)$$

denote the "odds ratio" between p-coins and α-coins. Observe that $0 < r < 1$ because $p < \alpha < 1$, and that all of the quantities $\alpha, \mathcal{H}$, and r are allowed to vary depending on k and n. Observe also that r and $\mathcal{H}$ are related: $\mathcal{H}'(a,p) = -\log(r)$, where the derivative is taken with respect to α, for fixed p. As always, we write $\sim$ to denote asymptotic equality, that is, that the ratio of two quantities tends to the limit 1.

Equation (11.5), which is the result of dividing Equation (11.7) into Equation (11.8), says that, conditional on the number of successes being at least $k = \alpha n$; the excess over k has, asymptotically, a geometric distribution with parameter $1 - r$. This geometric distribution is the distribution of the number of heads before the first tail when tossing an r-coin and has mean $r/(1-r)$.

$$\mathbb{P}(Y_n = \alpha n + i | Y_n \geq \alpha n) \to r^i(1-r) \text{ for } i = 0, 1, 2, \ldots \text{ as } n \to \infty. \quad (11.5)$$

Some useful facts about large deviations are collected in the next theorem.

Theorem 11.8 *For $Y_n \sim \mathcal{B}(n,p), p < \alpha < 1$, and r the odds ratio defined in Equation 11.4 above, as $n \to \infty$.*

$$\log \mathbb{P}(Y_n \geq n\alpha) \sim -n\mathcal{H}, \qquad (11.6)$$

$$\mathbb{P}(Y_n \geq \alpha n) \sim \frac{1}{1-r}\frac{1}{\sqrt{2\pi\alpha(1-\alpha)n}}e^{-n\mathcal{H}}, \qquad (11.7)$$

and

$$\mathbb{P}(Y_n = \alpha n + i) \sim \frac{1}{\sqrt{2\pi\alpha(1-\alpha)n}} r^i e^{-n\mathcal{H}}, \ i = 0, 1, 2, \ldots. \qquad (11.8)$$

11.2 Local Alignment

Now we turn to local comparisons, for which the algorithm H was devised. The local alignment problem is to find the best matching segments or intervals of sequence in comparisons of long genetic sequences. Recall

$$H(\mathbf{A}, \mathbf{B}) = \max\{S(I, J) : I \subset \mathbf{A}, J \subset \mathbf{B}\},$$

where S is similarity measured by the substitution function $s(\cdot, \cdot)$ and indel function $g(\cdot)$. $H(\mathbf{A}, \mathbf{B})$ is an appropriate random variable for the case where the alignment is unknown. It is sometimes the case that the alignment is known or given. Then the sequences are in a fixed alignment and the local comparison algorithm problem is to locate the highest scoring interval. In each situation, the statistical problem is to find the probability distribution of the alignment score when the sequences $\mathbf{A}$ and $\mathbf{B}$ are random. We wish to be able to detect statistically significant values of the random variables.

There are several levels of approach to these problems. In this section, we will study strong laws of large numbers to determine asymptotic behavior of the growth of the scores.

11.2.1 Laws of Large Numbers

In this section, we present laws of large numbers for the asymptotic behavior of the longest match between two random sequences. Random here means either independent and identically distributed or Markov, although similar laws for m-dependent processes can be obtained. Whereas the laws of large numbers only give order of magnitude results, these estimates are surprisingly good and give excellent rules of thumb for the expected magnitude of matches between random sequences. Later, we will give much more precise results that allow comparison with the more easily obtained results of this section.

Exact Matching

The first problem considered here is the length R_n of the longest match between two sequences of length n with iid letters which are in a fixed alignment. Let

$$p = \mathbb{P}(\text{Match}) = \sum_{i \in \mathcal{A}} \xi_i^2,$$

where ξ_i is the probability of letter i. The random variable R_n is only of interest in the cases $p \in (0, 1)$. The problem posed here can be restated as the longest run of heads in n coin tosses where $p = \mathbb{P}(\text{Heads})$. Erdös and Rényi (1970) presented results which contain the following theorem. The intuition is as follows. A headrun of length m has probability p^m. There are approximately n possible headruns so

$$\mathbb{E}(\text{\# headruns of length } m) \cong np^m.$$

If the largest run is unique, its length R_n should satisfy $1 = np^{R_n}$, which has solution $R_n = \log_{1/p}(n)$. This heuristic is a guide to this entire section and most of the remainder of this chapter. Before we make rigorous the heuristic, we give the Borel-Cantelli theorem, necessary for the proof of Theorem 11.10.

Theorem 11.9 (Borel-Cantelli)

(i) If $\sum_n \mathbb{P}(C_n) < \infty$, then $\mathbb{P}(C_n \text{ occur infinitely often}) = 0$.

(ii) Assume C_n are independent and $\sum_n \mathbb{P}(C_n) = \infty$. Then $\mathbb{P}(C_n \text{ occur infinitely often}) = 1$.

Although we frame the next theorem in terms of fixed alignment, it is really about n independent tosses of a coin with success probability p.

Theorem 11.10 Let $A_1, A_2, \ldots, B_1, B_2 \ldots$ be independent and identically distributed and let $0 < p \equiv \mathbb{P}(A_1 = B_1) < 1$. Define $R_n = \max\{m : A_{i+k} = B_{i+k} \text{ for } k = 1 \text{ to } m, 0 \leq i \leq n - m\}$. Then

$$\mathbb{P}\left(\lim_{n \to \infty} \frac{R_n}{\log_{1/p}(n)} = 1\right) = 1.$$

Proof. The idea of the proof is to exploit the heuristic $np^m = 1$ implies $m = \log_{1/p}(n)$. If $m = (1 + \epsilon)\log_{1/p}(n)$, we are able to show, using the Borel-Cantelli theorem, that runs of length m will occur only finitely often. This requires setting a "skeleton" of n_k's as $n_k \to \infty$ to make the calculations easy. The lower bound shows that $(1 - \epsilon)\log_{1/p}(n)$ occurs infinitely often. To exploit Borel-Cantelli for this problem, the sets must be independent, making the lower bound a little harder to get.

We begin with the upper bound. Let $D_i = \{A_{i+k} = B_{i+k} \text{ for } k = 1 \text{ to } m\}$, where $0 \leq i \leq n - m$. Now, for $\epsilon > 0$, let $m = \lceil(1 + \epsilon)\log_{1/p}(n)\rceil$. Because $\mathbb{P}(D_i) = p^m$, we have

$$n^{-(1+\epsilon)} \leq \mathbb{P}(D_i) = p^m \leq \frac{1}{p}n^{-(1+\epsilon)}.$$

For $n = n_k = \lceil(1/p)^k\rceil$, define

$$E(n_k) = \bigcup_{i=1}^{n_k - m_k} \{A_{i+k} = B_{i+k} \text{ for } k = 1 \text{ to } m_k\} = \bigcup_{i=1}^{n_k - m_k} D_i(n_k),$$

where $\lceil(1+\epsilon)\log_{1/p}(n_k)\rceil = m_k$ and the dependence of D_i on n_k is made explicit:

$$\mathbb{P}\left(\cup_k E(n_k)\right) \leq \sum_{k=1}^{\infty} \sum_{i=0}^{n_k-m_k} \mathbb{P}(D_i(n_k))$$

$$\leq \sum_{k=1}^{\infty} n_k n_k^{-(1+\epsilon)}$$

$$< \sum_{k=1}^{\infty} p^{\epsilon k} < \infty.$$

Therefore, by Theorem 11.9,

$$\mathbb{P}(E(n_k) \text{ occur infinitely often}) = 0.$$

Because longest match length increases with n, it follows that

$$\mathbb{P}(\overline{\lim_n} R_n / \log_{1/p}(n) \leq 1) = 1.$$

This establishes half of the result.

Theorem 11.9 can be recast is a slightly different form. Let $X_n = \mathbb{I}(C_n)$. Then $\mathbb{P}(C_n) = \mathbb{E}(X_n)$ and the theorem concerns the sums $\sum_n \mathbb{E}(X_n)$.

To obtain a corresponding lower bound, note that the Theorem 11.9 has a converse if the events are independent. To create nonoverlapping headruns, let

$$D_{m,i} = \{A_{mi+k} = B_{mi+k} \text{ for } k = 1, \ldots, m\} = E_i$$

and

$$T_{m_n} = \sum_i \mathbb{I}(E_i).$$

This time, with $m = \lceil(1-\epsilon)\log_{1/p}(n)\rceil$,

$$\mathbb{E}(T_{m_n}) \geq \left(\frac{n}{m} - 1\right) p^m \geq \left(\frac{n}{m} - 1\right) pn^{-(1-\epsilon)} = \frac{p}{m} n^\epsilon - pn^{-(1-\epsilon)}. \quad (11.9)$$

Therefore,

$$\lim_n \mathbb{E}(T_{m_n}) = \infty$$

and

$$\mathbb{P}(D_{m_n,i} \text{ occur infinitely often}) = 1.$$

It follows that

$$\mathbb{P}\left(\underline{\lim}_{n \to \infty}(R_n / \log_{1/p}(n)) \geq 1\right) = 1.$$

■

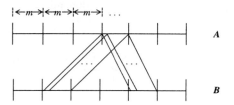

Figure 11.1: *Dependence created by blocking and shifting*

Next, we take up a problem of more direct interest to molecular biology, the length H_n of the longest match between two sequences when shifts are allowed. To be explicit, this is the local alignment score H_n when $\mu = -\delta = -\infty$ and no indels or mismatches are allowed. Allowing shifts gives approximately n^2 choices for (i,j), the starting position of a match run. Above, there were only n starting positions. This naive approach suggests that H_n grows like $\log_{1/p}(n^2) = 2\log_{1/p}(n)$. The heuristic again turns out to be correct and is formalized in the next theorem.

Theorem 11.11 *Let $A_1, A_2, \ldots, B_1, B_2, \ldots$ be independent and identically distributed with $0 < p \equiv \mathbb{P}(X_1 = Y_1) < 1$. Define*

$$H_n = \max\{m : A_{i+k} = B_{j+k} \text{ for } k = 1 \text{ to } m,\ 0 \leq i,j \leq n-m\}.$$

Then

$$\mathbb{P}\left(\lim_{n\to\infty} \frac{H_n}{\log_{1/p}(n)} = 2\right) = 1.$$

Proof. The upper bound is established just as in Theorem 11.9 with n^2 replacing n. As might be expected, the lower bound is more difficult due to a more complex dependence structure. We divide each sequence into nonoverlapping blocks of length m and consider matching between the blocks. See Figure 11.1. Define $D_{i,j} = \{A_{i+k} = B_{j+k} \text{ for } k = 1 \text{ to } m\}$ and let $m = \lceil (2-\epsilon)\log_{1/p}(n) \rceil$. Let $E_{i,j} = D_{mi,mj}$ and $T_n = \sum_{i,j} \mathbb{I}(E_{i,j})$. As in Theorem 11.9, $\mathbb{E}(T_n) \cong m^{-2}n^\epsilon \to \infty$. To handle the dependence introduced by shifting, it is sufficient to show

$$\operatorname{Var}(T_n)/(\mathbb{E}(T_n))^2 \to 0$$

to establish $\mathbb{P}(T_n \text{ occurs infinitely often}) = 1$. See Problem 11.6.

Using the above definition of T_n,

$$\operatorname{Var}(T_n) = \sum_{\substack{i,j \\ k,l}} \operatorname{cov}(\mathbb{I}(E_{i,j}), \mathbb{I}(E_{k,l}))$$

$$= \sum_{\substack{i=k \\ j=l}} + \sum_{\substack{i \neq k \\ j \neq l}} + 2 \sum_{\substack{i=k \\ j \neq l}} \text{cov}(\mathbb{I}(E_{i,j}), \mathbb{I}(E_{k,l})).$$

In this expansion, there are $\approx (n/m)^2$ diagonal terms $i = k, j = l$:

$$\sum_{i,j} \text{cov}(\mathbb{I}(E_{i,j}), \mathbb{I}(E_{i,j})) = \sum_{i,j} \mathbb{E}(\mathbb{I}(E_{i,j}) - \mathbb{P}(E_{i,j}))^2$$

$$= \sum_{i,j} \mathbb{P}(E_{i,j})(1 - \mathbb{P}(E_{i,j}))$$

$$\leq \sum_{i,j} \mathbb{P}(E_{i,j}) = \mathbb{E}(T_n).$$

In the second sum of $\approx (n/m)^4$ terms, $E_{i,j}$ and $E_{k,l}$ are independent, so the total contribution is 0. Finally, the third sum has approximately $2(n/m)^3$ terms where each term has the form indicated in Figure 11.1. Let $p = \sum_a (\mathbb{P}(A = a))^2 = \sum_a \xi_a^2$, that is, ξ_a is the probability distribution on the atoms of A. Now, we need a version of Hölder's inequality.

Lemma 11.4 (**Hölder**) *Let* $p_r = \mathbb{P}(r$ *independent letters are equal*) *for the distribution* $\xi_a = \mathbb{P}(A = a)$. *Then* $p_r = \sum_{a \in \mathcal{A}} (\xi_a)^r$, *and for* $0 < r < s$,

$$(p_s)^{1/s} < (p_r)^{1/r}.$$

Now

$$\text{cov}(\mathbb{I}(E_{i,j}), \mathbb{I}(E_{i,l})) = \mathbb{P}(E_{i,j} \cap E_{il}) - \mathbb{P}(E_{i,j})\mathbb{P}(E_{i,l}) \leq \mathbb{P}(E_{i,j} \cap E_{i,l}))$$

$$= \left[\sum_a (\xi_a)^3 \right]^m \leq \left[\sum_a \xi_a^2 \right]^{3m/2} = (\mathbb{P}(E_{i,j}))^{3/2}.$$

Summing on i, j and noting that $\mathbb{P}(E_{i,j}) = \mathbb{P}(E_{1,1})$ for all i, j,

$$2 \sum_{\substack{i=k \\ j \neq l}} \mathbb{P}(E_{i,j}))^{3/2} = 2 \left(\frac{n}{m}\right)^3 (\mathbb{P}(E_{1,1}))^{3/2}$$

$$= 2 \left(\left(\frac{n}{m}\right)^2 \mathbb{P}(E_{1,1}) \right)^{3/2} = 2(\mathbb{E}T_n)^{3/2}.$$

Combining these estimates yields

$$\text{Var}(T_n) < \mathbb{E}(T_n) + 2(\mathbb{E}(T_n))^{3/2},$$

and

$$\frac{\text{Var}(T_n)}{(\mathbb{E}(T_n))^2} < \frac{1}{\mathbb{E}(T_n)} + \frac{2}{(\mathbb{E}T_n)^{1/2}} \cong m^2 n^{-\epsilon} + 2mn^{-\epsilon/2} \to 0.$$

This completes the proof of the theorem. ∎

Note that the effects introduced by shifting make the theorem more difficult to prove. For the remainder of this section we will give some relevant generalizations where the proofs are too lengthy to give here. First, we give results for Markov chains. The logic of the theorem is based on the idea that for chains to match, they must take the same steps or transitions in each chain. Another mathematical result is needed.

Proposition 11.1 (Perron-Frobenius) *Let $Q = (q_{ij})$ be a substochastic matrix, that is, $q_{ij} \geq 0, \sum_j q_{ij} \leq 1$, that is irreducible and aperiodic. The largest $|\lambda|$ eigenvalue is real and $\lambda = 1$ if Q is stochastic, and $0 < \lambda < 1$ otherwise. The corresponding eigenvectors $\alpha Q = \lambda \alpha$ and $Q\beta = \lambda\beta$ are normalized so $\sum \alpha_i = 1$ and $\sum \alpha_i \beta_i = 1$. Then*

$$q_{ij}^{(n)} \sim \lambda^n \beta_i \alpha_j \text{ as } n \to \infty.$$

This tells us that remaining in Q's states is like a coin tossing experiment with $\mathbb{P}(H) = \lambda$. If we consider Q as the transitions between states in a subset S of a Markov chain, then if X_i is the state of the Markov chain and π is the equilibrium distribution

$$\mathbb{P}(X_1 X_2 \cdots X_n \in S) = \sum_{j \in S} \sum_{i \in S} \mathbb{P}\left(X_n = j \,\middle|\, \begin{array}{l}\text{stay in } S \\ \text{start at } i\end{array}\right) \pi_i$$

$$\sim \sum_{j \in S} \sum_{i \in S} \lambda^{n-1} \beta_i \alpha_j \pi_i$$

$$= \left(\sum_{i \in S} \pi_i \beta_i\right) \lambda^{n-1}.$$

If our sequences $A_1 A_2 \cdots$ and $B_1 B_2 \cdots$ are both Markov chains, they match when the transitions are identical. Form the "product" Markov chain (A_i, B_i) and consider the diagonal (p_{ij}^2). Runs on the diagonal keeping the chains identical behave like coin tossing with probability λ, the largest eigenvalue of the substochastic matrix (p_{ij}^2). Our interest in establishing the next theorem is the fact that (first-order) nearest neighbor effects in DNA and protein sequences are often statistically significant.

Theorem 11.12 *Let $A_1, A_2, \ldots$, and $B_1, B_2, \ldots$ be two independent Markov chains on a finite alphabet $\mathcal{A}$ which are irreducible, aperiodic, and have transition probabilities (p_{ij}), $i, j \in \mathcal{A}$. Let $\lambda \in (0, 1)$ be the largest eigenvalue of the substochastic matrix $((p_{ij})^2)$, $i, j \in \mathcal{A}$. Then*

$$\mathbb{P}\left(\lim_n \frac{H_n}{\log_{1/\lambda}(n)} = 2\right) = 1.$$

Probability and Statistics for Sequence Alignment

Another feature of biological sequences is that all sequences do not have the same distribution. There are results for Markov as well as iid sequences, but only the iid results are discussed here. The surprising discovery is that H_n can still have $2\log_{1/p}(n)$ behavior even when the marginal distributions are quite different.

Theorem 11.13 *Let $A_1, A_2, \ldots$ be distributed as ξ, $B_1, B_2, \ldots$ be distributed as ν with all letters independent and $p = \mathbb{P}(X_1 = Y_1) \in (0,1)$. Then, there is a constant $C(\xi, \nu) \in [1, 2]$ such that*

$$\mathbb{P}\left(\lim_{n\to\infty} H_n/\log_{1/p}(n) = C(\xi,\nu)\right) = 1.$$

In addition

$$C(\xi,\nu) = \sup_{\gamma \in Pr(\mathcal{A})} \min\left\{\frac{\log(1/p)}{\mathcal{H}(\gamma,\xi)}, \frac{\log(1/p)}{\mathcal{H}(\gamma,\nu)}, \frac{2\log(1/p)}{\log(1/p)+\mathcal{H}(\gamma,\beta)}\right\},$$

where $\beta_a \equiv \xi_a \nu_a / p$, $\mathcal{H}(\beta,\nu) = \sum_a \beta_a \log(\beta_a/\nu_a)$, and γ ranges over the probability distributions on the alphabet $\mathcal{A}$. Here, log can be to any base. Also, $C(\xi,\gamma) = 2$ if and only if

$$\max\{\mathcal{H}(\beta,\nu), \mathcal{H}(\beta,\xi)\} \leq \left(\frac{1}{2}\right)\log\frac{1}{p}.$$

The set of ξ such that $C(\xi,\nu) = 2$ for a fixed ν has positive diameter. Of course, $C(\nu,\nu) = 2$ by Theorems 11.11. That a large set of ξ satisfies $C(\xi,\nu) = 2$ is another indication of the strength of the "$2\log(n)$" law.

To illustrate this phenomenon, we give a parametric example where we can explicitly determine when shifts double the length of the longest match. Let $A_1, A_2, \ldots$ have $\mathbb{P}(A_i = H) = \mathbb{P}(A_i = T) = 1/2$ and $B_1, B_2, \ldots$ have $\mathbb{P}(B_i = H) = \theta = 1 - \mathbb{P}(B_i = T)$, where $\theta = [0,1]$. Then

$$p = \mathbb{P}(A_i = B_i) = 1/2\theta + 1/2(1-\theta) = 1/2.$$

So R_n, which is governed only by p, has growth $R_n \sim \log_2(n)$. In the case $\theta = 1$, the value $H_n =$ length of the longest run of H's in $A_1 A_2 \cdots A_n$. Therefore, H_n has the same distribution as $R_n \sim \log_2(n)$. The same holds for $\theta = 0$ where $H_n =$ length of the longest run of T's in $A_1 A_2 \cdots A_n$. However, if $\theta = 1/2$, A_i and B_i have the same distribution and $H_n \sim 2\log_2(n)$. The theorem tells us

$$H_n \sim C \log_2(n), \quad C \in [1,2],$$

and $C = 2$ if and only if

$$\max\{\mathcal{H}(\beta,\nu), \mathcal{H}(\beta,\xi)\} \leq 1/2 \log(2).$$

Because $\nu = (1/2, 1/2)$ and $\xi = (\theta, 1-\theta)$, $\beta = \left(\frac{1/2\theta}{1/2}, \frac{1/2(1-\theta)}{1/2}\right) = \xi$, $\mathcal{H}(\beta,\xi) = 0$ and $C = 2$ if and only if

$$\theta \log\theta + (1-\theta)\log(1-\theta) \leq 1/2\log(2).$$

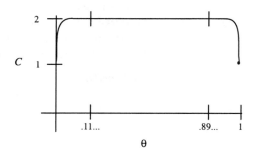

Figure 11.2: C vs θ for an example

The inequality holds in the interval $[0.11002786\ldots, 0.88997214\ldots]$. See the graph in Figure 11.2.

Approximate Matching

Another feature of sequence evolution that concerns us is substitutions, insertions, and deletions of letters as well as inversions. We ask, How many letters can be "removed" from the sequences to lengthen the match and still retain the $2\log_{1/p}(n)$ behavior? The result is surprisingly strong.

Theorem 11.14 *Let $A_1, A_2, \ldots, B_1, B_2, \ldots$ and p be as in Theorems 11.11 or 11.12. Let $H_n^*(k)$ be the longest match between $X_1 \ldots X_n$ and $Y_1 \ldots Y_n$, allowing shifts and removal of l single letters, that is,*

$$H_n^*(l) \equiv \max\{m : A_{i+k} = B_{j+k} \text{ for } k = 1 \text{ to } m, \text{ except for at most } l \text{ letters}\}$$

Then, for any constant l or any deterministic sequence $l = l(n)$ where

$$l = o(\log(n)/\log\log(n)),$$

it follows that

$$\lim_n M_n^*(l)/\log_{1/p}(n) = 2$$

in probability.

Now, we turn to matching with a fraction of mismatches. For fixed alignment, set

$$R_n^\alpha = \max\left\{t : \alpha t \leq \sum_{1 \leq k \leq t} C_{i+k},\ 0 \leq i \leq n - t\right\},$$

where $C_i = \mathbb{I}(X_i = Y_i)$. The result due to Erdös-Rényi is

Theorem 11.15 *For $C_1, C_2, \ldots$ independent Bernoulli $p \in (0, 1)$ and $1 \geq \alpha > p$,*

$$\mathbb{P}\left(\frac{R_n^\alpha}{\log(n)} \to \frac{1}{\mathcal{H}(\alpha, p)}\right) = 1.$$

The heuristic for this result is consistent with what was given for the easier case of pure headruns. The large deviations results in Theorem 11.10 say that a rich headrun of length t occurs with probability approximately $e^{-t\mathcal{H}(\alpha,p)}$. Then because a run has approximately n starting places, we solve

$$1 = n e^{-t\mathcal{H}(a,p)}$$

to obtain

$$R_n^\alpha \equiv t = \frac{\log(n)}{\mathcal{H}(a, p)}.$$

In exactly the same manner, this heuristic gives the correct result for sequence matching:

$$H_n^\alpha \equiv \max_{\substack{0 \leq i \leq n-t, \\ 0 \leq j \leq m-t}} \left\{ t : \alpha t \leq \sum_{1 \leq k \leq t} \mathbb{I}\{A_{i+k} = B_{j+k}\} \right\}.$$

Although it is routine to make Theorem 11.10 for pure headruns rigorous and provide a proof of Theorem 11.15, the extension from Theorem 11.11 for exact matching between sequences to the next theorem does not follow as easily.

Theorem 11.16 *Let $A_1, A_2 \ldots, B_1, B_2, \ldots$ be independent and identically distributed with $0 < p \equiv \mathbb{P}(A_1 = B_1) < 1$. Then*

$$\mathbb{P}\left(\frac{H_n^\alpha}{\log n} \to \frac{2}{\mathcal{H}(\alpha, p)}\right) = 1.$$

Matching with Scores

We state one more important generalization. Whereas our alignment algorithms made use of general scoring $s(a, b)$ for $a, b \in \mathcal{A}$, the above results have restricted scoring. To relax some of these restrictions, we now require $\mathbb{E}(s(A, B)) < 0$, $s^* = \max s(a, b) > 0$, and do not allow deletions. This is to assure that the expected segment match has negative score, but that positive scores are possible. Although clearly a context for large deviations, the results are surprisingly nice.

Furthermore, the content of the letters in maximal matches have a distribution that is often different from the initial $\xi_a, a \in \mathcal{A}$. To see that, let the letter l have $s(l, l) = +1$ and $s(a, b) = -\infty$ otherwise. Then any positive scoring match has 100% l's. Remarkably, there is a formula for this distribution.

Theorem 11.17 *Let* $\mathbf{A} = A_1 A_2 \cdots$ *and* $\mathbf{B} = B_1 B_2 \cdots$ *have iid letters with distribution* ξ. *Assume* $s(a,b)$ *is a scoring function with* $\mathbb{E}(s(A,B)) < 0$ *and* $s^* = \max\{s(a,b) : a,b \in \mathcal{A} \text{ and } \xi_a \xi_b > 0\} > 0$. *Let* $p > 0$ *be the largest real root of*

$$f(\lambda) = 1 - \mathbb{E}[\lambda^{-s(A,B)}] = 0. \tag{11.10}$$

Then

$$\mathbb{P}\left(\lim_{n \to \infty} \frac{H_n}{\log_{1/p}(n)} = 2\right) = 1,$$

and the proportion of letter a *aligned with letter* b *in the best matching interval converges to*

$$\xi_a \xi_b p^{-s(a,b)}. \tag{11.11}$$

Heuristics for proof. Recall that we are studying alignment segments without indels. For convenience of notation, let $C_i = (A_i, B_i)$ be iid with distribution $\xi \times \xi = \mu$.

Our interest is in motivating an Erdös-Rényi type result for scoring $A_1 \cdots A_n$ against $B_1 \cdots B_n$. There are about n^2 ways of beginning alignments $c_1 c_2 \cdots c_k$ with score $\sum_{j=1}^{k} s(c_j)$. These segments of length k with composition γ (a probability measure on $\mathcal{A} \times \mathcal{A} = \mathcal{A}^2$) have the following properties:

(i) They score $\sum_{c \in \mathcal{A}^2} s(c) \gamma_c$ per letter

and

(ii) they occur with probability approximately $e^{-k\mathcal{H}(\gamma, \mu)}$.

The Erdös-Rényi heuristic tells us to solve

$$n^2 e^{-k\mathcal{H}(\gamma, \mu)} = 1,$$

which yields

$$k = \frac{\log(n^2)}{\mathcal{H}(\gamma, \mu)}.$$

Therefore, we have the average score per letter and the average length. This implies the best γ composition segment scores

$$\frac{\sum \gamma_c s(c)}{\mathcal{H}(\gamma, \mu)} \log(n^2) = r(\gamma) \log(n^2).$$

Finally, we maximize over all compositions γ:

$$M_n \sim \log(n^2) \cdot \max\{r(\gamma) : \gamma \text{ probability distribution on } \mathcal{A}^2\}.$$

Next we characterize the distribution γ that maximizes $r(\gamma)$. Because $\sum \gamma_c = 1$, we employ Lagrange multipliers. Define

$$\varphi(\gamma) = \frac{\sum \gamma_c s(c)}{\mathcal{H}(\gamma, \mu)} + \lambda \left(1 - \sum \gamma_c\right).$$

Then taking $\partial \varphi(\gamma)/\partial \gamma_c = 0$,

$$\mathcal{H}(\gamma, \mu) s(c) - \left(\sum \gamma_i s(i)\right) \left(1 + \log \frac{\gamma_c}{\mu_c}\right) = \lambda \mathcal{H}^2(\gamma, \mu).$$

Solving for γ_c,

$$\log \frac{\gamma_c}{\mu_c} = \delta s(c) + \delta_1$$

and

$$\gamma_c = \frac{e^{\delta s(c)}}{\delta^*} \mu_c = \frac{\mu_c e^{\delta s(c)}}{\sum_c \mu_c e^{\delta s(c)}}.$$

To finish the derivation, we show that when $r(\gamma_c(\delta))$ is maximal, then $M(\delta) = 1$, where

$$M(\delta) = \sum \mu_c e^{\delta s(c)}.$$

Setting $\gamma_c = \gamma_c(\delta)$,

$$\sum \gamma_c(\delta) s(c) = \frac{\sum \mu_c e^{\delta s(c)} s(c)}{M(\delta)}$$

$$= \frac{M'(\delta)}{M(\delta)} = K'(\delta),$$

where $K(\delta) = \log M(\delta)$. Also,

$$\mathcal{H}(\gamma(\delta), \mu) = \sum \gamma_c \log \frac{\gamma_c}{\mu_c}$$

$$= \sum \gamma_c (\delta s(c) - K(\delta)).$$

From the definition of $r(\gamma)$,

$$r(\gamma(\delta)) = \frac{K'(\delta)}{\delta K'(\delta) - K(\delta)},$$

and by solving $\frac{dr(\gamma(\delta))}{d\delta} = 0$, we find $K(\delta) K''(\delta) = 0$. The Cauchy-Schwartz inequality tells us $K'' M^2 = MM'' - (M')^2 \geq 0$ with equality only when $s(c)$ is constant. Therefore, $K'' > 0$ and $K(\delta) = 0$, and it follows that $M(\delta) = 1$. ∎

Now we can check our intuition for $s(a, a) = +1$ and $s(a, b) = -\infty$ if $a \neq b$. Then

$$f(\lambda) = 1 - \xi_a^2 \lambda^{-1},$$

which has root $f(p) = 0$ with $p = \xi_a^2$. The probability of a aligned with a is

$$\mu_{a,a} = \xi_a \xi_a \left(\xi_a^2\right)^{-1} = 1.$$

This theorem also gives us a rational way to assign scores. Suppose we know the statistical description of the alignments that we wish to find. These statistics might be obtained from a study of known sequence relationships. We summarize this description as $\mu_{a,b}$, the probability that a and b will be aligned. Of course, $\sum_{a,b} \mu_{a,b} = 1$. It is natural to go to Equation (11.11) and solve for $s(a,b)$:

$$\mu_{a,b} = \xi_a \xi_b p^{-s(a,b)},$$

which implies

$$s(a,b) = \log_{1/p} \left(\frac{\mu_{a,b}}{\xi_a \xi_b}\right). \tag{11.12}$$

Certainly, this assignment of $s(a,b)$ is intuitive. We now give a likelihood ratio interpretation. Let $A_1 \cdots A_k$ and $B_1 \cdots B_k$ be random. Then the alignment

$$\mathbb{A} = \frac{A_1 A_2 \cdots A_k}{B_1 B_2 \cdots B_k}$$

has under distribution $\mu_{a,b}$ probability

$$\prod_{i=1}^{k} \mu_{A_i, B_i}$$

and under distribution $\xi_a \xi_b$, it has probability

$$\prod_{i=1}^{k} \xi_{A_i} \xi_{B_i}.$$

The Neyman-Pearson likelihood ratio for determining if the alignment is from $\mu_{a,b}$ versus $\xi_a \xi_b$ is

$$\left(\prod_{i=1}^{k} \mu_{A_i, B_i}\right) / \prod_{i=1}^{k} \xi_{A_i} \xi_{B_i} = \prod_{i=1}^{k} \frac{\mu_{A_i, B_i}}{\xi_{A_i} \xi_{B_i}}.$$

This is the statistic for the best hypothesis test of the hypothesis of distribution $\{\mu_{a,b}\}$ versus $\{\xi_a \xi_b\}$. We decide in favor of $\{\mu_{a,b}\}$ when the ratio is large. Of course, with $s(a,b)$ defined in Equation (11.12),

$$s(\mathbb{A}) = \sum_{i=1}^{k} s(A_i, B_i)$$

$$= \log_{1/p} \prod_{i=1}^{k} \frac{\mu_{A_i, B_i}}{\xi_{A_i} \xi_{B_i}}.$$

Therefore, large scores $s(\mathbb{A})$ under s determined by Equation (11.12) correspond to best hypothesis tests for the composition of the alignment. When all $\binom{n}{2}$ intervals are maximized over, the intuition remains in that we are choosing the largest of all possible likelihood ratios.

11.3 Extreme Value Distributions

If we have a Poisson distribution of arrivals in time with intensity $\lambda > 0$, we mean that we have an arrival distribution with

$$\mathbb{P}(k \text{ arrivals in } [s, s+t)) = \frac{e^{-\lambda t}(\lambda t)^k}{k!}.$$

No arrivals in $[0, t)$ has probability $e^{-\lambda t}$, and if $W=$ time until the first arrival, then

$$\mathbb{P}(W \geq t) = e^{-\lambda t}$$

or

$$\mathbb{P}(W \in A) = \int_A \lambda e^{-\lambda t} dt.$$

The mean time until an arrival is

$$\mathbb{E}(W) = 1/\lambda.$$

The distribution W is known as the exponential distribution with mean $1/\lambda$.

In our studies of the longest run of heads and the longest run of matches, our interest is in the maximum of the collection of random variables. There is a classical theorem for this problem.

Theorem 11.18 *Let $X_1, X_2, \ldots, X_n, \ldots$ be iid with distribution function F. Let $Y_n = \max\{X_1, X_2, \ldots, X_n\}$. Assume there are sequences $a_n > 0, b_n > 0$ such that, for all y,*

$$\lim_{n \to \infty} n\{1 - F(a_n + b_n y)\} = u(y)$$

exists. Then

$$\lim_{n \to \infty} \mathbb{P}(Y_n < a_n + b_n y) = e^{-u(y)}.$$

In our problem, $F(t) = 1 - \mathbb{P}(W \geq t) = 1 - e^{-\lambda t}$ and

$$n\{1 - F(a_n + b_n y)\} = n e^{-\lambda a_n} e^{-\lambda b_n y},$$

so if $a_n = \log(n)/\lambda$ and $b_n = 1/\lambda$,

$$n\{1 - F(a_n + b_n y)\} = e^{-y}$$

and
$$\lim_{n\to\infty} \mathbb{P}\left(Y_n < \frac{\log(n)}{\lambda} + \frac{y}{\lambda}\right) = e^{-e^{-y}}.$$

This is a so-called Type I extreme value distribution, and Y_n has the same distribution as
$$\log(n)/\lambda + V/\lambda,$$
where
$$\mathbb{P}(V \leq t) = e^{-e^{-t}}.$$

Remarkably, the mean
$$\mathbb{E}(Y_n) = \log(m)/\lambda + \mathbb{E}(V)/\lambda = \log(m)/\lambda + \gamma/\lambda,$$
where $\gamma = 0.5722\ldots$ is the Euler-Macheroni constant. On the other hand,
$$\operatorname{Var}(\max_{1 \leq i \leq m} W_i) = \operatorname{Var}(V)/\lambda^2$$
$$= (\pi^2/6)/\lambda^2$$
is independent of m. This is very different from the Central Limit Theorem where $\sum_{i=1}^m W_i$ has mean $m\mathbb{E}(W)$ and variance $m\operatorname{Var}(W)$.

We return to the longest run problem. R_n is the length of the longest headrun, where $p = \mathbb{P}(\text{Heads})$. Each headrun is preceded by a tail and has length m with probability qp^m, $q = 1 - p$. There are approximately nq tails in n trials so that
$$R_n \approx \max_{1 \leq i \leq nq} Z_i,$$
where Z_i is geometric and $\mathbb{P}(Z_i = m) = qp^m$. Also, it is an exercise that $Z_i = \lfloor W_i \rfloor$ where W_i are iid exponential random variables with mean $1/\lambda$, $\lambda = \log(1/p)$. Therefore it follows that
$$R_n \approx \left\lfloor \max_{1 \leq i \leq nq} W_i \right\rfloor.$$

As noted above, the maximum of iid exponential random variables is an extreme value random variable. Letting V denote a random variable such that $\mathbb{P}(V \leq t) = \exp(-e^{-t})$, R_n should satisfy
$$R_n \approx \lfloor \log(nq)/\lambda + V/\lambda \rfloor.$$

Hence,
$$\mathbb{E}(R_n) \approx \frac{\log(nq)}{\lambda} + \frac{\mathbb{E}(V)}{\lambda} - \frac{1}{2}$$
$$= \frac{\log(nq)}{\lambda} + \frac{\gamma}{\lambda} - \frac{1}{2},$$
$$= \log_{1/p}(n) + \log_{1/p}(q) + \gamma/\lambda - \frac{1}{2},$$

Probability and Statistics for Sequence Alignment

where $1/2$ is *Sheppard's continuity correction*. Integerizing continuous random variables decreases the mean. For $p = q = 1/2$, the approximation is $(\log(n) + \gamma)/\lambda - 3/2$. Applying the same approach to the variance, $\text{Var}(R_n) \approx \pi^2/6\lambda^2 + 1/12$. Here the $1/12$ is *Sheppard's correction for variance* and $\text{Var}(V) = \pi^2/6$. Integerizing continuous random variables increases the variance.

Although it is not entirely routine to rigorously establish this heuristic, it can be done for a more general case.

Theorem 11.19 *Let $A_1, A_2, \ldots B_1, B_2, \ldots$ have iid letters and let $0 < p = \mathbb{P}(A_1 = B_1) < 1$. Let*

$$R_n(k) = \max\{m : A_{i+l} = B_{i+l} \text{ for } l = 1 \text{ to } m \text{ except for at}$$

$$\text{most } k \text{ failures, } 0 \le i \le n - m\}.$$

Then for $\lambda = \ln(1/p)$,

$$\mathbb{E}(R_n(k)) = \log_{1/p}(qn) + k \log_{1/p} \log_{1/p}(qn)$$
$$+ (k) \log_{1/p}(q) - \log_{1/p}(k!) + k + \gamma/\lambda - 1/2 + r_1(n) + o(1)$$

and

$$\text{Var}(R_n(k)) = \pi^2/6\lambda^2 + 1/12 + r_2(n) + o(1),$$

where, for $\theta = \pi^2/\lambda$,

$$|r_1(n)| < (2\pi)^{-1} \theta^{1/2} e^{-\theta} (1 - e^{-\theta})^{-2}$$

and

$$|r_2(n)| < (1.1 + 0.7\theta)(2\theta^{1/2} e^{-\theta}(1 - e^{-\theta})^{-3}).$$

Note that the bounds are about equal to 1.6×10^{-6} (or 3.45×10^{-4}) for the mean and 6×10^{-5} (or 2.64×10^{-2}) for the variance when $p = 1/2$ (or $1/4$). The striking feature of the variance being approximately constant with n is derived from the extreme value distribution.

The next question of interest to DNA sequence analysis is whether these results carry over to matching with shifts. This is answered in the affirmative by the next theorem.

Theorem 11.20 *Let $A_1, A_2, \ldots, B_1, B_2 \ldots$ have iid letters and let $0 < p = \mathbb{P}(A_1 = B_1) < 1$. Let*

$$H_n(k) = \max\{m : A_{i+l} = B_{j+l} \text{ for } l = 1 \text{ to } m \text{ fails at most}$$

$$k \text{ times, for some } 0 \le i, j \le n - m\}.$$

Then

$$\mathbb{E}(H_n(k)) = \log_{1/p}(qn^2) + k\log_{1/p}\log_{1/p}(qn^2) + k\log_{1/p}(q)$$
$$- \log_{1/p}(k!) + k + \frac{\gamma}{\lambda} - \frac{1}{2} + r_1(n) + o(1),$$

and

$$\mathrm{Var}(H_n(k)) = \frac{\pi^2}{6\lambda^2} + \frac{1}{12} + r_2(n) + o(1).$$

The functions $r_1(n)$ and $r_2(n)$ are bounded by the corresponding functions of θ in the statement of Theorem 11.18. DNA sequences do not always have equal lengths, and a more general theorem exists with n^2 replaced by $n_1 n_2$, the product of the lengths of the sequences. A necessary condition is $\log(n_1)/\log(n_2) \to 1$. It is also possible to present these results in the case of Markov chains or even m-dependence.

11.4 The Chen-Stein Method of Poisson Approximation

There is a connection between the Poisson distribution and the exponentials of the last section. In fact, the extreme value distribution V has tails $\mathbb{P}(V > t) = 1 - e^{-e^{-t}} \approx 1 - (1 - e^{-t}) = e^{-t}$, which is $\mathbb{P}(Z = 0)$ for Z distributed Poisson mean 1. As we will see, this is no accident. Just how the Poisson is connected to long headruns will be detailed later, but it should be clear that whereas the Central Limit Theorem and its generalizations are widely known, only the simplest limit theorem is known for the Poisson. The random variable of interest is the sum of Bernoulli random variables.

Theorem 11.21 *Let $D_{1,n}, D_{2,n}, \ldots, D_{n,n}$ be independent random variables with $\mathbb{P}(D_{i,n} = 1) = 1 - \mathbb{P}(D_{i,n} = 0) = p_{i,n}$. If*

(i)

$$\lim_{n \to \infty} \max\{p_{i,n} : 1 \leq i \leq \infty\} = 0$$

and

(ii)

$$\lim_{n \to \infty} \sum_{i=1}^{n} p_{i,n} = \lambda > 0,$$

then

$$W_n = \sum_{i=1}^{n} D_{i,n} \text{ converges in distribution to } Z,$$

where Z is Poisson with mean λ.

Probability and Statistics for Sequence Alignment

This is the so-called *law of rare events*, which is restricted to the sum of indicator events. A fruitful direction for generalization lies in relaxing the assumption of independence. The origin of the theorem we will present lies in work of Stein, who gave the Central Limit Theorem based on a functional-differential equation. Chen later applied Stein's idea to Poisson approximation. A difference equation characterizes the Poisson in the following manner. Define the operator L on functions f by

$$(Lf)(x) = \lambda f(x+1) - xf(x).$$

Then $\mathbb{E}((Lf)(W)) = 0$ for all f such that $\mathbb{E}(Zf(Z)) < \infty$ if and only if W is distributed as Poisson(λ). The beauty of this approach is that it allows explicit error bounds.

Let I be an index set, and for each $i \in I$, let X_i be an indicator random variable. Because X_i indicates whether or not some event occurs, the total number of occurrences of events is

$$W = \sum_{i \in I} X_i.$$

It seems intuitive that if $\mathbb{E}(X_i) = \mathbb{P}(X_i = 1)$ is small and $|I|$, the size of the index set is large, then W should have an approximate Poisson distribution as is the case when all the $X_i, i \in I$, are independent. In the case of dependence, it seems plausible that the same approximation should hold when dependence is somewhat confined. For each i, we let J_i be the set of dependence for i; that is, for each $i \in I$, we assume we are given a set $J_i \subset I$ such that

$$X_i \text{ is independent of } \{X_j\},\ j \notin J_i. \qquad \text{(Condition I)}$$

This assumption will be referred to as Condition I. Define

$$b_1 \equiv \sum_{i \in I} \sum_{j \in J_i} \mathbb{E}(X_i)\mathbb{E}(X_j)$$

and

$$b_2 \equiv \sum_{i \in I} \sum_{i \neq j \in J_i} \mathbb{E}(X_i X_j).$$

Let Z denote a Poisson random variable with mean λ, so that for $k = 0, 1, 2, \ldots$,

$$\mathbb{P}(Z = k) = e^{-\lambda} \frac{\lambda^k}{k!}.$$

Let $h : Z^+ \to R$, where $Z^+ = \{0, 1, 2, \ldots\}$, and write $\|h\| \equiv \sup_{k \geq 0} |h(k)|$. We denote the total variation distance between the distributions of W and Z by

$$\|W - Z\| \equiv \sup_{\|h\|=1} |\mathbb{E}h(W) - \mathbb{E}h(Z)| = 2 \sup_{A \subset Z^+} |\mathbb{P}(W \in A) - \mathbb{P}(Z \in A)|.$$

A more general version of the following theorems can be proven. The general approach is known as the Chen-Stein method.

Theorem 11.22 *Let W be the number of occurrences of dependent events, and let Z be a Poisson random variable with $\mathbb{E}Z = \mathbb{E}W = \lambda$. Then under Condition I*

$$\|W - Z\| \leq 2(b_1 + b_2)\frac{1 - e^{-\lambda}}{\lambda} \leq 2(b_1 + b_2),$$

and in particular

$$|\mathbb{P}(W = 0) - e^{-\lambda}| \leq (b_1 + b_2)(1 - e^{-\lambda})/\lambda.$$

The second theorem is a process version of the Poisson approximation which is useful when we have to use the entire process of indicators $\{X_\alpha\}_{\alpha \in I}$.

Theorem 11.23 *For $\{Z_\alpha\}_{\alpha \in I}$, let Z_α be independent Poisson random variables with means $p_\alpha = \mathbb{E}(X_\alpha)$. Assuming Condition I, the total variation distance between $\mathbf{X} = \{X_\alpha\}_{\alpha \in I}$ and $\mathbf{Z} = \{Z_\alpha\}_{\alpha \in I}$ satisfies*

$$\frac{1}{2}\|\mathbf{X} - \mathbf{Z}\| \leq 2b_1 + 2b_2.$$

This material is organized around the rigorous Theorems 11.22 and 11.23 guided by the Poisson clumping heuristic. The idea of the clumping heuristic is that extreme events, such as large alignment scores, often occur in clumps approximately at random, according to Poisson process. In the case of alignments, the clumps are intersecting alignments. We can model this by a random number of clumps with locations laid down by a Poisson process. The clump sizes are modeled by another independent random variable. The analysis of pure headruns that we give next will illustrate these ideas.

11.5 Poisson Approximation and Long Matches

11.5.1 Headruns

Already, two analyses of long headruns have been given. R_n = length of the longest headrun was shown to satisfy

$$\lim_{n \to \infty} \frac{R_n}{\log_{1/p}(n)} = 1 \text{ with probability 1,}$$

and

$$\mathbb{E}(R_n) \approx \log_{1/p}(n) + \log_{1/p}(q) + \gamma/\lambda - 1/2$$

was given in an unproven theorem. We may well want even more. For example, if we observe a long headrun with length in excess of $\mathbb{E}(R_n)$, knowing the p-value is essential to sequence analysis. In other words, we need reliable estimates of $\mathbb{P}(R_n \geq t)$, the probability of observing a run of length t or larger.

The problem is now approached by Poisson approximation. Long headruns, those of length t, in this case, are rare events, that is, they have small probability. Certainly, such a headrun can begin at $n - t + 1$ places, so for $t \ll n$, there is a large number of places for the rare event to occur.

Applying the Poisson approximation to the events "a headrun of length t begins at i" fails because such runs occur in "clumps." If a headrun of length t begins at i, then with probability p, another headrun of length t begins at $i + 1$, and so on. On the average with fair coins such runs occur "every" 2^t positions, but because they are clumped there are gaps that average much more than 2^t between the clumps. As we are interested in $\mathbb{P}(R_n \geq t)$, we are justified in counting only the first such run in a clump and in this way "declumping." The clumping heuristic formulated by Aldous says the number of clumps will be Poisson in the limit. The idea is that locations of clumps are Poisson and that their size can by assigned by another random variable for clump size. In the case of headruns, clump size is the geometric distribution and of size k with probability $p^{k-1}(1 - p)$. Now the Chen-Stein theorem can be applied to give us the limit with error bounds. If in applying the heuristic or in setting up Condition I we choose too large a dependence set, then the rigorous theorem will give correct but useless results. (See Problem 11.10.)

For our fixed alignment or coin tossing problem, $D_i = \mathbb{I}(A_i = B_i)$, or $\mathbb{P}(D_i = 1) = 1 - \mathbb{P}(D_i = 0) = p$. Define X_i to be the event that a clump begins at i. Then

$$X_1 = \prod_{i=1}^{t} D_i$$

and

$$X_i = (1 - D_{i-1}) \prod_{j=0}^{t-1} D_{i+j}, \ i \geq 2.$$

Then the index set $I = \{1, 2, \ldots, n - t + 1\}$ and the dependence set $J_i = \{j \in I : |i - j| \leq t\}$. This gives

$$\lambda = \mathbb{E}(W) = \sum_{i \in I} \mathbb{E}(X_i) = p^t + (n - t)(1 - p)p^t.$$

Due to the factor of $(1 - D_{i-1})$ for clumping, X_i and X_j cannot both be 1 for $i \neq j$ where $i, j \in J_i$. Therefore,

$$b_2 = \sum_{i \in I} \sum_{i \neq j \in J_i} \mathbb{E}(X_i X_j) = 0.$$

It remains to calculate a bound for b_1

$$b_1 = \sum_{i \in I} \sum_{j \in J_i} \mathbb{E}(X_i)\mathbb{E}(X_j)$$

$$= p^t \sum_{j \in J_1} \mathbb{E}(X_j) + \sum_{i=2}^{n-t+1} (1-p)p^t \sum_{j \in J_i} \mathbb{E}(X_j)$$
$$< p^t(2t(1-p)p^t + p^t) + (n-t)(2t+1)((1-p)p^t)^2$$
$$< (2t+1)((1-p)p^t)^2 \{n-t+(1-p)^{-1} + p\{(1-p)^2(2t+1)\}^{-1}\}.$$

The Chen-Stein theorem says that

$$|\mathbb{P}(W=0) - e^{-\lambda}| \leq b_1 \min\{1, 1/\lambda\}.$$

Now, $W = 0$ if and only if there are no headruns $R_n \geq t$ or

$$\{W = 0\} = \{R_n < t\}$$

and

$$|\mathbb{P}(R_n < t) - e^{-\lambda}| \leq b_1 \min\{1, 1/\lambda\}.$$

Also, this equation can be restated as

$$|\mathbb{P}(R_n \geq t) - (1 - e^{-\lambda})| \leq b_1 \min\{1, 1/\lambda\},$$

and in this form, the relevance to computing p-values is most evident. For interesting probabilities, we want $\lambda = \lambda(t)$ to be bounded away from 0 and ∞. The term of the mean λ to worry about is np^t, so having $t - \log_{1/p}(n)$ bounded is required. For $t = \log_{1/p}(n(1-p)) + c$, we obtain

$$\lambda = p^t + (n-t)(1-p)p^t$$
$$= \frac{p^c}{n(1-p)} + (1 - \frac{t}{n})p^c \approx p^c$$

and

$$|\mathbb{P}(R_n < \log_{1/p}(n(1-p)) + c) - e^{-\lambda}|$$
$$< \frac{(2t+1)}{n}p^{2c}\left(1 - \frac{t}{n} + \frac{1}{n(1-p)} + \frac{p}{n(1-p)^2(2t+1)}\right)$$
$$= O\left(\frac{\log(n)}{n}\right). \tag{11.13}$$

11.5.2 Exact Matching Between Sequences

For the longest pure match between two sequences, $\mathbf{A} = A_1 A_2 \cdots A_n$ and $\mathbf{B} = B_1 B_2 \cdots B_m$, we set the test value t. Then

$$I = \{(i, j) : 1 \leq i \leq n - t + 1, 1 \leq j \leq m - t + 1\}.$$

Probability and Statistics for Sequence Alignment

The match indicator is $C_{i,j} = \mathbb{I}\{A_i = B_j\}$, so

$$Y_{i,j} = C_{i,j} C_{i+1,j+1} \cdots C_{i+t-1,j+t-1}$$

and, declumping,

$$X_{i,j} = Y_{i,j} \text{ if } i = 1 \text{ or } j = 1,$$

otherwise

$$X_{ij} = (1 - C_{i-1,j-1}) Y_{ij}.$$

The usual definition is made:

$$W = \sum_{\nu \in I} X_\nu$$

so that

$$\mathbb{E}(W) = [(n + m - 2t + 1) + (n - t)(m - t)(1 - p)] p^t.$$

The sets of dependence for $\nu = (i, j)$ are

$$J_\nu = \{\tau = (i', j') \in I : |i - i'| \le t \text{ or } |j - j'| \le t\}.$$

Clearly, X_ν and X_τ are independent when $\tau \notin J_\nu$. The next proposition gives an easy result when the distribution is uniform. For that case, $b_2 = 0$ due to independence.

Proposition 11.2 *The events $\{A_i = B_j\}$ and $\{A_i = B_k\}, j \ne k$, are positively correlated unless $\xi_a = 1/|\mathcal{A}|$ when they are independent.*

Proof. We have

$$\mathbb{P}(A_i = B_j \text{ and } A_i = B_k) = \mathbb{P}(A_i = B_j = B_k)$$
$$= \sum_{a \in \mathcal{A}} \xi_a^3 = p_3$$

and

$$\mathbb{P}(A_i = B_j) = \sum_{a \in \mathcal{A}} \xi_a^2 = p_2.$$

By definition, the correlation is positive if $p_3 - p_2^2 > 0$.

The following technique is very useful. Define

$$f(x) = \xi_x \text{ for } x \in \mathcal{A}$$

Then

$$\mathbb{E}(f(X)) = \sum_{a \in \mathcal{A}} \xi_a^2 = p_2$$

and

$$\mathbb{E}(f^2(X)) = \sum_{a \in \mathcal{A}} \xi_a^3 = p_3.$$

Using the Cauchy-Schwartz inequality,

$$p_2^2 = (\mathbb{E}f(X))^2 \leq \mathbb{E}(f^2(X)) = p_3.$$

Equality holds if and only if $f(x)$ is constant or $\xi_a = 1/|\mathcal{A}|$. ∎

As usual, b_1 is straightforward to bound. We will now study the dependencies between X_β and X_α when $\beta \in J_\alpha$. To represent Y_α, the identities in X_α without the declumping mismatch, draw t parallel lines between A_i and B_j, A_{i+1} and $B_{j+1}, \ldots, A_{i+t-1}$ and B_{j+t-1}. Do the same for Y_β. The resulting graph separates into connected components. We define the size of a component to be the number of verticies in the component. When $\beta \notin J_\alpha$, there are $2t$ components each connecting A_x with B_y. Otherwise, the A_x's or B_y's can intersect between Y_α and Y_β. There are two cases, one for the four "arms" of Figure 11.3 and one for the center.

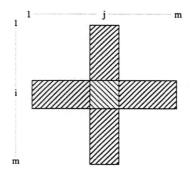

Figure 11.3: *The dependence set $J_{(i,j)}$*

The cases are easy to describe. The "arms" of Figure 11.3 can be described by $|i - i'| \leq t$ and $|j - j'| > t$ or $|i - i'| > t$ and $|j - j'| \leq t$. This case is called *crabgrass* because of the alignment graphs that result. The next graph gives two examples of crabgrass.

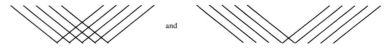

The first example has five components of size 3 whereas the second has six components of size 2, and two of size 3.

Probability and Statistics for Sequence Alignment

The center of Figure 11.3 can be described by $|i - i'| < t$ and $|j - j'| < t$. This case is called *accordions* because of the folding in the graphs. Here, for clarity, we represent Y_α with solid lines and Y_β with dashed lines. The most compact example is

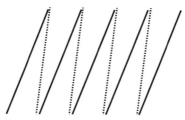

This graph has one component of size $2t + 1 = 11$. Rather more complex graphs are possible. The next graph has four components of size 2, one component of size 3, and one component of size 5:

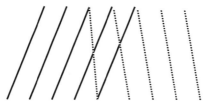

Although there are dependencies within a component, the components, or rather the event of identities between the verticies of a component, are independent from one another. That means we can multiply the probabilities p_k of obtaining identities in a component. Now we derive bounds for $\mathbb{E}(X_\nu X_\tau) \leq \mathbb{E}(Y_\nu Y_\tau)$ in preparation for bounding b_2.

Before we derive our bounds we need another Lemma. In Section 11.2.1, we used Hölder's inequality: $0 < r < s$ implies $p_s^{1/s} < p_r^{1/r}$. We are now required to obtain more precise information.

Lemma 11.5 *For $r \geq 3$ and $p = p_2$, we have*

$$p_r^{1/r} = p^{\delta(r)} p^{1/2},$$

where $\delta(r) \in (0, 1/2 - 1/r]$.

Proof. Because $p_r^{1/r} < p_2^{1/2} = p^{1/2}$, define $\delta(r) = \delta$ by

$$p_r^{1/r} = p^\delta p^{1/2}.$$

By our earlier device,

$$p^{r-1} = \left(\sum_a \xi_a^2\right)^{r-1} = (\mathbb{E}(\xi_X))^{r-1}$$
$$\leq \mathbb{E}(\xi_X^{r-1}) = p_r.$$

Therefore,
$$p^{(r-1)/r} \leq p_r^{1/r} < p^{1/2}$$
and
$$1/2 < 1/2 + \delta \leq 1 - 1/r$$
or $\delta \in (0, 1/2 - 1/r]$. ∎

First, we consider crabgrass where there are from 1 to t components of size 3. If there are x components of size 3 and y components of size 2, we have $(3+1)x + 2y = 4t$, since there are $2t$ letters in each Y_ν, Y_τ. This means $y = 2t - 2x$. Finally,
$$\mathbb{E}(Y_\nu Y_\tau) = p_2^{2t-2x} p_3^x$$
when $1 \leq x \leq t$. Using Lemma 11.5 with $\delta(3) = \delta \in (0, 1/6]$,
$$\mathbb{E}(Y_\nu Y_\tau) = p^{2t-2x} p^{(3/2+3\delta)x} = p^{2t} p^{(3\delta-1/2)x}.$$

We note that $3\delta - 1/2 \leq 1/2 - 1/2 = 0$, so to bound $\mathbb{E}(Y_\nu Y_\tau)$ we replace $(3\delta - 1/2)x$ by its smallest value $(3\delta - 1/2)t$:
$$\mathbb{E}(Y_\nu Y_\tau) \leq p^{3/2t} p^{3\delta t} = (\mathbb{E} Y_\nu)^{3/2} p^{3\delta t}. \tag{11.14}$$

Next, we consider accordions. Here, there are x_i components of size i, where $2 \leq i \leq 2t+1$. As before, there is a conservation equation: $\sum_{i=2}^{2t+1} x_i(2(i-1)) = 4t$. This equation can be used as above with an additional inequality $p_s \leq p^{s/2} p^{s\delta(3)}$ to produce a bound. Instead, we give a better bound using $s = 2t+1$ in Lemma 11.5:
$$\mathbb{E}(Y_\nu Y_\tau) = p_{k_1}^{x_{k_1}} p_{k_2}^{x_{k_2}} \cdots p_{k_l}^{x_{k_l}},$$
where there are l components. Define
$$q_k = \mathbb{P}(A_{k+1} = A_k | A_1 = \cdots = A_k).$$
If we set $p_1 = 1$, then
$$q_k = \frac{p_{k+1}}{p_k}.$$
Observe that q_k are increasing:
$$\frac{q_{k+1}}{q_k} = \frac{p_k p_{k+2}}{(p_{k+1})^2} \geq 1$$
is equivalent to
$$\mathbb{E}(\xi_X^{k-1}) \mathbb{E}(\xi_X^{k+1}) \geq (\mathbb{E}(\xi_X^k))^2.$$

To establish this inequality, use $\mathbb{E}Z^2\mathbb{E}Y^2 \geq (\mathbb{E}ZY)^2$ with $Z = \xi_X^{(k-1)/2}$ and $Y = \xi_X^{(k+1)/2}$. Returning to

$$\mathbb{E}(Y_\nu Y_\tau) = p_{k_1}^{x_{k_1}} p_{k_2}^{x_{k_2}} \cdots p_{k_l}^{x_{k_l}}$$

$$= (q_1 q_2 \cdots q_{k_1-1})^{x_{k_1}} (q_1 q_2 \cdots q_{k_2-1})^{x_{k_2}} \cdots$$

$$\leq q_1 q_2 \cdots q_{\sum x_i(i-1)}$$

$$= q_1 q_2 \cdots q_{2t} = p_{2t+1}.$$

Applying Lemma 11.5 with $\delta(2t+1) = \gamma \in (0, \frac{1}{2} - \frac{1}{2t+1}]$,

$$\mathbb{E}(Y_\nu Y_\tau) \leq p_{2t+1} \leq p^{\gamma(2t+1)} p^{t+1/2}. \tag{11.15}$$

Now we compute the Chen-Stein bounds. First, we handle b_1 ($\nu = (i,j), \tau = (i',j')$):

$$b_1 = \sum_{\nu \in I} \sum_{\tau \in J_\nu} \mathbb{E}(X_\nu)\mathbb{E}(X_\tau).$$

First of all,

$$\sum_{\tau \in J_\nu} \mathbb{E}(X_\tau) < 2(2t+1)(n \vee m)p^t.$$

Therefore,

$$b_1 < 2\lambda(2t+1)(n \vee m)p^t. \tag{11.16}$$

Next, we bound b_2 by breaking the sum into crabgrass and accordions and applying Equations (11.14) and (11.15):

$$b_2 = \sum_{\nu \in I} \sum_{\nu \neq \tau \in J_\nu} \mathbb{E}(X_\nu X_\tau)$$

$$= \sum_{(i,j) \in I} \left(\sum_{\substack{|i-i'| \leq t \\ |j-j'| \leq t}} + 2 \sum_{\substack{|i-i'| \leq t \\ |j-j'| > t}} \right) \mathbb{E}(X_{(i,j)} X_{(i',j')})$$

$$< (n-t+1)(m-t+1)\{(2t+1)^2 p^{t+1/2} p^{\gamma(2t+1)}$$
$$+ 2(2t+1)(n \vee m) p^{3t/2} p^{3\delta t}\}. \tag{11.17}$$

Recall that $\lambda = \mathbb{E}(W)$ satisfies

$$\lambda = p^t[(n+m-2t+1) + (n-t)(m-t)(1-p)].$$

To control λ, we set

$$t = \log_{1/p}(nm(1-p)) + c$$

so that
$$\lambda = \left(\frac{n+m-2t+1}{nm(1-p)} + (1-\frac{t}{n})(1-\frac{t}{m})\right) p^c,$$
which, when $n = m$, is approximately p^c.

From Equations (11.16) and (11.17),
$$b_1 + b_2 < nm \left\{ \frac{2\lambda(2t+1)(n \vee m)p^t}{nm} + (2t+1)^2 p^{t(1+2\gamma)} p^{\gamma+1/2} \right.$$
$$\left. + 2(2t+1)(n \vee m)p^{t(3/2+3\delta)} \right\}$$
$$= nm \left\{ \frac{2\lambda(2t+1)(n \vee m)p^t}{nm} + \frac{(2t+1)^2 p^{c(1+2\gamma)+\gamma+1/2}}{(nm(1-p))^{1+2\gamma}} \right.$$
$$\left. + \frac{2(2t+1)(n \vee m)p^{c(3/2+3\delta)}}{(nm(1-p))^{3/2+3\delta}} \right\}.$$

If $n = m$, then
$$b_1 + b_2 = O\left(\frac{\log(n)}{n}\right) + O\left(\left(\frac{\log(n)}{n^{2\gamma}}\right)^2\right) + O\left(\frac{\log(n)}{n^{6\delta}}\right).$$

11.5.3 Approximate Matching

In Chapter 8 we studied k-tuple matches between sequences. These k-tuple matches make a nice setting for extending the applications of Poisson approximation. We will cover several random variables of interest: the number of clumps, the sum of the clump sizes, and the maximum clump size. In addition, pure k-tuple matching is covered for all k and for $k = O(\log(n))$ approximate matching results are available.

For ease of exposition, the disscussion is restricted to the diagonal $1 \leq i = j \leq n$ of the comparison. Some cases of interest are:

(i) matching k-tuples $\{A_1 A_2 \cdots A_k = B_1 B_2 \cdots B_k\}$,

(ii) approximate matching k-tuples where at least q of $A_i = B_i$, $i = 1, 2, \ldots, k$, hold (for small k),

and

(iii) approximate matching k-tuples where at least q of $A_i = B_i$, $i = 1, \ldots, k$, hold and where $q/k > p = \mathbb{P}(A_1 = B_1)$ (for large k).

Let $D_i = \mathbb{I}(A_i = B_i)$. The k-tuple beginning at position i has score
$$V_i = D_i + D_{i+1} + \cdots + D_{i+k-1}.$$

A quality q/k match begins at i if

Probability and Statistics for Sequence Alignment

$$Y_i = \mathbb{I}(V_i \geq q) = 1.$$

As in earlier problems, the i where $Y_i = 1$ usually occur in clumps. To declump, define

$$X_i = Y_i(1 - Y_{i-1}) \cdots (1 - Y_{i-s}),$$

where s depends on q and k. Of course, for $q = k$, $s = 1$ is appropriate. For large k and $q/k > p$, $s = k$ is a correct choice. In any case where $s > 1$, the analysis is more complex than the examples worked out above. For example, clump size is not simply the number of consecutive $Y_\alpha = 1$. The clump that begins at i ends at j defined by

$$j = \min\{l \geq i : Y_l = 1, Y_{l+1} = Y_{l+2} = \cdots = Y_{l+s} = 0\}.$$

Therefore, clump size is defined for $X_i = 1$ by

$$C = \sum_{l=i}^{j} Y_l$$

and

$$\mathbb{P}(C = m) = \mathbb{P}\left(\sum_{l=i}^{j} Y_l = m \mid X_i = 1\right).$$

The Number of Clumps

The number of clumps on the diagonal is

$$W = \sum_{i=1}^{n-k+1} X_i.$$

The independence assumptions of the Chen-Stein theorems are met with $J_i = \{j : |j - i| \leq k + s - 1\}$.

Theorem 11.24 *Let Z be Poisson with mean $\lambda = \mathbb{E}(W)$. Then*

$$\frac{1}{2}\|W - Z\| \leq (b_1 + b_2)(1 - e^{-\lambda})/\lambda.$$

As usual, the task is to provide bounds for the approximation. Neglecting boundary effects, $\lambda = (n - k + 1)\mathbb{E}(X_i)$. Now

$$\mathbb{E}X_i = \mathbb{E}(X_i = 1 | Y_i = 1)\mathbb{P}(Y_i = 1).$$

By definition of Y_i,

$$\mathbb{P}(Y_i = 1) = \mathbb{P}(\mathcal{B}(k, p) \stackrel{d}{=} U \geq q)$$

and this quantity can accurately be approximated using large deviations when k is large and $1 \geq q/k > p$. A little examination produces a formula for $\mathbb{P}(X_i = 1 | Y_i = 1)$, which is the average fraction of the time a q/k match begins a clump. Therefore, this quantity is equal to the average number of clumps per match or

$$\mathbb{P}(X_i = 1 | Y_i = 1) = \frac{1}{\mathbb{E}(C)}.$$

So

$$\lambda = (n - k + 1)\mathbb{E}(X_i) = \frac{(n - k + 1)\mathbb{P}(\mathcal{B}(k, p) \geq q)}{\mathbb{E}(C)}.$$

To continue, we take $s = k$, which is not the best choice for $q = k$ but where the error bound $(b_1 + b_2) \to 0$ as $n \to \infty$.

For many cases, $q/k - p$ is a good approximation to $\mathbb{P}(X_i = 1 | Y_i = 1)$. For $s = k$, it can be proved using the ballot theorem that

$$\alpha - p \leq \mathbb{P}(X_i = 1 | Y_i = 1) \leq \alpha - p + 2(1 - \alpha)e^{-k\mathcal{H}(\alpha, p)},$$

where $\alpha = q/k$. For small k or α close to p, $\alpha - p$ might not be a good approximation to $\mathbb{P}(X_i = 1 | Y_i = 1) = \frac{1}{\mathbb{E}(C)}$.

Now we give bounds on b_1 and b_2.

$$b_1 = \sum_{i \in I} \sum_{j \in J_i} \mathbb{E}(X_i)\mathbb{E}(X_j) = |I||J_i|(\mathbb{E}(X_i))^2$$

$$= \lambda^2 \frac{|J_i|}{|I|}.$$

For b_2, we first note that $X_i = 0$ implies $X_i X_j = 0$. Otherwise, $X_i = 1$ and $X_j = 0$ for $j = i + 1$ to $i + k$. Otherwise, we have $X_i X_j \leq X_i Y_j = 0$ for $j = i - k$ to $i - 1$. For the remaining $2k$ elements of J_i,

$$\mathbb{E}(X_i X_j) \leq \mathbb{E}(X_i Y_j)$$
$$= \mathbb{E}(X_i)\mathbb{E}(Y_j) = \mathbb{E}(X_i)\mathbb{P}(Y_j = 1)$$
$$= (\mathbb{E}(X_i))^2 \mathbb{E}(C).$$

Therefore,

$$b_1 + b_2 < |I||J_i|(1 + \mathbb{E}(C))(\mathbb{E}X_i)^2$$
$$= b_1(1 + \mathbb{E}(C)).$$

Sum of Clump Sizes

Recall that the algorithms of Section 8.3 for approximate matching give the sum of the k-tuple matches on a diagonal. The idea of $\alpha = q/k$ matches will be

Probability and Statistics for Sequence Alignment

maintained in this section. The sum is denoted as S:

$$S = \sum_{i=1}^{n-k+1} Y_i.$$

It is easy to anticipate our approach. The number of clumps is approximately Poisson($\lambda = \mathbb{E}(W)$). If the clump distribution C is iid at each clump and $C_1, C_2, \ldots$ are iid as C, then the convolution

$$C^{*j} = C_1 + C_2 + \cdots + C_j, \quad j = 1, 2, \ldots,$$

will give the distribution of the sum of j clumps on a diagonal. If j is Poisson(λ), then the sum of a random number of random variables should approximate S. First the convolution:

$$\mathbb{P}(C^{*j} = m) = \begin{cases} \mathbb{P}(C = m) & \text{if } j = 1, \\ \sum_{i=1}^{m-1} \mathbb{P}(C^{*(j-1)} = m - i)\mathbb{P}(C = i), & \text{otherwise,} \end{cases}$$

using the fact that $\mathbb{P}(C = 0) = 0$ in our setting.

Returning to our approximation, using Z in place of j should approximate S:

$$\hat{S} = \sum_{i=1}^{Z} C_i.$$

$\hat{S}$ is called a *compound Poisson distribution* and note that $\mathbb{P}(Z = 0) > 0$, so $\hat{S} = 0$ is possible. For $m > 0$,

$$\mathbb{P}(\hat{S} = m) = \sum_{j=1}^{m} \mathbb{P}(\hat{S} = m | Z = j)\mathbb{P}(Z = j)$$

$$= \sum_{j=1}^{m} \mathbb{P}(C^{*j} = m)\mathbb{P}(Z = j).$$

For the first time in this chapter, we need the full Poisson process of clump distribution and we apply the process version of the Chen-Stein theorem 11.23. The next theorem justifies our above approximation, but before that, we sketch the general idea that allows the approximation.

The idea is that we begin with an index set I, events X_ν and a neighborhood structure J_ν where Condition I holds and where b_1 and b_2 are small. Then each event X_ν is further identified by a "type" $i \in T$. The new process has index set $I^* = I \times T$ and for each $(\nu, i) \in I^*$,

$$X_{\nu,i} = X_\nu \cdot \mathbb{I}\{\text{the occurrence is type } i\},$$

so that
$$X_\nu = \sum_{i \in T} X_{\nu,i}.$$
The new neighborhood structure is $J_{\nu,i} = J_\nu \times T$. The decomposition allows us to conclude
$$b_1^* = b_1 \text{ and } b_2^* = b_2. \tag{11.18}$$
If $\{X_{\nu,i}\}_{\nu \in I, i \in T} = \mathbf{X}^*$ and $\mathbf{Y}^* = \{Y_{\nu,i}\}_{\nu \in I, i \in T}$ is the Poisson process with $\lambda_{\nu,i} = \mathbb{E}(X_{\nu,i})$, then the process Chen-Stein theorem says
$$\|\mathbf{X}^* - \mathbf{Y}^*\| \leq 4(b_1 + b_2).$$

Theorem 11.25 *The total variation distance between S and $\hat{S}$ satisfies*
$$\frac{1}{2}\|S - \hat{S}\| \leq 2(b_1 + b_2) + 2\lambda \mathbb{P}(C > k),$$
where the neighborhood structure is
$$J_\nu = \{j : |j - i| \leq ks - 1\}.$$

Proof. The type is defined as clump size (the number of 1's in a clump) if the size is $i < k$ and defined as k if the clump size is at least k. Obviously,
$$X_\nu = \sum_{i=1}^{k} X_{\nu,i}.$$

Let us check Condition I. $X_{\nu,i}$ is a function of $D_{\nu-s} \cdots D_{\nu-1}$ for declumping. The most extended dependence comes when i 1's are obtained in the configuration $1\,0_{s-1}\,1\cdots 0_{s-1}\,1\,0_s$, which involves $D_\nu \cdots D_{\nu+s} \cdots D_{\nu+(i-1)s} \cdots D_{\nu+is}$. To be independent of X_ν, X_τ must have $\tau = \nu + (i+1)s + 1$, due to the s declumpling indicators for τ. In other words, $|\tau - \nu| \geq (i+1)s + 2$ ensures independence, so we need $|\tau - \nu| \leq (i+1)s + 1$ for dependence, $i = i, 2, \ldots, k-1$. Note that we will find many X_ν of size $k, k+1$, etc., but we are not guaranteed to find them all.

As claimed above, b_1 and b_2 for the unpartitioned $\{X_\nu\}_{\nu \in I}$ give b_1 and b_2 for the partitioned random variables $\{X_{\nu,i}\}_{(\nu,i)\in I}$. To capture the clump sum, note that
$$\sum_{\nu=1}^{n-k+1} C_\nu \mathbb{I}\{C_\nu < k\} = \sum_{\nu=1}^{n-k+1} \sum_{i=1}^{k-1} i X_{\nu,i}$$
counts all clumps of size less then k. Now
$$\mathbb{P}\left(\bigcup_{\nu=1}^{n-k+1} \{X_\nu = 1, C_\nu \geq k\}\right) \geq \sum_{\nu=1}^{n-k+1} \mathbb{P}\{X_\nu = 1, C_\nu \geq k\}$$

Probability and Statistics for Sequence Alignment

$$= \sum_{\nu=1}^{n-k+1} \mathbb{P}(C_\nu \geq k | X_\nu = 1) \mathbb{P}(X_\nu = 1)$$

$$= \mathbb{P}(C \geq k) \sum_{\nu=1}^{n-k+1} \mathbb{P}(X_\nu = 1)$$

$$= \lambda \mathbb{P}(C \geq k).$$

A similar argument for Z completes the proof. ∎

Order Statistics of Clump Sizes

Now return to the largest clump size and ask a natural question: What is the size of the second largest clump? The process setting is natural here. To review the setup, $W = \sum_i X_i$ and for each $X_i = 1$,

$$C_i = \sum_{l=i}^{j} Y_l$$

when $j = \min\{l : Y_l = 1, Y_{l+1} = Y_{l+2} = \cdots = Y_{l+s} = 0\}$.

For all i such that $X_i = 1$, we write the order statistics of C_i by

$$C_{(1)} \geq C_{(2)} \geq \cdots \geq C_{(W)},$$

where, for example,

$$C_{(1)} = \max\{C_i : X_i = 1\}.$$

Our approximation says that W is approximately Poisson($\lambda = \mathbb{E}(W)$), so we approximate $C_{(1)}, C_{(2)}, \ldots$ by

$$\tilde{C}_{(1)} \geq \tilde{C}_{(2)} \geq \cdots \geq \tilde{C}_{(Z)},$$

the order statistics of the iid $C_1, C_2, \ldots, C_Z$. If we "thin" $C_1, C_2, \ldots$ by requiring $C_i > x$, then

$$U = \sum_{j=1}^{Z} \mathbb{I}(C_j > x)$$

is Poisson($\lambda \mathbb{P}(C > x)$) (See Problem 11.15.) Note that

$$\{\tilde{C}_{(j)} \leq x\} = \{U \leq j - 1\}$$

because $x \geq \tilde{C}_{(j)} \geq \tilde{C}_{(j+1)} \geq \cdots \geq \tilde{C}_{(Z)}$ means that no more than $j-1$ C's are $> x$. This gives the remarkable formula

$$\mathbb{P}(\tilde{C}_{(j)} \leq x) = e^{-\lambda \mathbb{P}(C > x)} \sum_{i=0}^{j-1} \frac{(\lambda \mathbb{P}(C > x))^i}{i!}. \tag{11.19}$$

Error bounds can be obtained, as in the last section, by partitioning X_ν into types $X_{\nu,i}$. We have for $x < k$

$$\{C(j) \leq x\} = \Big\{ \sum_{\substack{\nu \in I \\ i > x}} X_{\nu,i} \leq j - 1 \Big\},$$

and we use $Z_{\nu,i}$, Poisson mean $\mathbb{E}(X_{\nu,i})$, to approximate $X_{\nu,i}$. The corresponding theorem is:

Theorem 11.26 *The total variation distance between $C_{(j)}$ and $\tilde{C}_{(j)}$ satisfies*

$$\frac{1}{2}\|C_{(j)} - \tilde{C}_{(j)}\| \leq 2(b_1 + b_2) + 2\lambda \mathbb{P}(C \geq k),$$

where b_1 and b_2 are determined by

$$J_\nu = \{\tau : |\nu - \tau| \leq ks + 1\}.$$

11.6 Sequence Alignment with Scores

Long perfect matchings or long matchings without indels are not the results of most sequence comparisons. Instead, we have $s(a, b)$ defined on $\mathcal{A}^2$ and subadditive deletion function $g(k)$, $k \geq 1$. In Section 11.1, we had some results for scoring for global alignment, which include linear growth of score and the large deviation inequality Theorem 11.4. If $s(a,b) = -\infty$ for $a \neq b$ and $g(k) = \infty$ for all k, then we have the $2\log_{1/p}(n)$ centering term for H_n, the local alignment score. It turns out that the linear growth/logarithmic growth of score is a general phenomenon, and when $g(k) = \alpha + \beta k$, there is a sharp phase transition between these regimes. After we sketch the proof of this surprising result, we will give a prescription that successfully provides a Poisson approximation for the entire logarithmic region. For the linear region, the best result is the Azuma-Hoeffding large deviations result Theorem 11.4 which does not provide useful bounds for protein sequence comparisons where n is usually in the range 200 to 600. It may be that asymptotically as $n \to \infty$ the bound is useful or it may be the case that the bound is not close to the real decay rate. Practically speaking, it is not important which of these situations holds. Theoretically, it would be very interesting to know the answer.

In Section 11.6.1, we sketch the proof of the phase transition between linear and logarithmic growth. Part of the proof has already been given in Section 11.1. Then, in Section 11.6.2, we show how to give a useful Poisson approximation in the logarithmic region.

11.6.1 A Phase Transition

In this section, we study the local alignment score

$$H(A_1 \cdots A_n, B_1 \cdots B_n) \equiv H(\mathbf{A}, \mathbf{B}) \equiv H_n = \max\{S(I,J) : I \subset \mathbf{A}, J \subset \mathbf{B}\}.$$

The theorem is true for more general scoring, but for simplicity we study

$$s(a,b) = \begin{cases} +1, & a = b, \\ -\mu, & a \neq b, \end{cases}$$

and $g(k) = \delta k$ where $(\mu, \delta) \in [0, \infty]^2$.

From Section 11.1 (and subadditivity) we know that the following limit exists:

$$\rho = \rho(\mu, \delta) = \lim_{k \to \infty} \frac{\mathbb{E}(S_k)}{k} = \sup_{k \geq 1} \frac{\mathbb{E}(S_k)}{k},$$

where

$$S_k = S(A_1 \cdots A_k, B_1 \cdots B_k).$$

The (μ, δ) with $\rho > 0$ clearly have linear growth of S_n as $n \to \infty$. Intuitively, H_n should also have linear growth for such (μ, δ). If we can show that $\{(\mu, \delta) : \rho < 0\}$ implies logarithmic growth for H_n as $n \to \infty$, this almost establishes the phase transition. The next theorem treats the set where $\rho = 0$.

Theorem 11.27 *The set $\{(\mu, \delta) : \rho(\mu, \delta) = 0\}$ defines a line in the parameter space $[0, \infty]^2$, separating the negative and positive regions $\{\rho < 0\}$ and $\{\rho > 0\}$.*

Proof. First, ρ is obviously nonincreasing in each of its parameters, and we have the global inequality $\rho(\mu + \epsilon, \delta + \epsilon/2) \geq \rho(\mu, \delta) - \epsilon$, because the corresponding inequality is satisfied by each possible alignment of two sequences of length n, and taking maxima, expectation, and limit preserves this inequality. This shows that ρ is continuous. In detail, with $Q = (\mu, \delta)$ and $Q' = (\mu', \delta')$, we have $|\rho(Q) - \rho(Q')| \leq \epsilon \equiv |\mu - \mu'| + 2|\delta - \delta'|$, because with $R = (\mu_0, \delta_0) = (\mu \wedge \mu', \delta \wedge \delta')$ and $S = (\mu_0 + \epsilon, \delta_0 + \epsilon/2)$, monotonicity and the global inequality gives $\rho(R) \geq \rho(Q) \geq \rho(S) \geq \rho(R) - \epsilon$, and similarly for $\rho(Q')$. Second, although ρ is not strictly monotone in each parameter everywhere in the parameter space, it is strictly monotone in the $(1, 1)$ direction, in a neighborhood of the line $\rho = 0$. To se! e this, let $\gamma \equiv \max(\mu, 2\delta)$ and observe that in alignments which score g or less per pair of letters, the proportion x of nonmatching pairs satisfies $-\gamma x + (1 - x) \leq g$, so that $x \geq (1 - g)/(\gamma + 1)$. For such alignments, increasing each of the penalty parameters by $\epsilon > 0$ must decrease the score by at least ϵx. It follows that $\rho(\mu + \epsilon, \delta + \epsilon) \leq \rho(\mu, \delta) - \epsilon(1 - \rho(\mu, \delta))/(1 + \mu + 2\delta)$ for all $\epsilon, \mu, \delta > 0$. The cases where $\mu = \infty$ or $\delta = \infty$ require a separate but similar argument. ∎

Next, we define the decay rate for S_n exceeding qn, $q \geq 0$:

$$r(q) = \lim_{n \to \infty} \frac{-\log \mathbb{P}(S_n \geq qn)}{n} = \inf \left(-\frac{\log \mathbb{P}(S_n \geq qn)}{n} \right).$$

Again, this limit exists and equals the inf because of subadditivity: $\mathbb{P}(S_{n+m} \geq q(n+m)) \geq \mathbb{P}(S_n \geq qn)\mathbb{P}(S_m \geq qm)$. The next result is a corollary of the Azuma-Hoeffding lemma.

Proposition 11.3 *If $\rho(\mu, \delta) < 0$ and $q \geq 0$, then $r(q) > 0$.*

Proof. Corollary 11.1 states that

$$\mathbb{P}(S_n \geq (\gamma + \rho)n) \leq e^{\frac{-\gamma^2 n}{2c^2}}.$$

Now

$$\mathbb{P}(S_n \geq qn) = \mathbb{P}(S_n - \rho n \geq (q - \rho)n) \leq e^{\frac{-(q-\rho)^2 n}{2c^2}}$$

so

$$r(q) = \frac{-\log \mathbb{P}(S_n \geq qn)}{n} \geq \frac{(q-p)^2}{2c^2} > 0.$$

∎

When $\rho(\mu, \delta) < 0$, we define

$$b = b(\mu, \delta) = \max_{q \geq 0} \frac{q}{r(q)}.$$

Because $r(1) = -\log \mathbb{P}(A_1 = B_1) < \infty$, $b > 0$. This constant b is the multiplier of $\log(n)$ when $\rho < 0$, which is the content of the next theorem. Although it is only shown that $H_n/\log(n)$ is in $[b, 2b]$, the limit is conjectured to be $2b$. The motivation for b is related to the heuristic for Theorem 11.17. Alignments of length t and score $S_t \geq qt$ can be extended by

$$1 = n^2 e^{-tr(q)}$$

or

$$t = \log(n^2)/r(q).$$

The maximum score should be

$$\max_{q \geq 0} qt = \max \frac{q}{r(q)} \log(n^2) = 2b \log(n).$$

This heuristic is now made rigorous.

Theorem 11.28 *For all $(\mu, \delta) \in [0, \infty]^2$ with $\rho(\mu, \delta) < 0$,*

$$\mathbb{P}\left((1-\epsilon)b < \frac{H_n}{\log(n)} < (2+\epsilon)b\right) \to 1$$

for all $\epsilon > 0$.

Proof. The proof is in two parts, the lower and the upper bounds. The upper bound is not too difficult but is lengthy and ommitted.

The lower bound is $\mathbb{P}((1-\epsilon)b \log(n) < H_n) \to 1$ as $n \to \infty$. Given $\epsilon > 0$, take $\gamma > 0$ small and $q > 0$ so that $q/(r(q))$ approximates b and

$$(1-\epsilon)b \left(\frac{r(q) + \gamma}{q}\right) < 1 - \frac{\epsilon}{2}.$$

Probability and Statistics for Sequence Alignment

As usual, we take $t = (1-\epsilon)b\log(n)$. With $k = \lfloor t/q \rfloor$, k is of order $\log(n)$, $k \approx c \log n$.

For n large enough, k will satisfy

$$-\frac{1}{k}\log \mathbb{P}(S_k \geq qk) \leq r(q) + \gamma,$$

and

$$\mathbb{P}(S_k \geq qk) \geq \exp\{-k(r(q) + \gamma)\}$$
$$\geq \exp\left\{-t\left(\frac{r(q)+\gamma}{q}\right)\right\}$$
$$\geq \exp\{-(1-\frac{\epsilon}{2})\log(n)\} = n^{-1+\epsilon/2}.$$

The proof is completed by taking nonoverlapping blocks of length k, so there are about $\frac{n}{k} \sim \frac{n}{c \log n}$ independent chances to get a large score. Each block has probability at least $n^{-1+\epsilon/2}$ to get a large score, so the expected number of large scores goes to infinity like $\frac{n^{\epsilon/2}}{(c \log(n))}$. By the Borel-Cantelli lemma, the lower bound holds. ∎

Next, we show that the growth of score is linear when $\rho > 0$.

Theorem 11.29 *If $\rho = \rho(\mu, \delta) > 0$, then*

$$S_n/n \to \rho \quad \text{in probability}$$

and

$$H_n/n \to \rho \quad \text{in probability.}$$

Proof. Note that $H_n \geq S_n$, so it is required that

$$\mathbb{P}(H_n > (1+\epsilon)n\rho) \to 0$$

and

$$\mathbb{P}(S_n < (1-\epsilon)n\rho) \to 0.$$

The second half follows from subadditivity and $S_n/n \to \rho$ with probability 1.

A lemma is needed. The motivation is that $S_k = S(A_1 \cdots A_k, B_1 \cdots B_k)$ uses sequences of length k, whereas H_n uses sequences of any length for the arguments.

Lemma 11.6 *Define $S_{i,j} = S(A_1 \cdots A_i, B_1 \cdots B_j)$ and*

$$r'(q) = \lim -\frac{1}{k}\log \max_{i+j=2k} \mathbb{P}(S_{i,j} \geq qk).$$

Then $r(q) = r'(q)$.

Proof. Subadditivity shows that $r'(q)$ exists. Clearly, $r' \leq r$, as the maximum for r' is over a larger set. We will show $r' \geq r - \epsilon$ for all $\epsilon > 0$. Pick i, j and $k = (i+j)/2$ large enough so that

$$-\frac{1}{k} \log \mathbb{P}(S_{i,j} \geq qk) < r' + \epsilon,$$

Now

$$\mathbb{P}(S_{2k} = S_{k,k} \geq q(2k)) \geq \mathbb{P}(S_{i,j} \geq qk, S(A_{i+1} \cdots A_{i+j}, B_{j+1} \cdots B_{i+j}) \geq qk)$$
$$= (\mathbb{P}(S_{i,j} \geq qk))^2.$$

This equality depends on A_i and B_i having identical distribution so that $S_{i,j}$ and $S_{j,i}$ have the same distribution. Therefore,

$$r = r(q) \leq -\frac{1}{2k} \log \mathbb{P}(S_{2k} \geq 2qk)$$
$$\leq -\frac{1}{2k} \log \left(\mathbb{P}(S_{i,j} \geq qk)\right)^2 < r' + \epsilon.$$

∎

Applying Lemma 11.6,

$$r = \dot{r}((1+\epsilon)\rho) = \lim \left(-\frac{1}{k} \log \mathbb{P}(S_k \geq (1+\epsilon)\rho k)\right)$$

$$= \inf \left(-\frac{1}{k} \max_{i+j=2k} \log \mathbb{P}(S_{i,j} \geq (1+\epsilon)\rho k)\right).$$

Application of the Azuma-Hoeffding lemma says $r > 0$. Therefore,

$$\mathbb{P}(S_{i,j} \geq (1+\epsilon)\rho k) \leq e^{-rk}$$

for all $i, j, k = (i+j)/2$. For $i, j \leq n, k = (i+j)/2 \leq n$ and

$$\mathbb{P}(S_{i,j} \geq (1+\epsilon)n\rho) \leq \mathbb{P}(S_{i,j} \geq (1+\epsilon)k\rho) \leq e^{-rk}.$$

Now $S_{i,j} \geq (1+\epsilon)n\rho$ requires $k = (i+j)/2 \geq (1+\epsilon)n\rho$ because identities score $+1$, and, therefore,

$$\mathbb{P}(S_{i,j} \geq (1+\epsilon)n\rho) \leq e^{-rk} \leq e^{-r(1+\epsilon)n\rho}.$$

Finally,

$$\mathbb{P}(H_n \geq (1+\epsilon)n\rho)$$

$$= \mathbb{P}\left(\bigcup_{\substack{i,j \\ k,l}} \{S(A_{i+1} \cdots A_{i+k}, B_{j+1} \cdots B_{j+l}) \geq (1+\epsilon)n\rho\}\right)$$

$$\leq n^4 e^{-r(1+\epsilon)n\rho} \to 0.$$

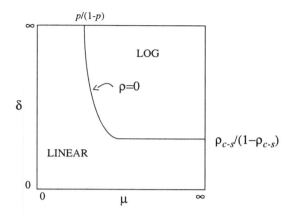

Figure 11.4: *Phase transition curve*

This completes the proof. ∎

Now we explore the shape of the curve $\rho = 0$ in $(\mu, \delta) \in [0, \infty]^2$. First of all, note that

$$H_n(2\delta, \delta) = H_n(\mu, \delta), \quad 2\delta \leq \mu,$$

since when $\mu > 2\delta$, there will be no non-identities or mismatches in alignments. The curve $\rho = 0$ will hit the (μ, ∞) boundary at the μ where the average score per letter is 0:

$$p \cdot 1 + (1-p)(-\mu) = 0$$

or $\mu = p/(1-p)$. With a uniform distribution on DNA, this value is $\mu = 1/3$. Similarly let the Chvátalátal-Sankoff longest common subsequence constant be ρ_{c-s}. On the boundary (∞, δ) the equation is

$$\rho_{c-s} \cdot 1 + (1 - \rho_{c-s})(-\delta) = 0$$

or $\delta = \rho_{c-s}/(1-\rho_{c-s})$. As with a uniform two-letter alphabet, $\rho_{c-s} \approx 0.82$, $\delta \approx 4.6$.

11.6.2 Practical p-Values

The ultimate goal is to provide a rapid and accurate estimates of p-values for values of $H(\mathbf{A}, \mathbf{B})$. As pointed out in the Introduction, there are not yet practical results for the linear region of (μ, δ). Simulation is too time-consuming for large numbers of comparisons of varying sequence lengths and compositions but can be used for specific comparisons of interest. However, a blend of theory and simulation gives a solution for the logarithmic region even for general scoring schemes.

To set the stage, recall that we presented a strong law for scoring without deletions ($g(k) = \infty$ for all k). This requires $\mathbb{E}(s(A, B)) < 0$ and $s^* = \max s(a, b) > 0$, which puts us in the logarithmic region. For p the largest root of $1 = \mathbb{E}(\lambda^{-s(A,B)})$, $\mathbb{P}(\lim_{n\to\infty} H_n / \log_{1/p}(n^2) = 1)$. Therefore, for the case of no indels, we know the center of the score is $\log_{1/p}(nm)$. A Chen-Stein style theorem can be proved.

Theorem 11.30 *Under the above assumptions, there is a constant $\gamma > 0$ such that*
$$\mathbb{P}(H(\mathbf{A}, \mathbf{B}) > \log_{1/p}(nm) + c) \approx 1 - e^{-\gamma p^c}.$$

Although this result has not been proved for the usual alignment problem with indels, it is another indication of the range of validity of Poisson approximation.

Next, we sketch the general set up for Poisson approximation in sequence matching. Recall that the local algorithm for $H(\mathbf{A}, \mathbf{B})$ has a computational algorithm for declumping. One alignment from the clump with largest score is used to declump, so that the second largest clump can be found, and so on. Let $H_{(1)} \geq H_{(2)} \cdots$ denote these clump sizes. The random variables $Y_\nu = Y_{(i,j)}$ are indicators of the event that some alignment ending at (i, j) scores at least t:
$$Y_{(i,j)} = \mathbb{I}\{\max\{S(A_k \cdots A_i, B_l \cdots B_j) : k \leq i, l \leq j\} \geq t\}.$$

$X_\nu = X_{(i,j)}$ are the declumped random variables. The model we develop for this process is the *Poisson clumping heuristic*. In this model alignment clumps are laid down according to a Poisson process on $[1, n] \times [1, m]$, and for each alignment clump, there is an independent clump size. The clumps have scores $H_{i,j}$. We model the number $W(t)$ of clumps with scores larger than a test value $t = \text{center} + c$ by a Poisson distribution with mean $\lambda = \lambda_{n,m}(t)$. This is applied in the form $\mathbb{P}(\text{at least one score exceeds } t) = 1 - \mathbb{P}(\text{no score exceeds } t) = 1 - e^{-\lambda}$ or
$$\mathbb{P}(W = 0) = \mathbb{P}(H(A, B) < t)$$
$$\approx 1 - e^{-\lambda}.$$

When
$$t = \log_{1/\xi}(nm) + c,$$
$$\mathbb{P}(W = 0) \approx 1 - e^{-\gamma \xi^c}$$

There are several assumptions that this model makes that can be numerically tested. First of all, we can test to see that $W(t)$ is approximately Poisson distributed with mean $\lambda = \mathbb{E}(W(t))$. Second, we can check to see that λ has the exponential for $\mathbb{E}(W(t)) = \hat{\gamma} p^t$. Finally, we can verify that $\hat{\gamma}$ has the form $\hat{\gamma} = \gamma mn$,

Probability and Statistics for Sequence Alignment

proportional to $nm =$ product of sequence lengths. Fortunately, although we do not have a theorem that establishes these results, numerical tests show the assumptions hold throughout the logarithmic region.

The problem of estimating γ and ξ remains. The distribution function of H is $\mathbb{P}(H \leq t) = \mathbb{P}(W(t) = 0) = e^{-\gamma mn\xi^t}$, so we can collect a sample of H by simulation and take $\log(-\log(\text{empirical cdf}))$ which can then be fit by simple regression. The theoretical curve is $\log(\gamma mn) + t\log(\xi)$ so that estimates of γ and ξ are then available. This is a reliable although computationally expensive method of optaining γ and ξ.

Another method, *declumping estimation*, illustrates the power of the Poisson approximation. Simulation of many sequence comparisons is quite expensive because each costs $O(nm)$ time. However, in Section 11.5.3 on approximate matching, we decomposed the indicators X_ν of clump at ν into $X_{\nu,i}$, indicators of a clump at ν of size i. The process approximation that was developed to handle the compound Poisson process says that the indicators $X_{\nu,i}$ give a random sample of clump sizes H, truncated so that only $H_i > t$ are seen; that is, we have $H_{(1)} \geq H_{(2)} \geq \cdots \geq H_{(N)}$, where $H_{(N)} > t$ and $H_{(N+1)} \leq t$. This sample can be analyzed numerically to find a good estimate of ξ.

Problems

Problem 11.1 Apply the Azuma-Hoeffding lemma to $Y_n = C_1 + C_2 + \cdots + C_n$, where C_i are independent Bernoullis with $\mathbb{P}(C_i = 1) = 1 - \mathbb{P}(C_i = 0) = p$.

Problem 11.2 Show $\min_{\beta>0}\{e^{-\alpha\beta}(1-p+pe^\beta)\} = e^{-\mathcal{H}(\alpha,p)}$, where $0 < \alpha < 1$.

Problem 11.3 Prove that $\mathbb{P}(|S_n - E(S_n)| > \gamma n) \leq 2e^{-\frac{\gamma^2 n}{2c^2}}$.

Problem 11.4 Use Corollary 11.1 to show $\text{Var}(S_n) \leq 4nc^2$.

Problem 11.5 Prove that $\mathbb{E}|S_n|^k < \infty$ for all $k \geq 1$.

Problem 11.6 (Chung) If $\sum_n \mathbb{P}(C_n) = \infty$ and

$$\lim_n \frac{\sum_{i=1}^n \sum_{j=1}^n \mathbb{P}(C_i C_j)}{(\sum_{i=1}^n \mathbb{P}(C_i))^2} = 1,$$

then show that $\mathbb{P}(C_n \text{ occur infinitely often}) = 1$.

Problem 11.7 For the Bernoulli random variables of Theorem 11.20, find $g_n(t)$, the generating function of $X_{1,n} + \cdots + X_{n,n}$. Show that $\log g_n(t) \to -\lambda(1-s)$. (The generating function of Poisson mean λ is $e^{-\lambda(1-s)}$.)

Problem 11.8 *Open reading frames.* Recall the genetic code reads mRNA in nonoverlapping triplet or codons. There are three stop codons: UAA, UGA, and UAG. A reading frame is one of the six ways of assigning codons. An open reading frame is a stretch of sequence in a reading frame without stop codons. (i) For a fixed reading frame F_1 in a sequence of n iid letters, estimate the probability that the longest open reading frame has length $L_1 \geq t$. (ii) Use (i) to bound $\mathbb{P}(\max\{L_i : i = 1,\ldots,6\} \geq t)$. (iii) In bacterial genes, the gene is uninterrupted and (usually) starts with AUG. An open reading frame F_1 is the maximum length between AUG and a stop codon. Estimate the probability that the longest open reading frame F_1 has length $L_1 \geq t$.

Problem 11.9 Let $s(a,b) = 1$ if $a = b$, and $-\infty$ if $a \neq b$ and $g(k) = -\infty$ for all $k \geq 1$. Let $\mathbf{A} = A_1 A_2 \cdots \mathbf{B} = B_1 B_2 \cdots$ have iid letters with distribution $\{\xi_a : a \in \mathcal{A}\}$. Suppose $\mu_{a,b}$ is the probability a is aligned with b in the optimal local alignment. (i) If $\xi_a = 1/|\mathcal{A}|$ for all $a \in \mathcal{A}$, find $\mu_{a,a}$. (ii) Derive $s(a,b)$ so that $\mu_{a,a} = 1/|\mathcal{A}|$.

Problem 11.10 In this problem, we look again at long headruns. $D_1, D_2, \ldots, D_n$ are iid as $\mathbb{P}(D_i = 1) = 1 - \mathbb{P}(D_i = 0) = p \in (0,1)$. We define X_i **without** declumping: for $i = 1, 2, \ldots, n - t + 1$

$$X_i = \prod_{j=i}^{i+t-1} D_j \text{ and } W = \sum_{i \in I} X_i.$$

As usual $I = \{1, 2, \ldots, n - t + 1\}$ and $J_i = \{j \in I : |i - j| < t\}$. (i) Compute $\lambda = \mathbb{E}(W)$. (ii) Find t so that $\lambda = \lambda(n)$ stays in $(0, \infty)$ as $n \to \infty$. (iii) Compute $b_1 = \sum_{i \in I} \sum_{j \in J_i} \mathbb{E}(X_i)\mathbb{E}(X_j)$, and $b_2 = \sum_{i \in I} \sum_{i \neq j \in J_i} \mathbb{E}(X_i X_j)$. (iv) For t in (ii), what is the behavior of b_1 and b_2 as $n \to \infty$?

Problem 11.11 Take N sequences of length n where each letter is iid uniform (probability $1/|\mathcal{A}| = 1/a$). Use the heuristic to find the value $t = t(n, N)$, where $H = H(n, N)$, the length of the longest sequence, common to all N sequences, satisfies $\mathbb{P}(\lim_{n \to \infty} H/t = 1)$. For $N = 2$, $H(n,2) \sim 2\log_a(n) = t(n,2)$. What happens to t as $N \to \infty$?

Problem 11.12 This problem is to test intuition and understanding of concepts, not mathematical manipulative skills. Suppose the $n \times n$ lattice, $L_n = \{(i,j) : 1 \leq i, j \leq n\}$, is colored with red (R) or black (B), iid at each lattice point by

$$\mathbb{P}(Z(i,j) = R) = 1 - \mathbb{P}(Z(i,j) = B) = p \in (0,1).$$

Our interest is in red squares, where the $d \times d$ square at (k, l) is

$$S = S(k,l)$$
$$= \{(i,j) : 1 \leq k \leq i \leq k + d - 1 \leq n \text{ and } 1 \leq l \leq j \leq l + d - 1 \leq n\}.$$

The square is red if $\nu \in S$ implies $Z(\nu) = R$. Define $X_n = $ side length of the largest red square.

(i) As a function of n, find $d(n)$, the asymptotic side length of the largest red square in L_n; that is, find $d(n)$ (no proof, just heuristic) so that

$$\mathbb{P}\left(\lim_{n\to\infty} \frac{X_n}{d(n)} = 1\right) = 1.$$

(ii) Repeat (i) for d_n^α, the side length of the largest red square with fraction $\alpha > p$ lattice points colored red.

Problem 11.13 Show that $\mathbb{E}((Lf)Z) = 0$ for all f satisfying $\mathbb{E}(Zf(Z)) < \infty$ if and only if Z is distributed as Poisson(λ).

Problem 11.14 Let X_α be iid Bernoulli(p), $0 < p < 1$, and $I = \{1, 2, \ldots, n\}$. For $W = \sum_{\alpha \in I} X_\alpha$ and Z Poisson mean $= \mathbb{E}(W)$, bound $\|W - Z\|$.

Problem 11.15 Let Y be a Poisson process on $[0, \infty)$ with intensity λ. Let X be derived from Y by removing Y points (arrivals) with probability $1 - p$. Show that X is a Poisson process on $[0, \infty)$ with intensity λp.

Problem 11.16 Show that in the general process construction for the decomposition of X_ν into $X_{\nu,i}$, $i \in T$, that $b_1^* = b_1$ (Equation 11.17).

Chapter 12

Probability and Statistics for Sequence Patterns

The last chapter was concerned with the statistical distribution of alignment scores. Due to the fact that alignment scores are obtained from optimizing over a huge number of potential alignments, the score distribution involves large deviations. The chapter could have been included under the heading "extremal statistics." Not all statistical features of sequences involve extremes. The first statistical studies of sequences involved counting the frequencies of k-tuples. The statistical analyses were naive but very natural. If $p_\mathbf{w}$ is the probability of the k-tuple $\mathbf{w}$ and $N_\mathbf{w}$ is its frequency in a sequence of length n, then we expect $(n - k + 1)p_\mathbf{w}$ to be the mean of $N_\mathbf{w}$, and $(N_\mathbf{w} - (n - k + 1)p_\mathbf{w})/\sqrt{(n - k + 1)p_\mathbf{w}(1 - p_\mathbf{w})}$ should be approximately normally distributed for large n. The simplest case is that of uniformly distributed letters where in DNA sequences $p_\mathbf{w} = 4^{-k}$. Although the above expectation is correct, the count distribution is strongly dependent on the word $\mathbf{w}$ when $p_\mathbf{w} \equiv 4^{-k}$ for all k-tuples $\mathbf{w}$. If the $n - k + 1$, k-tuples were not from a sequence but generated iid, this would not be the case. One of the goals of this chapter is to present the reasons for this surprising feature of sequence k-tuple counts. We will also discuss the generalization of these ideas to Markov sequences. One reason for performing these statistical studies is to characterize features of genomes or organisms.

To set the stage for this chapter, consider a sequence $\mathbf{A} = A_1 A_2 \cdots$ with iid letters from $\mathcal{A} = \{R, Y\}$, $\mathbb{P}(R) = 1 - \mathbb{P}(Y) = 1/2$. Let $\mathbf{w} = w_1 w_2 \cdots w_k$ be an arbitrary word with letters $w_i \in \mathcal{A}$. Then

$$\mathbb{P}(A_{i+1} A_{i+2} \cdots A_{i+k} = \mathbf{w}) = 2^{-k}$$

for all $\mathbf{w}$. In fact,

$$f(\mathbf{w}, n) = \frac{1}{n - k + 1} \sum_{i=1}^{n-k+1} \mathbb{I}(A_i A_{i+1} \cdots A_{i+k-1} = \mathbf{w}) \to 2^{-k}$$

by the strong law of large numbers. It is also true that

$$\mathbb{E}(f(\mathbf{w}, n)) = \frac{1}{n-k+1} \sum_{i=1}^{n-k+1} \mathbb{E}\left((\mathbb{I}(A_i A_{i+1} \cdots A_{i+k-1} = \mathbf{w})\right) = 2^{-k}.$$

So all k-tuples $\mathbf{w}$ occur about $(n - k + 1)2^{-k}$ times in a sequence of length n. Where does the difference between k-tuples come in?

Now, let $\mathbf{w}_R = RR \cdots R = R^k$ be the word with $w_i \equiv R$. Then, conditional on finding $\mathbf{w}_R$ starting at position $i + 1$, the probability of finding $\mathbf{w}_R$ at $i + 2, \ldots, i + k + 1$ is $1/2$, at $i + 3, \ldots, i + k + 2$ is $(1/2)^2$ and so on; that is,

$$\mathbb{P}(A_{i+1} \cdots A_{i+k+1} = R^{k+1} | A_{i+1} \cdots A_{i+k} = \mathbf{w}_R)$$
$$= \mathbb{P}(A_{i+k+1} = R) = 1/2,$$

$$\mathbb{P}(A_{i+1} \cdots A_{i+k+2} = R^{k+2} | A_{i+1} \cdots A_{i+k} = \mathbf{w}_R)$$
$$= \mathbb{P}(A_{i+k+1} = A_{i+k+2} = R) = (1/2)^2,$$

$$\mathbb{P}(A_{i+1} \cdots A_{i+k+j-1} = R^{k+j-1} | A_{i+1} \cdots A_{i+k} = \mathbf{w}_R)$$
$$= (1/2)^{j-1}, \; j \geq 2.$$

If we require that $i + 1$ starts the leftmost occurrence of $\mathbf{w}_R$, then $A_i = Y$. Therefore, the number of overlapping k-tuples $\mathbf{w}_R$ beginning at $i + 1$ is geometric with mean 2:

$$Y \underbrace{RR \cdots R}_{k} \underbrace{RR \cdots R}_{j-1} Y.$$

Therefore, the pattern $RR \cdots R = \mathbf{w}_R$ occurs in clumps, where the clump size J has a geometric distribution $\mathbb{P}(J = j) = \left(\frac{1}{2}\right)^j$, $j \geq 1$. There is an average of two occurrences of $\mathbf{w}_R$ per clump. From the work on Poisson approximation in Chapter 11, the number of clumps is approximately Poisson with mean $\lambda = (n - k + 1)(1/2)^{k+1}$, where n =sequence length.

When we move to a pattern without self-overlap, the situation dramatically changes. Let $\mathbf{w}_1 = RR \cdots RY$. Now $\mathbf{w}_1$ has no self-overlap and with one occurrence of $\mathbf{w}_1$ per "clump," it must be much more evenly spaced along the sequence than $\mathbf{w}_R$, which occurs in clumps that average two patterns per clump. The fact that both $\mathbf{w}_1$ and $\mathbf{w}_R$ have the same average number of occurrences motivates these observations.

At least two ways to count words are relevant in biology. Often the frequencies $N_\mathbf{w}$ of the word $\mathbf{w}$ ($N_\mathbf{w} = nf(\mathbf{w}, n)$ in the above notation) in a sequence is tabulated for $\mathbf{w}$ in some set $\mathcal{W}$. The count distribution depends on the self-overlap and pairwise overlaps of words in $\mathcal{W}$. In Section 12.1, we derive a

Central Limit Theorem for these counts based on a Markov chain analysis. These distributional results can be used for genome and sequence comparisons.

In other circumstances, nonoverlapping occurrences of $\mathbf{w} \in \mathcal{W}$ are the objects of interest. Here we proceed down the sequence, and when $\mathbf{w} \in \mathcal{W}$ is encountered, we jump ahead to the end of $\mathbf{w}$ and begin the search anew. Now, in the simple two-letter, uniform iid case, the mean frequencies of all length k patterns are not identical. Restriction site frequencies should be counted in this way. Renewal theory and Li's method are used to give a statistical analysis of counts in Section 12.2. Later in the chapter we use Poisson approximation from Chapter 11 as another approach to the analysis of pattern counts.

Another question that arises frequently is the spacings of certain patterns along a sequence or genome. These patterns might be restriction sites or gene locations. The biologist is often interested in whether these sites are randomly (uniformly) distributed or not. Statistical distributions and tests are required to answer these questions.

12.1 A Central Limit Theorem

The goal of this section is to derive a Central Limit Theorem for the joint distribution of $(N_{\mathbf{w}_1}, N_{\mathbf{w}_2}, \ldots)$ for $\mathbf{w}_i \in \mathcal{W}$. As usual, the limiting variance/covariance matrix is required. This is the reason we will study $\text{cov}(N_{\mathbf{u}}, N_{\mathbf{v}})$ for two words $\mathbf{u}$ and $\mathbf{v}$.

One approach to word counts is to observe that even when the sequence $A_1 A_2 \cdots A_n$ is generated iid, the transition from $\mathbf{w}$, $A_i A_{i+1} \cdots A_{i+k-1} = w_1 w_2 \cdots w_k$, to $\mathbf{v}$, $A_{i+1} A_{i+2} \cdots A_{i+k} = v_1 v_2 \cdots v_k$, is a Markov chain. Then for long sequences, the counts for length k words obey the central limit theorem for Markov chains. Not much difficulty is introduced by allowing the sequence $A_1 A_2 \cdots$ itself to be generated by a Markov chain. To reduce the required notation, we will take the order of the chain to be 1.

Assume the sequence is generated by a Markov chain $\{A_i\}$ with state space $\mathcal{A}$. The chain has order 1:

$$p_{a,b} = \mathbb{P}(A_m = b | A_1 = a_1, \ldots, A_{m-2} = a_{m-2}, A_{m-1} = a)$$
$$= \mathbb{P}(A_m = b | A_{m-1} = a).$$

As usual,
$$p_{a,b}^{(k)} = \mathbb{P}(A_{m+k} = b | A_m = a)$$

is the probability of moving from a to b in k steps. We assume the chain is irreducible and aperiodic (see Chapter 4.4). Because $|\mathcal{A}| < \infty$, there is a stationary distribution π independent of the initial distribution $p^{(0)}$.

Lemma 12.1 *Under the above assumptions, there exists a unique stationary distribution and constants $0 \leq K, 0 \leq \rho < 1$ such that*

$$|p_{a,b}^{(n)} - \pi_b| \leq K \rho^n.$$

Let $\mathbf{u} = u_1 \cdots u_k$ and $\mathbf{v} = v_1 v_2 \cdots v_l$ be two words. In general, $\mathbb{I}\mathbf{u}(i) = \mathbb{I}(A_i A_{i+1} \cdots A_{i+k-1} = \mathbf{u})$ is the indicator that a finite word $\mathbf{u}$ occurs starting at position i in the sequence $A_1 \cdots A_n \cdots$. Then

$$N_{\mathbf{u}} = N_{\mathbf{u}}(n) = \sum_{i=1}^{n-k+1} \mathbb{I}\mathbf{u}(i).$$

We will derive a Central Limit Theorem for sets of finite words. Essentially, for each pair of words $(\mathbf{u}, \mathbf{v})$ we need the first two moments of $(N_{\mathbf{u}}, N_{\mathbf{v}})$. The first moments are easy to derive:

$$\mathbb{E}(\mathbb{I}\mathbf{u}(i)) = \sum_a p_a^{(0)} p_{a,u_1}^{(i-1)} P_{\mathbf{u}}(k-1), \tag{12.1}$$

where we define $P_{\mathbf{u}}(l)$ to be the probability of seeing $u_{k-l+1} \cdots u_k$ given that $u_1 \cdots u_{k-l}$ occurs; that is,

$$P_{\mathbf{u}}(l) = p_{u_{k-l}, u_{k-l+1}} p_{u_{k-l+1}, u_{k-l+2}} \cdots p_{u_{k-1}, u_k}$$

and

$$P_{\mathbf{u}}(k-1) = p_{u_1, u_2} p_{u_2, u_3} \cdots p_{u_{k-1}, u_k},$$

where $P_{\mathbf{u}}(l) = 1$ if $l \leq 0$. Let $\pi_{\mathbf{u}}$ be the equilibrium probability of $\mathbf{u}$: $\pi_{\mathbf{u}} = \pi_{u_1} p_{u_1, u_2} \cdots p_{u_{k-1}, u_k}$. Although $\mathbb{E}(N_{\mathbf{u}}(n))$ is a complicated function of n, it grows like $n\pi_{\mathbf{u}}$ in the limit.

Theorem 12.1
$$\lim_{n \to \infty} \frac{1}{n} \mathbb{E}(N_{\mathbf{u}}(n)) = \pi_{\mathbf{u}}.$$

Proof. Using the lemma,

$$|\mathbb{E}(\mathbb{I}\mathbf{u}(i)) - \pi_{\mathbf{u}}| \leq \sum_a p_a^{(0)} P_{\mathbf{u}}(k-1) |p_{a,u_1}^{(i-1)} - \pi_{u_1}|$$

$$\leq K\rho^{i-1} \sum_a p_a^{(0)} P_{\mathbf{u}}(k-1) \leq K\rho^{i-1}.$$

Then

$$n^{-1}|\mathbb{E}(N_{\mathbf{u}}(n)) - n\pi_{\mathbf{u}}| \leq n^{-1} \sum_i |\mathbb{E}(\mathbb{I}\mathbf{u}(i)) - \pi_{\mathbf{u}}|$$

$$\leq n^{-1} \sum_{i=1}^{n-k+1} K\rho^{i-1} \leq n^{-1} \frac{K}{1-\rho}.$$

∎

Probability and Statistics for Sequence Patterns

The next step to a Central Limit Theorem is to calculate the covariance of $(N_{\mathbf{u}}(n), N_{\mathbf{v}}(n))$. We will derive the limiting value of $\frac{1}{n}\text{cov}(N_{\mathbf{u}}(n), N_{\mathbf{v}}(n))$. This calculation is complicated by overlaps of $\mathbf{u}$ and $\mathbf{v}$ as well as by the dependence between nonoverlapping occurrences due to the Markov structure of the sequence $A_1 A_2 \cdots$.

First, we consider $\text{cov}(N_{\mathbf{u}}, N_{\mathbf{v}})$. We must consider the dependence introduced when the words $\mathbf{u}$ and $\mathbf{v}$ overlap. Shift $\mathbf{v}$ along $\mathbf{u}$ a distance $j \geq 0$. We define the *overlap bit* $\beta_{\mathbf{u},\mathbf{v}}(j) = 1$ if the shifted words allow this overlap (that is, if $u_{j+1} = v_1, \ldots, u_k = v_{k-j}$). Otherwise $\beta_{\mathbf{u},\mathbf{v}}(j) = 0$. Now we use the overlap bit to compute $\text{cov}(N_{\mathbf{u}}(n), N_{\mathbf{v}}(n))$. For $i \leq j$ and $j - i < k$,

$$\mathbb{E}(\mathbb{I}_{\mathbf{u}}(i)\mathbb{I}_{\mathbf{v}}(j)) = \mathbb{E}(\mathbb{I}_{\mathbf{u}}(i))\beta_{\mathbf{u},\mathbf{v}}(j-i)P_{\mathbf{v}}(j-i+l-k). \tag{12.2}$$

For the case of $k \leq j - i$, the words $\mathbf{u}$ and $\mathbf{v}$ do not overlap and $\mathbf{u}$ is followed by $j - i - k$ letters before $\mathbf{v}$ occurs. We need to collect v_1 to begin $\mathbf{v}$ so

$$\mathbb{E}(\mathbb{I}_{\mathbf{u}}(i)\mathbb{I}_{\mathbf{v}}(j)) = \mathbb{E}(\mathbb{I}_{\mathbf{u}}(i))p_{u_k,v_1}^{(j-i-k+1)}P_{\mathbf{v}}(l-1). \tag{12.3}$$

For use in this and later sections, we define the *overlap polynomial* $G_{\mathbf{u},\mathbf{v}}(s)$

$$G_{\mathbf{u},\mathbf{v}}(s) = \sum_{j=0}^{|\mathbf{u}|-1} s^j \beta_{\mathbf{u},\mathbf{v}}(j) P_{\mathbf{v}}(j + |\mathbf{v}| - |\mathbf{u}|). \tag{12.4}$$

Some straightforward but detailed calculation yields the value of $\text{cov}(N_{\mathbf{u}} N_{\mathbf{v}})$ in the next proposition. The proof is left as an exercise.

Proposition 12.1 *For the sequence $A_1 A_2 \cdots A_n$ and words $\mathbf{u}$ and $\mathbf{v}$, with $|\mathbf{u}| = k \leq |\mathbf{v}| = l$,*

$$\text{cov}(N_{\mathbf{u}}(n), N_{\mathbf{v}}(n))$$

$$= \sum_{j=0}^{k-1} \sum_{i=1}^{n-l-j+1} \mathbb{E}(\mathbb{I}_{\mathbf{u}}(i)) \{\beta_{\mathbf{u},\mathbf{v}}(j)P_{\mathbf{v}}(l-k+j) - \mathbb{E}(\mathbb{I}_{\mathbf{v}}(i+j))\}$$

$$+ \sum_{j=0}^{n-l-k} \sum_{i=1}^{n-l-j-k+1} \mathbb{E}(\mathbb{I}_{\mathbf{u}}(i)) \left\{ p_{u_k,v_1}^{(j+1)} P_{\mathbf{v}}(l-1) - \mathbb{E}(\mathbb{I}_{\mathbf{v}}(i+j+k)) \right\}$$

$$+ \sum_{j=0}^{l-1} \sum_{i=1}^{n-l-j+1} \mathbb{E}(\mathbb{I}_{\mathbf{v}}(i)) \{\beta_{\mathbf{v},\mathbf{u}}(j)P_{\mathbf{u}}(k-l+j) - \mathbb{E}(\mathbb{I}_{\mathbf{u}}(i+j))\}$$

$$+ \sum_{j=0}^{n-l-k} \sum_{i=1}^{n-l-j-k+1} \mathbb{E}(\mathbb{I}_{\mathbf{v}}(i)) \left\{ p_{v_l,u_1}^{(j+1)} P_{\mathbf{u}}(k-1) - \mathbb{E}(\mathbb{I}_{\mathbf{u}}(i+j+l)) \right\}$$

$$- \sum_{i=1}^{n-l+1} \text{cov}(\mathbb{I}_{\mathbf{u}}(i)\mathbb{I}_{\mathbf{v}}(i)) + \sum_{i=n-l+2}^{n-k+1} \sum_{j=1}^{n-l+1} \text{cov}(\mathbb{I}_{\mathbf{u}}(i)\mathbb{I}_{\mathbf{v}}(j)).$$

By using these formulas and an elementary result on Cesáro sums, we can derive the asymptotic covariance of the pattern counts.

Theorem 12.2 *Assume* $|\mathbf{u}| = k \leq l = |\mathbf{v}|$. *Then*

$$\lim_{n\to\infty} n^{-1} \operatorname{cov}(N_\mathbf{u}(n), N_\mathbf{v}(n)) = \pi_\mathbf{u} \sum_{i=0}^{k-1} \{\beta_{\mathbf{u},\mathbf{v}}(i) P_\mathbf{v}(l-k+i) - \pi_\mathbf{v}\}$$

$$+ \pi_\mathbf{u} P_\mathbf{v}(l-1) \sum_{j=0}^{\infty} \left\{ p^{(j+1)}_{u_k, v_1} - \pi_{v_1} \right\}$$

$$+ \pi_\mathbf{v} \sum_{i=0}^{l-1} \{\beta_{\mathbf{v},\mathbf{u}}(i) P_\mathbf{u}(k-l+i) - \pi_\mathbf{u}\}$$

$$+ \pi_\mathbf{v} P_\mathbf{u}(k-1) \sum_{j=0}^{\infty} \left\{ p^{(j+1)}_{v_1, u_1} - \pi_{u_1} \right\}$$

$$- \pi_\mathbf{u} \{\beta_{\mathbf{u},\mathbf{v}}(0) P_\mathbf{v}(l-k) - \pi_\mathbf{v}\}.$$

Proof. We find the limit of $n^{-1} \operatorname{cov}(N_\mathbf{u}(n), N_\mathbf{v}(n))$ term by term from Proposition 12.1. As by the proof of Theorem 12.1, $\lim_{i\to\infty} \mathbb{E}(\mathbb{I}\,\mathbf{u}(i)) = \pi_\mathbf{u}$, we have

$$\lim_{n\to\infty} \sum_{j=0}^{k-1} n^{-1} \sum_{i=1}^{n-l-j+1} \mathbb{E}(\mathbb{I}\,\mathbf{u}(i)) \{\beta_{\mathbf{u},\mathbf{v}}(j) P_\mathbf{v}(l-k+j) - \mathbb{E}(\mathbb{I}\,\mathbf{v}(i+j))\}$$

$$= \pi_\mathbf{u} \sum_{j=0}^{k-1} \{\beta_{\mathbf{u},\mathbf{v}}(j) P_\mathbf{v}(l-k+j) - \pi_\mathbf{v}\}.$$

The third term has a similar limit with the roles of $\mathbf{u}$ and $\mathbf{v}$ interchanged. The second term is considered next. By adding and subtracting $\pi_\mathbf{v}$ and applying Lemma 12.1,

$$\mathbb{E}(\mathbb{I}\,\mathbf{u}(i)) \left| \left\{ p^{(j+1)}_{u_k, v_1} P_\mathbf{v}(l-1) - \mathbb{E}(\mathbb{I}\,\mathbf{v}(i+j+k)) \right\} \right| \leq 2K\rho^j$$

and

$$n^{-1} \sum_{i=1}^{n-l-j-k+1} \mathbb{E}(\mathbb{I}\,\mathbf{u}(i)) \left| \left\{ p^{(j+1)}_{u_k, v_1} P_\mathbf{v}(l-1) - \mathbb{E}(\mathbb{I}\,\mathbf{v}(i+j+k)) \right\} \right| < 2K\rho^j.$$

Therefore, because the summand has a limit as $i \to \infty$, using the dominated convergence theorem,

$$\lim_{n\to\infty} n^{-1} \sum_{j=1}^{n-l-k} \sum_{i=1}^{n-l-j-k+1} \mathbb{E}(\mathbb{I}\,\mathbf{u}(i)) \left\{ p^{(j+1)}_{u_k, v_1} P_\mathbf{v}(l-1) - \mathbb{E}(\mathbb{I}\,\mathbf{v}(i+j+k)) \right\}$$

$$= \sum_{j=0}^{\infty} \pi_{\mathbf{u}} \left\{ p_{u_k,v_1}^{(j+1)} P_{\mathbf{v}}(l-1) - \pi_{\mathbf{v}} \right\}$$

$$= \pi_{\mathbf{u}} P_{\mathbf{v}}(l-1) \sum_{j=0}^{\infty} \left\{ p_{u_k,v_1}^{(j+1)} - \pi_{v_1} \right\}.$$

Note that by Lemma 12.1, this series converges geometrically. The fourth term has an analogous limit.

The fifth term of Proposition 12.1 is the negative of the first term of the first infinite sum derived above, so

$$\lim_{n \to \infty} n^{-1} \sum_{i=1}^{n-l-j+1} \operatorname{cov}(\mathbb{I}_{\mathbf{u}}(i)\mathbb{I}_{\mathbf{v}}(i)) = \pi_{\mathbf{u}} \left\{ \beta_{\mathbf{u},\mathbf{v}}(0) P_{\mathbf{v}}(l-k) - \pi_{\mathbf{v}} \right\}.$$

Finally,

$$n^{-1} \left| \sum_{i=n-l+2}^{n-k+1} \sum_{j=1}^{n-l+1} \operatorname{cov}(\mathbb{I}_{\mathbf{u}}(i)\mathbb{I}_{\mathbf{v}}(j)) \right| \leq n^{-1} \sum_{i=n-l+2}^{n-k+1} \sum_{j=1}^{\infty} |\operatorname{cov}(\mathbb{I}_{\mathbf{u}}(i)\mathbb{I}_{\mathbf{v}}(j))|.$$

But

$$\sum_{j=1}^{\infty} |\operatorname{cov}(\mathbb{I}_{\mathbf{u}}(i)\mathbb{I}_{\mathbf{v}}(j))| \leq k+l,$$

so this last limit is 0. ∎

The second and fourth terms of the right-hand side of Theorem 12.2 are to account for dependence. The theorem simplifies for independent sequences.

Corollary 12.1 *Assume* $|\mathbf{u}| = k \leq l = |\mathbf{v}|$ *and* $\{A_i\}_{i \geq 1}$ *has iid letters. Then*

$$\lim_{n \to \infty} n^{-1} \operatorname{cov}(N_{\mathbf{u}}(n), N_{\mathbf{v}}(n)) = \pi_{\mathbf{u}} G_{\mathbf{u},\mathbf{v}}(1) - k\pi_{\mathbf{u}}\pi_{\mathbf{v}} + \pi_{\mathbf{v}} G_{\mathbf{v},\mathbf{u}}(1)$$
$$- l\pi_{\mathbf{v}}\pi_{\mathbf{u}} - \pi_{\mathbf{u}} \left\{ \beta_{\mathbf{u},\mathbf{v}}(0) P_{\mathbf{v}}(l-k) - \pi_{\mathbf{v}} \right\}.$$

If $\mathbf{u} = RR \cdots R$ and $\mathbf{v} = RR \cdots RY$ with $|\mathbf{u}| = |\mathbf{v}| = k$ and $p = \mathbb{P}(R) = 1 - \mathbb{P}(Y) = q$, then $\pi_{\mathbf{u}} = p^k$ and $\pi_{\mathbf{v}} = p^{k-1}q$. Whereas $\pi_{\mathbf{u}} = \pi_{\mathbf{v}}$ when $p = q = 1/2$, the variances of $N_{\mathbf{u}}(n)$ and $N_{\mathbf{v}}(n)$ are quite different.

$$\frac{1}{n} \operatorname{Var}(N_{\mathbf{u}}(n)) \approx \left(\frac{1+p-2p^k}{q} \right) p^k - (2k-1)p^{2k}$$

and

$$\frac{1}{n} \operatorname{Var}(N_{\mathbf{v}}(n)) \approx \left(\frac{q}{p} \right) p^k - \left(\frac{q}{p} \right)^2 (2k-1)p^{2k}$$

and when $p = q = 1/2$, $\operatorname{Var} N_{\mathbf{u}}(n) \approx (2^{-k} - (2k+3)2^{-2k})n$ and $\operatorname{Var} N_{\mathbf{v}}(n) \approx (2^{-k} - (2k-1)2^{-2k})n$.

Two general theorems are needed to deduce a Central Limit Theorem. The first handles sums of real-valued random variables.

Theorem 12.3 (Ibragimov) *Suppose* $\{Z_i\}$, $-\infty < i < \infty$, *is a stationary sequence of random variables with* $\mathbb{E}(Z_i) = 0$ *and* $\mathbb{E}(|Z_i|^{2+\delta}) < \infty$ *for some* $\delta > 0$. *For* f *and* $g \in L^2$, *where* f *is measurable with respect to* $\{Z_i, i \leq 0\}$ *and* g *measurable with respect to* $\{Z_i, n \leq i\}$, *let*

$$\rho_n = \sup_{f,g} \text{cor}(f, g).$$

Define $S_n = \sum_{i=1}^n Z_i$ *and* $\sigma_n^2 = \text{Var}(S_n)$. *If* $\rho_n \to 0$ *and* $\sigma_n \to \infty$ *as* $n \to \infty$, *then* $S_n/\sigma_n \stackrel{d}{\Rightarrow} \mathcal{N}(0, 1)$.

The next theorem allows us to get a multivariate Central Limit Theorem by studying sums of the counts.

Theorem 12.4 (Billingsley) *Let* $\mathbf{X}_n$ *and* $\mathbf{Y}$ *be random vectors in* $\mathbb{R}^m$. *Then* $\mathbf{X}_n \stackrel{d}{\Rightarrow} \mathbf{Y}$ *if and only if* $\mathbf{X}_n \mathbf{t}^T \stackrel{d}{\Rightarrow} \mathbf{Y} \mathbf{t}^T$ *for every vector* $\mathbf{t}$.

Now let $\mathcal{W}$ be a set of m words and $\Sigma = (\sigma_{ij})$ be the limiting covariance matrix

$$\sigma_{ij} = \lim_{n \to \infty} n^{-1} \text{cov}(N_i, N_j),$$

where N_i and N_j are word counts for the i-th and j-th words. Also define

$$\boldsymbol{\mu} = \lim_{n \to \infty} n^{-1}(\mathbb{E}(N_1), \mathbb{E}(N_2), \ldots, \mathbb{E}(N_m)).$$

(This vector exists by Theorem 12.1.)

Theorem 12.5 *Let* $\{A_i\}_{i \geq 1}$ *be a stationary, irreducible, aperiodic first-order Markov chain. Let* $\mathcal{W} = \{\mathbf{w}_1, \ldots, \mathbf{w}_m\}$ *be a set of words and* $\mathbf{N} = (N_1(n), \ldots, N_m(n))$ *be the count vector. Then* $n^{-1}\mathbf{N}$ *is asymptotically normal with mean* $\boldsymbol{\mu}$ *and covariance matrix* $n^{-1}\Sigma$. *If* $\det(\Sigma) \neq 0$, *then*

$$n^{1/2} \Sigma^{-1/2} (\mathbf{N}/n - \boldsymbol{\mu}) \stackrel{d}{\Rightarrow} \mathcal{N}(\mathbf{0}, \mathbf{1}).$$

Proof. To apply Theorem 12.3, let $\mathbf{t}$ be a vector in $\mathbb{R}^m$ and

$$\mathbf{N}\mathbf{t}^T = \sum_{i=1}^m t_i N_i = \sum_{i=1}^m t_i \sum_{j=1}^n \mathbb{I}_{\mathbf{w}_i}(j) = \sum_{j=1}^n \sum_{i=1}^m t_i \mathbb{I}_{\mathbf{w}_i}(j) = \sum_{j=1}^n Z_j.$$

This defines Z_j, which has all finite moments and is stationary. As $\{A_i\}_{i \geq 1}$ is a Markov chain, $\rho_n \to 0$ as $n \to \infty$. Theorem 12.3 implies that $\mathbf{N}\mathbf{t}^T \stackrel{d}{\Rightarrow} \mathcal{N}(\boldsymbol{\mu}\mathbf{t}^T, \mathbf{t}\Sigma\mathbf{t}^T)$, and Theorem 12.4 gives our theorem. ∎

Theorem 12.5 is the basis of various hypothesis tests that can be made. For example,

Corollary 12.2 $n(\mathbf{N}/n - \boldsymbol{\mu})\Sigma^{-1}(\mathbf{N}/n - \boldsymbol{\mu})^T \stackrel{d}{\Rightarrow} \chi^2$ *with degrees of freedom* $k = \text{rank}(\Sigma)$.

12.1.1 Generalized Words

It is sometimes the case that more complex patterns are specified. For example, RYTR will be counted if any one of the eight words {A,G} {C,T} T {A,G} occurs. Let s and t represent such generalized words. The theory presented above holds if we compute $\text{cov}(N_\mathbf{s}(n), N_\mathbf{t}(n))$. This is easily done by summing over all words $\mathbf{u} \in \mathbf{s}$ and $\mathbf{v} \in \mathbf{t}$

$$\text{cov}(N_\mathbf{s}(n), N_\mathbf{t}(n)) = \sum_{\mathbf{u} \in \mathbf{s}} \sum_{\mathbf{v} \in \mathbf{t}} \text{cov}(N_\mathbf{u}(n), N_\mathbf{v}(n)).$$

12.1.2 Estimating Probabilities

It is a common practice to substitute parameter estimates for the corresponding parameters in distributional results. Sometimes this changes the distribution being studied. To illustrate this effect in our situation, we take $\mathbf{A}$ with iid letters where $p = \mathbb{P}(R) = 1 - \mathbb{P}(Y)$. Let $N_1(n)$ be the number of times $\mathbf{w}_1 = RR$ occurs in $A_1 \cdots A_n$. Then $\mu_1 = \mathbb{E}(N_1(n)) = (n-1)p^2$ and the asymptotic variance of $N_1(n)$ is given by Corollary 12.2. Define $\hat{q} = N_1(n)/n$. Theorem 12.5 then implies

$$\sqrt{n}(\hat{q} - p^2) \xrightarrow{d} \mathcal{N}(0, p^2 + 2p^3 - 3p^4).$$

If $p = \mathbb{P}(R)$ is unknown, it is natural to replace p^2 by $(\hat{p})^2$, where $\mathbf{w}_2 = R$ and

$$\hat{p} = \frac{N_2(n)}{n}.$$

Of course, $\hat{p}$ is the estimate of p. Next, we look at the asymptotic distribution of $\sqrt{n}(\hat{q} - \hat{p}^2)$. First of all, it is an exercise to use Corollary 12.2 to show that the joint distribution of $(\hat{q}, \hat{p})$ is asymptotically normal with covariance matrix

$$\frac{1}{n} \begin{pmatrix} p^2 + 2p^3 - 3p^4 & 2p^2 - 2p^3 \\ 2p^2 - 2p^3 & p - p^2 \end{pmatrix}.$$

The next theorem is used to obtain the distribution of $g(\hat{q}, \hat{p}) = \sqrt{n}(\hat{q} - \hat{p}^2)$.

Theorem 12.6 (Delta Method) *Let* $\mathbf{X}_n = (X_{n1}, X_{n2}, \ldots, X_{nk})$ *be a sequence of random vectors satisfying* $b_n(\mathbf{X}_n - \boldsymbol{\mu}) \xrightarrow{d} \mathcal{N}(\mathbf{0}, \Sigma)$ *with* $b_n \to \infty$. *The vector valued function* $\mathbf{g}(\mathbf{x}) = (g_1(\mathbf{x}), \ldots, g_l(\mathbf{x}))$ *has real valued* $g_i(\mathbf{x})$ *with non-zero differential*

$$\frac{\partial g_i}{\partial g_\mathbf{x}} = \left(\frac{\partial g_i}{\partial g_{x_1}}, \ldots, \frac{\partial g_i}{\partial g_{x_k}} \right).$$

Define $\mathbf{D} = (d_{i,j})$ *where* $d_{i,j} = \frac{\partial g_i}{\partial x_j}(\boldsymbol{\mu})$. *Then*

$$b_n(g(\mathbf{X}_n) - g(\boldsymbol{\mu})) \xrightarrow{d} \mathcal{N}(\mathbf{0}, \mathbf{D}\Sigma\mathbf{D}^T).$$

In other words, the delta method shows how to adjust the variance in a Central Limit Theorem. Applying the delta method to our problem shows that

$$\sqrt{n}(\hat{q} - \hat{p}^2) \overset{d}{\Rightarrow} \mathcal{N}(0, p^2 - 2p^3 + p^4).$$

Therefore, by using $\hat{p}$ instead of p, we change the asymptotic variance from $p^2 + 2p^3 - 3p^4$ to $p^2 - 2p^3 + p^4$. These subtle changes can make important differences when we are studying the statistics of patterns in sequences.

12.2 Nonoverlapping Pattern Counts

In the previous section, we studied counts of all occurrences of $\mathbf{w}$ in $A_1 A_2 \cdots A_n$. When the mean of $N_\mathbf{w}$ was simply $(n - k + 1)p_\mathbf{w}$ when $\mathbf{w} = w_1 \cdots w_k$, the variance depended on the overlap properties of $\mathbf{w}$. Now we modify our count statistics in a way, that we count only the *nonoverlapping* occurrences of patterns. To be explicit we move from A_1 along the pattern until $\mathbf{w}$ is first encountered at $A_i A_{i+1} \cdots A_{i+k-1} = \mathbf{w}$. Then we add +1 to the count $M_\mathbf{w}(n)$ and move to A_{i+k} to begin again. In this way, no overlapping patterns are counted in $M_\mathbf{w}(n)$. These counts are motivated by cuts of restriction enzymes. To be explicit, if $\mathbf{w} = RRR$,

$$\mathbf{A} = \underline{RRR}RY\,\underline{RRR}\,\underline{RRR}Y$$

has three occurrences of $\mathbf{w}$ (see the underlined RRR in $\mathbf{A}$) instead of six as in the last section.

This situation for one pattern belongs to renewal theory, and, below, we follow the treatment in Feller Vol. I, Chapter 8 (1971) in the case of iid sequences. We will give a Central Limit Theorem. For more general problems of $|\mathcal{W}| \geq 1$ and Markov sequences, a remarkable argument known as Li's method gives the mean times between occurrences of $\mathcal{W}$. Although not presented here, the general setting for these problems is Markov renewal processes.

12.2.1 Renewal Theory for One Pattern

In this section, we restrict attention to $\mathbf{w} = w_1 w_2 \cdots w_k$, where $A_1 A_2 \cdots$ is a sequence with iid letters. In contrast with the previous section, it will be convenient to index our counts at the right-hand end of an occurrence of $\mathbf{w}$. Let

$$u_i = \mathbb{P}(A_{i-k+1} \cdots A_i = \mathbf{w} \text{ and a renewal occurs at } i)$$

and

$$f_i = \mathbb{P}(\min\{l : A_{l-k+1} \cdots A_l = \mathbf{w}\} = i),$$

where we set $u_0 = 1$ and $f_0 = 0$. The associated generating functions are very useful:

$$U(s) = \sum_{i=0}^{\infty} u_i s^i$$

Probability and Statistics for Sequence Patterns

and
$$F(s) = \sum_{i=0}^{\infty} f_i s^i.$$

Whereas $U(1) = \infty$, $F(1) = \sum_{i=0}^{\infty} f_i = 1$. These generating functions are closely related. The probability of a renewal at i and that **w** first occurred at j has probability $f_j u_{i-j}$. Therefore,

$$u_i = f_1 u_{i-1} + f_2 u_{i-2} + \cdots + f_i u_0, \quad i \geq 1.$$

Thus, $\{u_i\}$ is essentially the convolution of $\{u_i\}$ and $\{f_i\}$. Multiply both sides by s^i and sum, we obtain $U(s) - 1$ on the left and $F(s)U(s)$ on the right.

Theorem 12.7 *The above generating functions satisfy*

$$U(s) = (1 - F(s))^{-1}.$$

In a more general setting, Feller proves the renewal theorem which is a nontrivial result.

Theorem 12.8 (Renewal) *Let $\mu = \sum_{i \geq 1} i f_i = F'(1)$. Then*

$$\lim_{n \to \infty} u_n = \mu^{-1}.$$

This theorem is the analog of the expected waiting time of $1/p$ for the first head when $\mathbb{P}(H) = p$.

Now we derive U and F for our pattern **w**. The ideas are quite straightforward. If **w** occurs ending at position i, it is a renewal or an earlier renewal overlaps this occurrence.

$$\mathbb{P}(\mathbf{w}) = \mathbb{P}(A_{i-k+1} \cdots A_i = \mathbf{w}) = \sum_{j=0}^{k-1} u_{i-j} \beta_{\mathbf{w},\mathbf{w}}(j) P_{\mathbf{w}}(j). \tag{12.5}$$

Multiply both sides by s^i for $i \geq |\mathbf{w}| = k$,

$$\mathbb{P}(\mathbf{w}) s^i = \sum_{j=0}^{k-1} u_{i-j} s^{i-j} s^j \beta_{\mathbf{w},\mathbf{w}}(j) P_{\mathbf{w}}(j).$$

Summing on $i \geq k$,

$$\frac{\mathbb{P}(\mathbf{w}) s^k}{1-s} = \sum_{i=k}^{\infty} \mathbb{P}(\mathbf{w}) s^i$$

$$= \sum_{j=0}^{k-1} \left(\sum_{i=k}^{\infty} u_{i-j} s^{i-j} \right) s^j \beta_{\mathbf{w},\mathbf{w}}(j) P_{\mathbf{w}}(j)$$

$$= (U(s) - 1) \sum_{j=0}^{k-1} s^j \beta_{\mathbf{w},\mathbf{w}}(j) P_{\mathbf{w}}(j).$$

Recall the definition of the overlap polynomial:

$$G_{\mathbf{w},\mathbf{w}}(s) = \sum_{j=0}^{k-1} s^j \beta_{\mathbf{w},\mathbf{w}}(j) P_{\mathbf{w}}(j). \tag{12.6}$$

Then

$$U(s) = \frac{\mathbb{P}(\mathbf{w})s^k + (1-s)G_{\mathbf{w},\mathbf{w}}(s)}{(1-s)G_{\mathbf{w},\mathbf{w}}(s)},$$

and by Theorem 12.7 we have proved the following theorem.

Theorem 12.9 *For sequences $A_1 A_2 \cdots$ of iid letters, the generating function for $\{f_i\}$ defined by $\mathbf{w} = w_1 \cdots w_k$ is*

$$F(s) = \frac{\mathbb{P}(\mathbf{w})s^k}{\mathbb{P}(\mathbf{w})s^k + (1-s)G_{\mathbf{w},\mathbf{w}}(s)}.$$

Corollary 12.3 *For sequences $A_1 A_2 \cdots$ of iid letters,*

$$\mu = F'(1) = G_{\mathbf{w},\mathbf{w}}(1)/\mathbb{P}(\mathbf{w}),$$

and

$$\sigma^2 = F^{(2)}(1) + F'(1) - (F'(1))^2$$
$$= \left(\frac{G_{\mathbf{w},\mathbf{w}}(1)}{\mathbb{P}(\mathbf{w})}\right)^2 + \frac{2G'_{\mathbf{w},\mathbf{w}}(1) - (2k-1)G_{\mathbf{w},\mathbf{w}}(1)}{\mathbb{P}(\mathbf{w})}.$$

The value of μ follows directly from Equation (12.5), but the generating function characterizes the distribution. To see the dependence on patterns, let $\mathbf{u} = RR \cdots RR$ and $\mathbf{v} = RR \cdots RY$ be words of length k. Then, if $p = \mathbb{P}(R) = 1 - q$,

$$\mu_{\mathbf{u}} = \frac{1 - p^k}{qp^k},$$

$$\sigma_{\mathbf{u}}^2 = \frac{1}{(qp^k)^2} - \frac{(2k-1)q + 2}{q^2 p^k} - \frac{p}{q^2},$$

whereas

$$\mu_{\mathbf{v}} = \frac{1}{qp^{k-1}}$$

and

$$\sigma_{\mathbf{v}}^2 = \frac{1}{(qp^{k-1})^2} - \frac{(2k-1)}{qp^{k-1}}.$$

Let $M_\mathbf{w}(n)$ be the number of renewals of $\mathbf{w}$ in $\mathbf{A} = A_1 A_2 \cdots A_n$. Let $T_\mathbf{w}^{(r)}$ be the number of letters up to and including the r-th renewal. Then

$$\mathbb{P}(M_\mathbf{w}(n) \geq r) = \mathbb{P}(T_\mathbf{w}^{(r)} \leq n). \tag{12.7}$$

If T_i is the number of trials associated with the i-th renewal,

$$T_\mathbf{w}^{(r)} = T_1 + T_2 + \cdots + T_r,$$

and T_i are iid $\mathbb{E}(T_i) = \mu$, $\text{Var}(T_i) = \sigma^2$. This immediately implies a Central Limit Theorem for $T_\mathbf{w}^{(r)}$ and, what is less obvious, a related result for $M_\mathbf{w}(n)$.

Theorem 12.10 (Central Limit Theorem) *For sequences $A_1 A_2 \cdots$ of iid letters and with the mean μ and variance σ^2 as in Corollary 12.3, we have*

$$(a) \quad \lim_{n \to \infty} \mathbb{P}\left(\frac{T_\mathbf{w}^{(r)} - r\mu}{\sigma \sqrt{r}} < x\right) = \frac{1}{\sqrt{2\pi}} \int_{-\infty}^{x} e^{-t^2/2} dt$$

and

$$(b) \quad \lim_{n \to \infty} \mathbb{P}\left(\frac{M_\mathbf{w}(n) - n/\mu}{\sqrt{\frac{\sigma^2}{\mu^3} n}} < x\right) = \frac{1}{\sqrt{2\pi}} \int_{-\infty}^{x} e^{-t^2/2} dt.$$

Proof. Clearly, for fixed x and $r \to \infty$

$$\mathbb{P}\left(\frac{T_\mathbf{w}^{(r)} - r\mu}{\sigma \sqrt{r}} < x\right) \to \frac{1}{\sqrt{2\pi}} \int_{-\infty}^{x} e^{-t^2/2} dt = \Phi(x).$$

Now let $n \to \infty$ as $r \to \infty$ so that

$$\frac{n - r\mu}{\sigma \sqrt{r}} \to x. \tag{12.8}$$

Using Equations (12.7) and (12.8), we have

$$\mathbb{P}(M_\mathbf{w}(n) \geq r) \to \Phi(x).$$

Now

$$\frac{M_\mathbf{w}(n) - n/\mu}{\sqrt{\frac{\sigma^2}{\mu^3} n}} \geq \frac{r - n/\mu}{\sqrt{\frac{\sigma^2}{\mu^3} n}} = \frac{r\mu - n}{\sqrt{\sigma^2 r}} \cdot \sqrt{\frac{r\mu}{n}}.$$

If we divide Equation (12.8) by $\sqrt{r}$, we see that $n/r \to \mu$ so that the right-hand side of the last equation tends to $-x$ or

$$\mathbb{P}\left(\frac{M_\mathbf{w}(n) - n/\mu}{\sqrt{\frac{\sigma^2}{\mu^3} n}} \geq -x\right) \to \Phi(x)$$

or

$$\mathbb{P}\left(\frac{M_{\mathbf{w}}(n)-n/\mu}{\sqrt{\frac{\sigma^2}{\mu^3}n}} < -x\right) \to 1 - \Phi(x) = \Phi(-x).$$

■

To return to our examples, we see from Theorem 12.10 that as $n \to \infty$,

$$\tfrac{1}{n}\mathbb{E}M_{\mathbf{u}}(n) \to \tfrac{1}{\mu_{\mathbf{u}}} = \tfrac{qp^k}{1-p^k};$$

$$\tfrac{1}{n}\operatorname{Var}M_{\mathbf{u}}(n) \to \tfrac{\sigma^2}{\mu^3} = (1-p^k)^{-3}\left(qp^k - q^2(2k+1)p^{2k} - pqp^{3k}\right);$$

$$\tfrac{1}{n}\mathbb{E}M_{\mathbf{v}}(n) \to \tfrac{1}{\mu_{\mathbf{v}}} = \tfrac{q}{p}p^k;$$

$$\tfrac{1}{n}\operatorname{Var}M_{\mathbf{v}}(n) \to \tfrac{\sigma^2}{\mu^3} = \left(\tfrac{q}{p}\right)^3\left(p^k - (2k-1)p^{2k}\right).$$

Contrast these results with the corresponding ones for $N_{\mathbf{u}}$ and $N_{\mathbf{v}}$.

12.2.2 Li's Method and Multiple Patterns

The idea of words corresponding to restriction sites motivates us to consider sets $\mathcal{W}$ of words when $|\mathcal{W}| \geq 1$. The generating function approach of Section 12.2.1 in the case of iid sequences can be generalized. See Problems 12.10 and 12.11. Markov renewal processes can be used to derive related results for Markov sequences. In this section, we apply a remarkable technique known as Li's method to analyze the mean time between renewals for the case of Markov and independent sequences. First, we rederive the formula for $\mu_{\mathbf{w}}$ in Corollary 12.3.

Li's Method for w

The method of Li is based on the concept of a fair game. Let $\mathbf{A} = A_1 A_2 \cdots$ be a sequence of iid letters. Suppose we want to find $\mu_{\mathbf{w}}$, the mean time until first occurrence, where w =CTC, $\mathbb{P}(C) = 1/3$, and $\mathbb{P}(T) = 1/4$. As each letter appears, a gambler joins the game betting \$1 on the occurrence of CTC in that and the following two letters. If C does not appear the gambler is broke and drops out. If C does appear, he receives \$$(x)$. To make the game fair requires $x\mathbb{P}(C) + 0(1 - \mathbb{P}(C)) = 1$ or $x = 1/\mathbb{P}(C) = 3$. He bets his stake, \$3, on the next letter where he loses or, by the same logic, if T appears, he receives \$3× 4=\$12. Again, the gambler bets his entire fortune on the next letter. If on the third letter a C appears to complete CTC he receives \$12× 3=\$36. On the average, CTC first appears at trial $\mu_{\mathbf{w}}$. Because a new gambler bets \$1 at each trial, the $\mu_{\mathbf{w}}$ gamblers have paid $-\$\mu_{\mathbf{w}}$ to play. One gambler wins \$36 for CTC and the last

Probability and Statistics for Sequence Patterns

gambler wins $3 for the last C. This means that for $\mu_\mathbf{w} = \mu_{CTC} = \mu$,

$$-\mu + 36 + 3 = 0,$$

or $\mu = 39$. To make this analysis more transparent,

$$\mu_{CTC} = \frac{1}{\mathbb{P}(CTC)} + \frac{1}{\mathbb{P}(C)}$$
$$= \frac{G_{CTC,CTC}(1)}{\mathbb{P}(CTC)}.$$

In fact, this simple argument implies

$$\mu_\mathbf{w} = \frac{G_{\mathbf{w},\mathbf{w}}(1)}{\mathbb{P}(\mathbf{w})},$$

as given in Corollary 12.3. This remarkable technique gives $\mu_\mathbf{w}$ without performing any summations or taking any derivatives.

Next, we generalize the method to stationary Markov sequences with stationary distribution π. Recall from Equation (12.1) that $P_\mathbf{w}(l)$ is the probability of seeing $w_{k-l+1} \cdots w_k$, given that $w_1 \cdots w_{k-l}$ just occurred with $P_\mathbf{w}(l) = 1$ if $l \le 0$. If we revisit our example of $\mathbf{w} =$ CTC, then the only change is to realize that the first C occurs with probability π_C and that successive letters are collected according to the transition probabilities. By the logic given above,

$$\mu_{CTC} = \frac{1}{\pi_C p_{CT} p_{TC}} + \frac{1}{\pi_C}$$
$$= \frac{G_{CTC,CTC}(1)}{\mathbb{P}(CTC)}.$$

where $\mathbb{P}(CTC) = \pi_C p_{CT} p_{TC}$ and $G_{\mathbf{w},\mathbf{w}}(1)$ is defined exactly as in Equation (12.4). This argument implies the next theorem.

Theorem 12.11 *Let* $\{A_i\}_{i \ge 1}$ *be a stationary irreducible, aperiodic Markov chain, with stationary measure* π. *Then the mean time* $\mu_\mathbf{w}$ *between renewals, of* $\mathbf{w}$ *is given by*

$$\mu_\mathbf{w} = \frac{G_{\mathbf{w},\mathbf{w}}(1)}{\mathbb{P}(\mathbf{w})},$$

where $\mathbb{P}(\mathbf{w}) = \mathbb{P}(w_1 \cdots w_k = A_1 A_2 \cdots A_k) = \pi_{w_1} \prod_{i=1}^{k-1} p_{w_i, w_{i+1}}$.

Multiple Patterns

The above arguments generalize to multiple patterns $\mathcal{W} = \{\mathbf{w}_1, \ldots, \mathbf{w}_m\}$, where no $\mathbf{w}_i$ is a subword of $\mathbf{w}_j, j \ne i$. Assume $\{A_i\}_{i \ge 1}$ is a stationary Markov chain. Then, the sequence of renewal types is itself a Markov chain and, hence, there is a stationary vector $\boldsymbol{\tau} = (\tau_1, \ldots, \tau_m)$, where τ_i is the probability that a renewal is by word $\mathbf{w}_i$. (Problems 12.10–12.12 treat the generating function approach for multiple patterns and sequences of independent letters.)

At each letter, we imagine m gamblers entering the game, the i-th betting on the word $\mathbf{w}_i$. Therefore, when $\mathbf{w}_i$ occurs to create a renewal, we must pay off the gamblers betting on $\mathbf{w}_j$ for $j = 1, 2, \ldots, m$. Define the payoff μ_{ij} to be that required to make a fair game for those gamblers betting on j when word $\mathbf{w}_i$ occurs. Define $G_{\mathbf{w}_i, \mathbf{w}_j}(s) = G_{i,j}(s)$. As before,

$$\mu_{i,i} = \frac{G_{i,i}(1)}{\mathbb{P}(\mathbf{w}_i)}.$$

The same argument yields

$$\mu_{i,j} = \frac{G_{i,j}(1)}{\mathbb{P}(\mathbf{w}_j)}.$$

Let us consider the example $\mathcal{W} = \{\text{CTC}, \text{TCT}\}$ with iid letters. Then

$$(G_{i,j}(1)) = \begin{pmatrix} 1 + p_T p_C & p_T \\ p_C & 1 + p_C p_T \end{pmatrix}$$

and define M by

$$M^T = (\mu_{i,j}) = \begin{pmatrix} \frac{1}{p_C^2 p_T} + \frac{1}{p_C} & \frac{1}{p_C p_T} \\ \frac{1}{p_C p_T} & \frac{1}{p_T^2 p_C} + \frac{1}{p_T} \end{pmatrix}.$$

Returning to Li's method, let μ be the expected distance between renewals. For the betters on j to be in a fair game, we have

$$\mu = \tau_1 \mu_{1,j} + \tau_2 \mu_{2,j} + \cdots + \tau_m \mu_{m,j}.$$

This hold for all j so that

$$\begin{pmatrix} \mu_{11} & \mu_{21} & \cdots & \mu_{m1} \\ \vdots & \vdots & & \vdots \\ \mu_{1m} & \mu_{2m} & \cdots & \mu_{mm} \end{pmatrix} \begin{pmatrix} \tau_1 \\ \tau_2 \\ \vdots \\ \tau_m \end{pmatrix} = \begin{pmatrix} \mu \\ \mu \\ \vdots \\ \mu \end{pmatrix}$$

or $M\boldsymbol{\tau}^T = \boldsymbol{\mu}^T$. Therefore,

$$\boldsymbol{\tau}^T = \mu M^{-1} \mathbf{1}^T.$$

Now $\boldsymbol{\tau} \mathbf{1}^T = 1$ and $\mathbf{1}\boldsymbol{\tau}^T = 1$, so that $1 = \mu \mathbf{1} M^{-1} \mathbf{1}^T$ and

$$\mu = \frac{1}{\mathbf{1} M^{-1} \mathbf{1}^T} \tag{12.9}$$

and

$$\boldsymbol{\tau}^T = \frac{M^{-1} \mathbf{1}^T}{\mathbf{1} M^{-1} \mathbf{1}^T}. \tag{12.10}$$

Probability and Statistics for Sequence Patterns 321

12.3 Poisson Approximation

So far, we have applied methods of Markov chains, renewal theory, and Li to pattern counts. When the word **w** has a small probability of occurrence, the method of Poisson approximation also applies. Recalling that even six-letter patterns have probability 4^{-6}, we see that Poisson approximation will often apply. In this section $\{A_i\}_{i \geq 1}$ has iid letters.

The method is easy to sketch. The number of overlapping clumps of words **w** is approximately Poisson. The self-overlaps of **w** determine a period which has probability p. The clumps are preceded by a "declumping" event with probability $1 - p$. If we count overlapping occurrences of **w** as in Section 12.1, the number of occurrences per clump is a geometric with mean $1/p$. If we count nonoverlapping occurrences, then the number of occurrences per clump is a geometric with mean $1/\mathbb{P}(\mathbf{w})$. In either case, the number of words in $A_1 \cdots A_n$, $N(n)$, or $M(n)$ has an approximate compound Poisson distribution. The Chen-Stein method from Chapter 11 allows us to compute bounds on the approximation.

We review the material of Section 11.4 by taking the example of the k-letter word $\mathbf{w} = RR \cdots R$. Even if we intend to count the overlapping occurrences $N_\mathbf{w}(n)$ of **w**, the first step in Poisson approximation is to count the **clumps** of occurrences. For occurrences $\mathbf{w} = RR \cdots R$ in $A_1 A_2 \cdots A_n$, this is easy. Define $D_i = \mathbb{I}(A_i = R)$ and

$$X_1 = \prod_{i=1}^{k} D_i,$$

$$X_i = (1 - D_i) \prod_{j=0}^{k-1} D_{i+j}, \; i \geq 2,$$

$$W = \sum_{i=1}^{n-k+1} X_i.$$

For the Chen-Stein method, we apply Theorem 11.22, with index set $I = \{1, 2, \ldots, n - k + 1\}$ and the dependence sets $J_i = \{j \in I : |i - j| \leq k\}$. The Poisson mean is

$$\lambda = \mathbb{E}(W) = p^k + (n - k)(1 - p)p^k,$$

where $\mathbb{P}(A_i = R) = p \in (0, 1)$. The random variables X_i and X_j cannot both be 1 for $i \neq j$ and $j \in J_i$. Therefore,

$$b_2 = \sum_{i \in I} \sum_{i \neq j \in J_i} \mathbb{E}(X_i X_j) = 0.$$

The constant b_1 was bounded by

$$b_1 < (2k + 1)\left((1 - p)p^k\right)^2 \left\{n - k + (1 - p)^{-1} + p\{(1 - p)^2(2k + 1)\}^{-1}\right\}.$$

We now refer to Chapter 11.5.3 on approximate matching. Let Y_i denote the occurrence of $\mathbf{w} = RR \cdots R$ in $A_1 A_2 \cdots A_n$ at position i. Then,

$$N_\mathbf{w}(n) = \sum_{i=1}^{n-k+1} Y_i.$$

The number of clumps is approximately $\mathcal{P}(\lambda)$, with $\lambda = \mathbb{E}(W)$. If the number of $\mathbf{w}$'s in each clump is C, then if C_i is iid C,

$$C^{*j} = C_1 + C_2 + \cdots + C_j, \quad j \geq 1,$$

is the random variable giving the number of $\mathbf{w}$'s in j clumps. As there is approximately a Poisson number Z of clumps,

$$\hat{N}_\mathbf{w}(n) = \sum_{i=1}^{Z} C_i$$

should approximate $N_\mathbf{w}(n)$.

Now C is geometric,

$$\mathbb{P}(C = l) = (1-p)p^{l-1}, \quad l \geq 1,$$

and, in fact,

$$\mathbb{P}(C^{*j} = l) = \binom{l-1}{j-1} (1-p)^j p^{l-j}, \quad l \geq j.$$

Using

$$\mathbb{P}(\hat{N}_\mathbf{w}(n) = l) = \begin{cases} \mathbb{P}(Z = l), & l = 0 \\ \displaystyle\sum_{j=1}^{l} \mathbb{P}(C^{*j} = l)\mathbb{P}(Z = j), & l \geq 1, \end{cases}$$

we obtain

$$\mathbb{P}(\hat{N}_\mathbf{w}(n) = l) = \begin{cases} e^{-\lambda}, & l = 0 \\ e^{-\lambda} \displaystyle\sum_{j=1}^{l} \binom{l-1}{j-1} \frac{(\lambda(1-p))^j}{j!} p^{l-j}, & \text{otherwise.} \end{cases}$$

Theorem 11.25 tells us that

$$\frac{1}{2}\|N_\mathbf{w}(n) - \hat{N}_\mathbf{w}(n)\| \leq 2b_1 + 2\lambda \mathbb{P}(C > k).$$

Now we study $M_\mathbf{w}(n)$ for $\mathbf{w} = RR \cdots R$. The number of clumps is identical to that for $N_\mathbf{w}(n)$. The geometric, however, has parameter p^k (instead of p)

because the occurrences of **w** must not overlap. Now $\mathbb{P}(C > k) = 1 - (1-p^k) = p^k$, so $M_\mathbf{w}(n) \approx Z$ when p^k is small, but the previous argument can be repeated to yield

$$\mathbb{P}(\hat{M}_\mathbf{w}(n) = l) = \begin{cases} e^{-\lambda}, & l = 0 \\ e^{-\lambda} \sum_{j=1}^{l} \binom{l-1}{j-1} \frac{(\lambda(1-p^k))^j}{j!} p^{k(l-j)}, & \text{otherwise.} \end{cases}$$

Finally, we turn to a general word **w**. The period of **w** is the minimum shift $s > 0$ such that $\beta_{\mathbf{w},\mathbf{w}}(s) = 1$. For $\mathbf{w} = RR \cdots R$, $s = 1$ and for $\mathbf{w} = HTT \cdots TH$, $s = k - 1$. The cases where **w** does not self-overlap, such as $\mathbf{w} = HH \cdots HT$, require a slightly different analysis given later. Let $p = \mathbb{P}(w_1 w_2 \cdots w_s = A_1 \cdots A_s)$. Then the declumping event is that $w_1 \cdots w_s$ does not precede $w_1 \cdots w_k$, so the mean of the Poisson is

$$\lambda = \mathbb{P}(\mathbf{w}) + (n-k)(1-p)\mathbb{P}(\mathbf{w})$$
$$= \mathbb{P}(\mathbf{w}) + (n-k)(1 - \mathbb{P}(w_1 \cdots w_s))\mathbb{P}(\mathbf{w}).$$

The mean of the geometric C for $N(n)$ is $1/p = 1/\mathbb{P}(w_1 \cdots w_s)$ whereas for $M(n)$ the mean remains $1/\mathbb{P}(\mathbf{w})$. When **w** does not self-overlap, there is no declumping necessary, so

$$\lambda = (n - k + 1)\mathbb{P}(\mathbf{w})$$

and the mean of the geometric C is $1/\mathbb{P}(\mathbf{w})$ for both $N(n)$ and $M(n)$.

12.4 Site Distributions

In biology, we often have the location of genes, restriction sites, or other features on an interval or sequence of DNA. The scale of these features relative to the DNA sequence varies greatly, but in this section, we will assume that it is appropriate to assume these locations are points on an interval. When this is not the case, as for genes that occupy a substantial fraction of the DNA under consideration, the continuous methods we present are inappropriate.

We assume then that there are n sites $s_1, s_2, \ldots, s_n$ all located in $[0, G]$. To make the discussion simpler, we will also consider the scaled locations $u_i = s_i/G \in (0, 1)$.

Tests for Poisson

One of the most often asked questions is whether the points $\{u_i\}$ are uniformly distributed in $(0, 1)$. If the points are laid down by a Poisson process, then it is easy to show that conditional on n points in $(0, 1)$, the locations $U_1, U_2, \ldots, U_n$

are iid uniform. Several tests are available. For large n, let

$$S_n^{(1)} = \sum_{i=1}^{n} U_i.$$

The Central Limit Theorem says

$$S_n \approx \mathcal{N}\left(\frac{n}{2}, \frac{n}{12}\right) \tag{12.11}$$

because $\mathbb{E}(U) = 1/2$ and $\text{Var}(U) = 1/12$.

Another test that is often used is to divide the interval into k intervals of equal length. It is recommended that $\frac{n}{k} \geq 5$. Let $N_i = \#$ of u's in interval i. We then perform a χ^2 test where

$$S_n^{(2)} = \sum_{i=1}^{k} \frac{(N_i - n/k)^2}{n/k}. \tag{12.12}$$

Under the null hypothesis, the statistic $S_n^{(2)}$ has asymptotically a χ^2 distribution with $k-1$ degrees of freedom. One difficulty with this test is the arbitrary location of the intervals.

Another test can be based on the Kolmogorov-Smirnov test which measures the largest deviation between the empirical distribution function and the distribution function $F(x) = x$, $x \in (0, 1)$, of the uniform. The statistic is

$$S_n^{(3)} = \max\left\{\max_{1 \leq i \leq n}\left\{\frac{i}{n} - u_i\right\}, \max_{1 \leq i \leq n}\left\{u_i - \frac{i-1}{n}\right\}\right\}. \tag{12.13}$$

The distribution of $\sqrt{n} S_n^{(3)}$ is tabulated in many places as the Kolmogorov-Smirnoff statistic.

12.4.1 Intersite Distances

We retain the hypothesis that the points are laid down by a Poisson process. The intersite distances are $x_i = s_i - s_{i-1}$, where $s_0 = 0$. Under the Poisson hypothesis, X_i are iid exponential random variables with mean μ. The density function is

$$f(x) = \frac{1}{\mu} e^{-x/\mu}, \quad x > 0.$$

To estimate μ from the values of X_i, use

$$\hat{\mu} = \frac{1}{n} \sum_{i=1}^{n} X_i.$$

Probability and Statistics for Sequence Patterns

$\hat{\mu}$ is the maximum likelihood estimate of μ. The values $X_1, X_2, \ldots, X_n$ are often examined to see if they are "too small" or "too large". To make rigorous such observations, let $X_{(1)} = \min_{1 \le i \le n} X_i$ and $X_{(n)} = \max_{1 \le i \le n} X_i$. It is routine to show $X_{(1)}$ is exponential with mean μ/n, so probability statements can be easily derived for $X_{(1)}$. In Chapter 11 we discussed $X_{(n)}$ which has an extreme value distribution:

$$\mathbb{P}(X_{(n)} \le \log(n) + t) = e^{e^{-t}}, \quad -\infty < t < \infty,$$

and

$$\mathbb{E}(X_{(n)}) = \mu(\gamma + \log(n)),$$

where $\gamma = 0.5772\ldots$ is Euler's constant.

Problems

Problem 12.1 For $\mathbf{u} = HHH$ and $\mathbf{v} \in \{HT, HH\}$, compute $\beta_{\mathbf{u},\mathbf{v}}(i)$, $i \ge 1$.

Problem 12.2 Establish Proposition 12.1.

Problem 12.3 Generalize Theorem 12.2 to the case of a Markov chain with memory $\nu \ge 1$.

Problem 12.4 In Section 12.2.1, show $F(1) = 1$.

Problem 12.5 In the proof of Theorem 12.2, show $\sum_{j=1}^{\infty} |\text{cov}(\mathbb{I}_\mathbf{u}(i)\mathbb{I}_\mathbf{v}(j))| \le k + l$.

Problem 12.6 Check the variances of $N_\mathbf{u}(n)$ and $N_\mathbf{v}(n)$ following Corollary 12.1 and derive the asymptotic value of $\frac{1}{n}\text{cov}(N_\mathbf{u}(n), N_\mathbf{v}(n))$.

Problem 12.7 Derive the asymptotic covariance matrix for $(\hat{q}, \hat{p})$ in Section 12.1.2. Check the result of applying the delta method.

Problem 12.8 Verify Corollary 12.3 and the values $\mu_\mathbf{u}, \sigma_\mathbf{u}^2, \mu_\mathbf{v}$, and $\sigma_\mathbf{v}^2$ that follow the corollary.

Problem 12.9 Show $M_\mathbf{w}(n) \sim n/\mu$.

Problem 12.10 A renewal is defined by the occurrence of **any** member of $\mathcal{W}$ $\{\mathbf{w}_1, \ldots, \mathbf{w}_m\}$ where no word is a subword of any other word. Let $G_{i,j}(s)$ $G_{\mathbf{w}_i, \mathbf{w}_j}(s)$ and $U_i(s) = U_{\mathbf{w}_i}(s)$. Show that for sequences of iid letters

$$\begin{pmatrix} G_{1,1}(s) & \cdots & G_{m,1}(s) \\ \vdots & \cdots & \vdots \\ G_{1,m}(s) & \cdots & G_{m,m}(s) \end{pmatrix} \begin{pmatrix} U_1(s) - 1 \\ \vdots \\ U_m(s) - 1 \end{pmatrix} = \frac{1}{1-s} \begin{pmatrix} s^{|\mathbf{w}_1|}\mathbb{P}(\mathbf{w}_1) \\ \vdots \\ s^{|\mathbf{w}_m|}\mathbb{P}(\mathbf{w}_m) \end{pmatrix}$$

Problem 12.11 In Problem 12.10, denoting the first matrix by $G(s)$ and assuming $G^{-1}(1)$ exists, show also that

$$\begin{pmatrix} 1/\mu_1 \\ \vdots \\ 1/\mu_m \end{pmatrix} = G^{-1}(1) \begin{pmatrix} \mathbb{P}(\mathbf{w}_1) \\ \vdots \\ \mathbb{P}(\mathbf{w}_m) \end{pmatrix},$$

where μ_i is the mean distance between renewals of $\mathbf{w}_i$.

Problem 12.12 In Problems 12.10 and 12.11 show $\mu = \left(1/\mu_1 + 1/\mu_2 + \cdots + 1/\mu_m\right)^{-1}$.

Problem 12.13 Let $X_1, X_2, \ldots, X_n$ be iid exponential mean λ. Show $X_{(1)} = \min_{1 \leq i \leq n} X_i$ is exponential with mean λ/n.

Problem 12.14 Consider independent tosses of a coin with $\mathbb{P}(H) = p = 1 - \mathbb{P}(T)$. Let $X = \#$ trials until a head. Use Li's method to show $\mathbb{E}(X) = 1/p$.

Chapter 13

RNA Secondary Structure

In Chapter 1, we emphasized the close relationship between structure and function in biology. The shape of macromolecules and of complexes of macromolecules determine the interactions that are allowed and, hence, the processes of life. In this section, we will study the possible shapes of singlestranded RNA molecules. Recall the shape of tRNA presented in the Introduction. Some of the problems are easy to describe: Given the linear sequence of an RNA, predict its structure in two and three dimensions. The three-dimensional structure of DNA is also important, but we will not study that in this chapter.

The singlestranded RNA is viewed as a linear sequence $\mathbf{a} = a_1 a_2 \ldots a_n$ of ribonucleotides. The sequence $\mathbf{a}$ is called the *primary structure*. Each a_i is identified with one of four bases or nucleotides: A (adenine), C (cytosine), G (guanine), and U (uracil). These bases can form basepairs, conventionally A pairs with U and C pairs with G. These are called Watson-Crick pairs. In addition, the pairing of G and U is frequently allowed. *Secondary structure* is a planar graph which satisfies the statement: If a_i pairs with a_j and a_k is paired with a_l with $i < k < j$, then $i < l < j$ also. The sugar phosphate backbone is represented as a solid line with the bases as dots. Hydrogen bonds or the basepairs are represented as lines between non-neighboring dots; basepair lines will not connect the bases in the usual biological representation of secondary structure. See Figure 13.1(a) for an example of a so-called cloverleaf structure for $\mathbf{a}$ = CAGCGUCACACCCGCGGGGUAAACGCU. In Figure 13.1(b), the same sequence and the same secondary structure is shown. The primary structure is written along the horizontal axis and the basepairs are shown as arcs.

13.1 Combinatorics

In this section, we will ignore the specific base sequence of a RNA molecule and allow basepairs between any a_i and a_j. In biological examples, the base

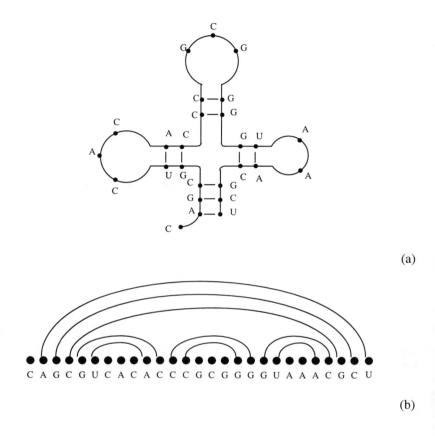

Figure 13.1: *Two representations of secondary structure*

sequence is essential in determining the secondary structure, and we return to those questions in Section 13.2. Here we are interested in counting possible topologies. Two secondary structures are distinct if the two sets of basepairs are not equal. There are 17 possible secondary structures for $a_1 a_2 a_3 a_4 a_5 a_6$. See Figure 13.2. Because we ignore the base sequence, we will write the sequence as $1-2-3-4-5-6$ or $[1,6]$.

Theorem 13.1 *Let $S(n)$ be the number of secondary structures for the sequence $[1, n]$. Then $S(0) = 0, S(1) = 1$, and for $n \geq 2$,*

$$S(n+1) = S(n) + S(n-1) + \sum_{k=1}^{n-2} S(k) S(n-k-1).$$

RNA Secondary Structure 329

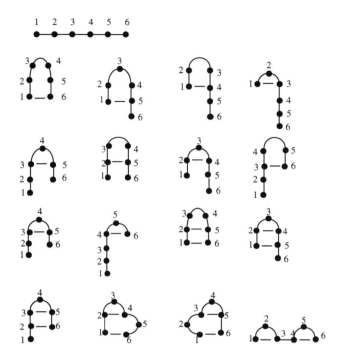

Figure 13.2: *The 17 secondary structures for* $[1,6]$

Proof. For $n = 1$ the only structure is ${}^1_\bullet$, while for $n = 2$ the only structure is ${}^1_\bullet \quad {}^2_\bullet$. Therefore $S(1) = S(2) = 1$.

Assume $S(k)$ is known for $1 \leq k \leq n$. Consider the sequence $[1, n+1]$. Either $n+1$ is not basepaired or $n+1$ is paired with j for $1 \leq j \leq n-1$. When $n+1$ is not basepaired there are $S(n)$ structures, since $[1, n]$ can form secondary structures.

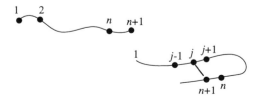

When $n+1$ is paired with j, $[1, j-1]$ and $[j+1, n]$ can each form secondary structures, $S(j-1)$ and $S(n-j)$, respectively. Therefore,

$$S(n+1) = S(n) + S(n-1) + S(1)S(n-2) + \cdots + S(n-2)S(1)$$

$$= S(n) + S(n-1) + \sum_{j=2}^{n-1} S(j-1)S(n-j)$$

∎

Lemma 13.1 *If $\varphi(x) = \sum_{k \geq 0} S(k)x^2$, then $\varphi(x)$ satisfies*

$$\varphi^2(x)x^2 - \varphi(x)(1 - x - x^2) + x = 0$$

Proof. Set $\varphi(x) = y$. Then

$$y^2 = \sum_{n=1}^{\infty} \left(\sum_{k=1}^{n-1} S(n-k)S(k) \right) x^n.$$

The recursion above states that $S(n+2) - S(n+1) - S(n) = \sum_{k=1}^{n-1} S(n-k)S(k)$. Therefore,

$$y^2 = \sum_{n \geq 1} \left(S(n+2) - S(n+1) - S(n) \right) x^n$$

$$= \sum_{n \geq 1} S(n+2)x^n - \sum_{n \geq 1} S(n+1)x^n - \sum_{n \geq 1} x^n$$

$$= \frac{(y - x^2 - x)}{x^2} - \frac{y - x}{x} - y$$

and the result follows. ∎

Theorem 13.2 *As $n \to \infty$,*

$$S(n) \sim \sqrt{\frac{15 + 7\sqrt{5}}{8\pi}} n^{-3/2} \left(\frac{3 + \sqrt{5}}{2} \right)^n.$$

Proof. Above, letting $y = \varphi(x)$, we derived

$$F(x, y) = x^2 y^2 - (1 - x - x^2)y + x = 0$$

There is a theorem that asserts when $F(x, y) = 0$, $r > 0$, $s > S(0)$ is the unique solution of

$$F(r, s) = 0, \quad F_y(r, s,) = 0,$$

RNA Secondary Structure

the n-th coefficient of $\varphi(x)$, $S(n)$, satisfies

$$S(n) \sim \sqrt{\frac{r^2 F_x(r,s)}{2\pi F_{yy}(r,s)}} n^{-1/2} r^{-n}.$$

The system becomes

$$r^2 s^2 - (1 - r - r^2)s + r = 0,$$

$$2sr^2 - (1 - r - r^2) = 0$$

or

$$s^2 = \frac{1}{r},$$

$$2r^{3/2} + r + r^2 = 1,$$

and $s^4 - s^2 - 2s - 1 = 0 = (s^2 - s - 1)(s^2 + s + 1)$. Then $s = \frac{1+\sqrt{5}}{2}$ and $\frac{1}{r} = \frac{3+\sqrt{5}}{2}$. Substitution into the asymptotic formula gives the result. ∎

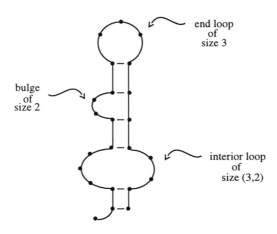

Figure 13.3: *Components of a hairpin*

The structural components of secondary structure are basepairs, bulges, interior loops, end loops, and multibranch loops. A hairpin is a structure with basepairs (at least one), bulges, interior loops, and exactly one end loop. See Figure 13.3. Multibranch loops are junctions, such as in the cloverleaf, where more than one hairpin or more complex secondary structures are appended. Figure 13.1(a) has one multibranch loop.

Theorem 13.3 *There are $2^{n-2} - 1$ hairpins for $[1, n]$.*

Proof. Let $L(n)$ be the number of secondary structures with at most one end loop. $L(1) = L(2) = 1$. Consider $[1, n+1]$. If $n+1$ is not in a basepair, there are $L(n)$ structures. Otherwise, $n+1$ basepairs j, for some j, $1 \leq j \leq n-1$. $[1, j-1]$ cannot have a basepair, as there is at most one end loop, so there are $L(n-(j+1)+1) = L(n-j)$ secondary structures with one loop:

$$L(n+1) = L(n) + L(n-1) + \cdots + L(1)$$

$$= 2L(n), \ n \geq 2.$$

Therefore, $L(n) = 2^{n-2}, n \geq 2$, and the result follows because all structures counted by $L(n)$ are hairpins except the structure with no basepairs. ∎

13.1.1 Counting More Shapes

It is of interest to further classify secondary structures. Let $\mathcal{S}_{n,k}$ be the set of secondary structures on $[1, n]$ with exactly k basepairs, and set $S(n, k) = |\mathcal{S}_{n,k}|$. The proof of the next theorem is left as an exercise.

Theorem 13.4 *Set $S(n, 0) = 1$ for all n and $S(n, k) = 0$ for $k \geq n/2$. Then for $n \geq 2$,*

$$S(n+1, k+1) = S(n, k+1) + \sum_{j=1}^{n-1} \left[\sum_{i=0}^{k} S(j-1, i) S(n-j, k-i) \right].$$

Corollary 13.1 $S(n) = \sum_{k=0}^{\lfloor n/2 \rfloor} S(n, k).$

Remarkably, $S(n, k)$ are much easier to compute than $S(n)$, and we will find an explicit, simple formula for $S(n, k)$. The combinatorics are accomplished by constructing a bijection of $\mathcal{S}_{n,k}$ onto a certain set of trees.

A *linear tree* is a rooted tree along with a linear ordering on the set of children of each vertex of the tree. We define $\mathcal{T}_{n,k}$ to be the set of unlabeled linear trees having n vertices, k of which are not terminal, that is, are of degree 1. Set $T(n, k) = |\mathcal{T}_{n,k}|$. In Figure 13.4, the set $\mathcal{T}_{5,3}$ is shown with roots at the top of the trees.

The next proposition gives a bijection between secondary structures and linear trees.

Proposition 13.1 *For all $n, k \geq 1$, there exists a bijection $\varphi : \mathcal{S}_{n+k-2, k-1} \to \mathcal{T}_{n,k}$.*

Proof. We illustrate the bijection with an example in Figure 13.5. For a member of $\mathcal{S}_{n+k-2, k-1}$ written in the loop form, put a node of $\mathcal{T}_{n,k}$ above the figure, outside all loops. Then insert nodes inside all loops visible from this node, and connect to the

RNA Secondary Structure 333

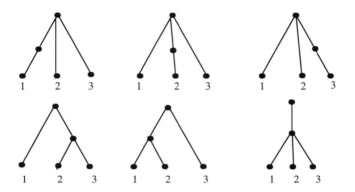

Figure 13.4: *The six members of $T_{5,3}$*

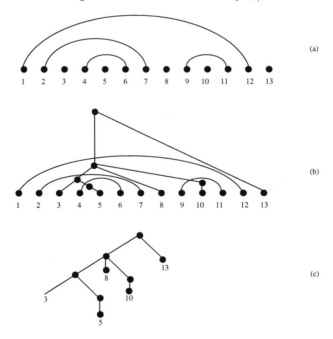

Figure 13.5: *Correspondence (b) between a member of $S_{13,4}$ (a) and a member of $T_{10,5}$.*

node. Also, connect all "visible" unpaired bases. Repeat this procedure until no loops remain "below" nodes. There will be one root node plus $k-1$ internal nodes (one for each basepair). In addition, there are $n+k-2-2(k-1) = n-k$ terminal nodes. The linear tree we have constructed is a member of $\mathcal{T}_{n,k}$. Evidently, this construction could be inverted. ∎

Theorem 13.5 *For $n, k \geq 0$,*

$$S(n,k) = \frac{1}{k} \binom{n-k}{k+1} \binom{n-k+1}{k-1}.$$

Proof. This result is obtained by counting $T(n,k) = |\mathcal{T}_{n,k}|$ which has solution

$$T(n,k) = \frac{1}{k-1} \binom{n-1}{k} \binom{n-2}{k-2}.$$

This can be obtained as an exercise in generating functions or by referring to more general results on tree combinatorics because $|S_{n,k}| = |\mathcal{T}_{n-k+1,k+1}|$. ∎

Corollary 13.2

$$\sum_{k=1}^{n} \frac{1}{k} \binom{n-k}{k+1} \binom{n-k+1}{k-1} \sim \sqrt{\frac{15+7\sqrt{5}}{8\pi}} n^{-3/2} \left(\frac{3+\sqrt{5}}{2}\right)^n.$$

Corollary 13.3 $S(n+k, k) = S(2n-k-1, n-k-1).$

These unexpected results show the value of looking beyond the straightforward view of mathematical objects. The mapping φ we utilize is known in other contexts as Poincaré duality.

13.2 Minimum Free-energy Structures

In the previous section, we studied the shapes of RNA molecules where all possible basepairings were allowed. RNA sequences have a specific base sequence, over the {A,U,G,C} alphabet. The basepairs A•U and G•C are called Watson-Crick basepairs. In many structural RNA molecules, G•U basepairs also appear. Define

$$\rho(a,b) = \begin{cases} 1 & \text{if } a \text{ and } b \text{ can basepair,} \\ 0 & \text{otherwise.} \end{cases}$$

First, let us search for structures that maximize the number of basepairs. Define $X(i,j) = $ maximum number of basepairs in the sequence $a_i a_{i+1} \cdots a_j, i \leq j$.

Theorem 13.6 *The following recursion holds where $X_{i,j} = 0$ if $|j - i| \leq 1$:*

$$X(i, j+1) = \max\{X(i,j), \max\{[X(i, l-1) + 1 + X(l+1, j)]\rho(a_l, a_{j+1}) : 1 \leq l \leq j-1\}\}.$$

RNA Secondary Structure

Proof. As usual, the proof of correctness of this dynamic programming recursion follows that of the combinatorial Theorem 13.1. ∎

The time complexity of the above algorithm is

$$\sum_{i<j}(j-i) = O(n^3).$$

In reality, this problem is much harder because the components all make a contribution to the stability of the overall structure. This stability is measured in free-energy.

Throughout, $h_{i,j}$ is the minimum free-energy (single hairpin) secondary structure on $a_i a_{i+1} \ldots a_j$, $i < j$, where a_i and a_j form a basepair and there is a single end loop. See Figure 13.6. If a_i and a_j cannot basepair, $h_{i,j} = +\infty$. The free-energy functions are assumed to be of the form

$\alpha(a, b)$ = free-energy of an a, b basepair,
$\xi(k)$ = destabilization free-energy of an end loop of k bases,
η = stacking energy of adjacent basepairs,
$\beta(k)$ = destabilization free-energy of bulge of k bases,
$\gamma(k)$ = destabilization free-energy of an interior loop of k bases.

It is possible to define more general energy functions. For example, ξ and η can depend on the adjacent basepair(s). The same complexity results derived below hold. Free-energy of a molecule without pairing is 0; the pairs A•U for example have $\alpha(A,U) < 0$ when $\xi(k) > 0$.

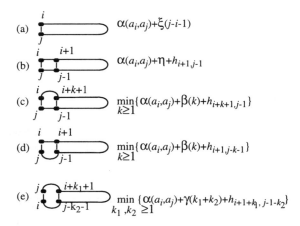

Figure 13.6: *Each of the five minimization situations with the corresponding formula: (a) end loop, (b) extension of helical region, (c) bulge at i, (d) bulge at j, and (e) interior loop*

There are exactly five ways to build $h_{i,j}$ from the basepair and these are presented in Figure 13.6, along with formulas which calculate their values in Theorem 13.7.

Theorem 13.7 *The minimum free-energy $h_{i,j}$ of a hairpin secondary structure on $a_i a_{i+1} \cdots a_j$ with $a_i a_j$ basepaired is the minimum of*

(a) loop: $\quad \alpha(a_i, a_j) + \xi(j - i - 1),$
(b) helix extension: $\alpha(a_i, a_j) + \eta + h_{i+1,j-1},$
(c) bulge: $\quad \min_{k \geq 1}\{\alpha(a_i, a_j) + \beta(k) + h_{i+k+1,j-1}\},$
(d) bulge: $\quad \min_{k \geq 1}\{\alpha(a_i, a_j) + \beta(k) + h_{i+1,j-k-1}\},$
(e) interior loop: $\min_{k_1,k_2 \geq 1}\{\alpha(a_i, a_j) + \gamma(k_1 + k_2) + h_{i+1+k_1, j-1-k_2}\}.$

Proof. See Figure 13.6. ∎

To estimate the computation complexity, each step of Figure 13.6 is treated. Only powers of n are, in the end, of interest for the rate of growth of computation. For purposes of estimating computational complexity, the various formulas take time proportional to

(a);(b) $$\sum_{1 \leq i \leq j \leq n} 1 = \sum_{i=1}^{n} \sum_{j=i}^{n} 1 = O(n^2),$$

(c);(d) $$\sum_{1 \leq i \leq j \leq n} (j - i) = \sum_{i=1}^{n} \sum_{j=i}^{n} (j - i),$$

$$= \sum_{i=1}^{n} \frac{(n-i)(n-i+1)}{2} = O(n^3),$$

(e) $$\sum_{1 \leq i \leq j \leq n} \left(\sum_{i \leq i' \leq j' \leq j} 1 \right) = \sum_{1 \leq i \leq j \leq n} C \cdot (j - i)^2 = O(n^4).$$

It is clear then, that previous algorithms for best hairpins basically have a time complexity of $O(n^4)$. The purpose of the next section is to reduce these algorithms to a time complexity of $O(n^3)$ in general.

13.2.1 Reduction of Computation Time for Hairpins

In this section, some details of the computation will be discussed. In the course of this, it will become clear that a major reduction in computation can be achieved. The dynamic programming calculations are stored in a matrix and, in fact, are sometimes referred to as "matrix methods." Organizing these calculations in a way which allows visualization of the associated RNA structures is a useful device.

RNA Secondary Structure

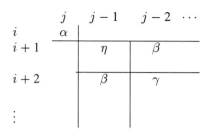

Figure 13.7: *Schematic of minimization*

As above, $h_{i,j}$ is the minimum free-energy of hairpins on $a_i a_{i+1} \ldots a_j$, $i < j$, satisfying a_i and a_j forming a basepair. If a_i and a_j cannot basepair, set $h_{i,j} = +\infty$. Organization of the matrix $(h_{i,j})$, $i = 1, 2, \ldots, n$ and $j = n, n-1, \ldots (i \leq j)$, with the base sequence written in reverse order along the columns is illustrated as

$$
\begin{array}{c|cccccc}
 & a_n & a_{n-1} & \cdots & a_2 & a_1 \\
\hline
a_1 & h_{1,n} & h_{1,n-1} & & h_{1,2} & h_{11} \\
a_2 & h_{2,n} & h_{2,n-1} & & h_{2,2} & \\
\vdots & & & \ddots & & \\
a_{n-1} & h_{n-1,n} & h_{n-1,n-1} & & & \\
a_n & h_{n,n} & & & &
\end{array}
$$

Next take $h_{i,j}$ where $h_{i,j} < +\infty$, that is, where a_i and a_j can form a basepair and where $j - i - 1 \geq m$ (the minimum end loop size). Of course, $h_{i,j}$ results from one of the situations discussed in Section 13.2, which will now be examined in greater detail.

If the basepair is at the "bottom" of an end loop, then $h_{i,j}$ equals

$$\alpha(a_i, a_j) + \xi(j - i - 1).$$

This is the only step in the minimization not indicated on the matrix in Figure 13.7.

If the basepair is part of a helical region, then $h_{i,j}$ equals

$$\alpha(a_i, a_j) + h_{i+1, j-1} + \eta,$$

which is indicated in the $(i+1, j-1)$ position of the matrix in Figure 13.7 by the letter η.

If $h_{i,j}$ results from a bulge, then $h_{i,j}$ equals

$$\alpha(a_i, a_j) + \beta(k) + h_{i+k+1, j-1}$$

or

$$\alpha(a_i, a_j) + \beta(k) + h_{i+1, j-k-1},$$

where $1 \leq k \leq j - i - 2 - m$. In the first situation, a minimization is performed down the $(j - 1)$-st column. This vertical region is indicated by the symbol β as is the horizontal region in the $(i + 1)$-st row.

The remaining possibility is interior loops, where $h_{i,j}$ equals

$$\alpha(a_i, a_j) + \gamma(k_1 + k_2) + h_{i+1+k_1, j-1-k_2},$$

where $j - i - 2 - m \geq k_1 + k_2 \geq 2$ (because $k_1 \geq 1$ and $k_2 \geq 1$). This region is indicated in Figure 13.7 by γ.

As shown in Theorem 13.7, the computational complexity of helix and end loop formation is $O(n^2)$, of bulge formation is $O(n^3)$, and of interior loop formation is $O(n^4)$. The remainder of this section will show that interior loop calculations can, in general, be reduced to $O(n^3)$.

For calculating interior loops from an $i - j$ basepair, the possible candidate positions are (k, l), where $h_{k,l} < \infty$ and (k, l) belongs to

$$\Gamma(i, j) = \{(k, l) : l - k - 1 \geq m, k \geq i + 2, j - 2 \geq l\}.$$

The size of the interior loop is

$$s = (j - i - 1) - (l - k + 1) = (j - i) - (l - k) - 2.$$

This equation implies that, along lines with $l - k = $ constant, the interior loop destabilization function $\gamma(s)$ is constant. The computation can now be organized to exploit this observation.

The idea is for each (i, j) to store values $h_{i,j}^*(s) = \min\{h_{k,l} : (k, l) \in \Gamma(i, j)$ and $s = (j - i) - (l - k) - 2)\}$, $s \geq 1$. Then when we move from $j - i = c$ to $j - i = c + 1$, each vector can be updated in time $O(n)$. Moreover, we find best interior loop free-energies by

$$\min\{\alpha(a_i, b_j) + \gamma((j - i) - (k - l)) + h_{i,j}^*(s)\}.$$

The above setup shows interior loops have an overall calculation time equivalent to that of bulges, $O(n^3)$. Moreover, this equivalence is established by showing that the interior loop problem can be given a data structure equivalent to that of bulges. The additional storage is $n^2/2$ while the computation time is bounded by $O(n^3)$.

13.2.2 Linear Destabilization Functions

As mentioned above, significant efficiencies can be achieved for linear destabilization functions. $O(n^3)$ sequence alignment algorithms can be reduced to $O(n^2)$ for linear deletion functions Here we give a proof of for both bulges and interior loops and frame the proof in a manner which indicates how to perform the computations.

Define the best bulges "down" the column by

$$\text{hdo}(i, j) = \min_{k \geq 1}\{\beta(k) + h_{i+k+1, j-1}\},$$

where $\beta(k) = \delta_1 + \delta_2(k-1)$. Then

$$\text{hdo}(i,j) = \min\{\delta_1 + h_{i+2,j-1}, \min_{k\geq 2}\{\beta(k) + h_{i+k+1,j-1}\}\}$$

$$= \min\{\delta_1 + h_{i+2,j-1}, \min_{l\geq 1}\{\beta(l+1) + h_{i+l+2,j-1}\}\}$$

$$= \min\{\delta_1 + h_{i+2,j-1} \min_{l\geq 1}\{\beta(l) + h_{i+l+2,j-1}\} + \delta_2\}$$

$$= \min\{\delta_1 + h_{i+2,j-1}, \text{hdo}(i+1,j) + \delta_2\}.$$

Similarly, if the bulges "over" a row are considered,

$$\text{hov}(i,j) = \min_{k\geq 1}\{\beta(k) + h_{i+1,j-k-1}\}$$

with $\beta(k) = \delta_1 + \delta_2(k-1)$, we obtain

$$\text{hov}(i,j) = \min\{\delta_1 + h_{i+1,j-2}, \text{hov}(i,j-1) + \delta_2\}.$$

If the computation is carried out on lines of $j - i = c$ for $c = m, m+1, \ldots$, it is easy to see that a single vector of length n suffices for each of $\text{hdo}(i,j)$ and $\text{hov}(i,j)$.

Next, for interior loops and $\gamma(k) = \lambda_1 + \lambda_2(k-2)$, define

$$\text{hil}(i,j) = \min_{1\leq k_1, k_2}\{\gamma(k_1 + k_2) + h_{i+1+k_1, j-1-k_2}\}.$$

Now

$$\min_{1< k_1, 1\leq k_2}\{\gamma(k_1 + k_2) + h_{i+1+k_1, j-1-k_2}\}$$

$$= \min_{1\leq l, 1\leq k_2}\{\gamma(1 + l + k_+2) + h_{i+2+l, j-1-k_2}\}$$

$$= \lambda_2 + \text{hil}(i+1, j).$$

For

$$\min_{k_1 = 1, k_2 \geq 1}\{\gamma(1 + k_2) + h_{i+2, j-1-k_2}\},$$

the problem is exactly equivalent to the bulge problem handled above.

Thus, when the energy functions $\beta(k)$ and $\gamma(k)$ are linear, computation of best hairpin can be accomplished in $O(n^2)$ time and space.

13.2.3 Multibranch Loops

In this section, we consider loops with more than one hairpin extending from them. The destabilization function $\gamma(\cdot)$ above assumes exactly one hairpin extends from the loop. As so little is known about the energetic properties of multibranch loops, we assume that $\rho(\cdot)$ is a single destabilization function which holds for all multibranch loops.

Theorem 13.8 *Minimum free-energy multibranch structures can be found in $O(n^4)$ time and $O(n^3)$ storage.*

Proof. Define $g(i,j)$ to be the minimum free-energy multibranch loop structures on $a_i a_{i+1} \ldots a_j$. The multibranch loop is to begin and/or end at i and j. Our convention is not to count the bases in the basepair at the beginning of helices when we determine loop size. Let $g(i,j;k)$ be the minimum free-energy of multibranch loop structures on $a_i a_{i+1} \ldots a_j$ with k unpaired bases in the multibranch loop. Finally, let $e(i,j)$ be the minimum free-energy of secondary structures on $a_i a_{i+1} \ldots a_j$, where a_i and a_j form a basepair.

The structure corresponding to $g(i, j+1; k)$ either has a_{j+1} basepaired or not. If a_{j+1} is not basepaired, then

$$g(i, j+1; k) = \rho(k) - \rho(k-1) + g(i, j; k-1).$$

If a_{j+1} is basepaired, then

$$g(i, j+1; k) = \min_{1 \leq j^* \leq j-1} \{g(i, j^*; k) + e(j^*+1, j+1)\}.$$

Now

$$g(i,j) = \min_k g(i,j;k),$$

and $e(i,j)$ is obtained by minimizing over multibranch loops, end loops, interior loops, bulges, and helix formation.

Storage for the multibranch loop algorithm described here is proportional to

$$\sum_{i=1}^{n} \sum_{j=i}^{n} (j-i) = O(n^3),$$

whereas time is proportional to

$$\sum_{i=1}^{n} \sum_{j=i}^{n} (j-i)^2 = O(n^4).$$

∎

Increasing storage to n^3 is not desirable, but earlier rigorous algorithms take time $O(n^{2L})$, where L = maximum number of arms helicies from a multibranch loop. Even cloverleaf structures took time $O(n^6)$.

13.3 Consensus folding

The last section gave an overview of dynamic programming methods for minimum free-energy folding of $\mathbf{a} = a_1 a_2 \cdots a_n$. These methods are not always reliable in practice, perhaps because of approximate energy functions or because of higher-order interaction with other molecules. An alternate approach has had excellent

RNA Secondary Structure

success in a number of important cases. Here, we call that method *consensus folding*, which is taken to mean that a folding common to a set of RNAs is the object. The idea is quite like consensus word analysis in the multiple sequence analysis in Section 10.6.

The sequences are placed into an initial alignment. Instead of one window as in multiple sequence consensus word analysis, we put two nonoverlapping windows onto the sequences. A helix size (k) must be specified as well as window width W. Then, the idea is to see if there are k-tuples in the left and right windows of each sequence (or most sequences) that are complementary, that is, that form helicies. Figure 13.8 shows a small example of this idea for $k = 5, W = 9$, and 4 sequences. Note that none of the helicies have any sequence identity.

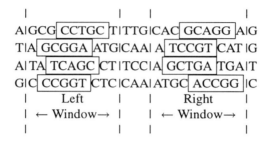

Figure 13.8: *Consensus helix*

This search takes $rO(n^2)$ time and presents very visual results. Our implementation proceeds in an interactive fashion. Once a helix is found and the ambiguities are resolved, it can be locked into place and the search continued. Ambiguities include alternate choices of helicies in a sequence.

To illustrate the results of the method, we show consensus folding for a set of 32 *E. coli* tRNA sequences They have limited sequence similarity, except for GTTCGA, the so-called $T\psi C$ loop. The folding can be seen in Figure 13.9 by $(xyz|\cdots|z^c y^c x^c)$ or $(\cdots|\cdots|\cdots)$ denoting helicies.

```
ALA1A  (ggggcalta(gctclagctggga  lgagc)g(cctgclttt gcaclgcagg)        aggtc(tgcgglttcgatclccgcg)lcgctccc)acca
ALA1B  (ggggctalta(gctclagctggga  lgagc)g(cctgclttt gcaclgcagg)        aggtc(tgcgglttcgatclccgca)ltagctcc)acca
CYS    (ggcgcgtlta(acaalagcggtt   latgt)a(gcggalttgcaaaltccgt)         ctag(tccgglttcgactlccgga)lacgcgcc)tcca
ASP    (ggagcgglta(gttclagtcggttalgaat)a(cctgclctgtcaclgcagg)          gggtc(gcgggltt cgagtlcccgt)lccgttcc)gcca
GLUE1  (gtcccctltc(gtctlagaggcccalggac)a(ccgcclcttt caclggcgg)         taac(agggg ltt cgaat lcccct)lgggggac)gcca
GLUE2  (gtcccctltc(gtctlagaggcccalggac)a(ccgcclctttcaclggcgg)          taac(aggggltt cgaat lcccct)laggggac)gcca
PHE    (gcccggalta(gctclagtcggta  lgagc)a(gggga lttgaaaaltcccc)        gtgtc(cttgglt tcgatt lccgag)lt ccgggc)acca
GLY    (gcgggcglta(gttclaatggta   lgaac)g(agagclt tcccaalgctct)        atac(gagggltt cgatt lccct t)lcgcccgc)tcca
GLY    (gcgggcaltc(gtatlaatggcta  lttac)c(tcagclcttccaalgctga)         tgat(gcgggltt cgatt lccccgc)lt gcccgc)tcca
GLY    (gcgggaalta(gctclagttggta  lgagc)a(cgacclttgccaalggtcg)         gggtc(gcgagltt cgagtlctcgt)lt tcccgc)tcca
HIS    g(gtggctalta(gctclagttggta lgagc)c(ctggalttgtgatltccag)         ttgtc(gtgggltt cgaatlcccat)ltagccac)ccca
ILE    (aggcttglta(gctclaggtggttalgagc)g(caccclctgataalgggtg)          aggtc(ggtgglttcaagtlccact)lcaggcct)acca
TRI2   (ggccccttlta(gctclagtggttta lgagc)a(agcgalctgataaltcgct)        tggtc(gctgglttcaagtlccagc)laggggcc)acca
LYS    (gggtcgtlta(gctclagttggta  lgagc)a(gttgalctttt aaltcaat)        tggtc(gcagglttcgaatlcctgc)lacgaccc)acca
LEU    (gcgaaggltg(gcgglaattggtaglacgc)g(ctagclttcaggtlgttag)          tgtcct tacggacgt(ggggglttcaagtlccccc)lccctcgc)acca
LEU    (gccgaggltg(gtgglaattggtaglacac)g(ctaccltt gaggtlggtag)         tgcccaataggggctt(acgggltt caagtlcccgt)lcctcggt)acca
LEU    (gcccggaltg(gtgglaatcggtaglacac)a(agggalttaaaaaltccct)          cggcgtt cgcgctgt(gcggglttcaagtlcccgc)lt ccgggt)acca
IMET   (cgcggggltg(gagclagcctggtalcgtc)g(tcggglctcataalcccga)          aggtc(gtcgglttcaaatlccggc)lccccgca)acca
MET    (ggctacgltagctclagttggtta lgagc)a(catcalctcataaltgatg)          gggtc(acagglttcaagtlcccgt)lcgtagcc)acca
ASN    (tcctctglta(gttclagtcggta  lgaac)g(gcggalctgttaaltccgt)         atgtc(actgglttcgagtlccagt)lcagagga)gcca
GLN    (tggggtaltc(gccalagcggta   laggc)a(ccggtltttt gatlaccgg)        cattc(cctgglttcgaatlcccagg)ltacccca)gcca
GLN    (tggggtaltc(gccalagcggta   laggc)a(ccggalttctgatlt ccgg)        cattc(cgagglttcgaatlcctcg)ltacccca)gcca
ARG    (gcatccglta(gctclagctggta  lgagt)a(ctcgglctgcgaalccgag)         cggtc(ggagglttcgaatlcctcc)lcggatgc)acca
ARG    (gcatccglta(gctclagctggatalgagt)a(ctcgglctgcgaalccgag)          cggtc(ggagglttcgaatlcctcc)lcggatgc)acca
SER    (ggaagtgltg(gccglagcggtt galaggc)a(ccggtlcttgaaalaccgg)         cgacccgaaagggttc(cagagltt cgaatlctctg)lcgcttcc)gcca
SER    (ggtgaggltg(gccglagagcctgalaggc)g(ctccclctgctaalgggag)tatgcggt caaaagctgcatc(cgggglttcgaatlccccg)lcctcacc)gcca
THR    (gctgatalta(gctclagttggta  lgagc)g(caccclttggtaalgggtg)         aggtc(ggcagltt cgaatlctgcc)lt atcagc)acca
VAL    (gggtgatltagctclagctggga   lgagc)a(cctcclcttacaalggagg)         gggtc(gcggglttcgatclccgtc)latcaccc)acca
VAL    (gcgtccglta(gctclagttggttalgagc)a(ccaccltt gacatlggtgg)         gggtc(ggtgglttcgagtlccact)lcggacgc)acca
VAL    (gcgttcalta(gctclagttggttalgagc)a(ccaccltt gacatlggtgg)         gggtc(gttgglttcgagtlccaat)ltgaacgc)acca
TRP    (aggggcglta(gttclaattggta  lgaac)a(ccggtlctccaaalaccgg)         gtgtt(ggggagltt cgagtlctctc)lcgccccct)gcca
TYR    (ggtggggltt(cccglagcggccaalaggg)a(gcagalctgtaaaltctgc)          cgtcatcgactt c(gaagglttcgaatlccttc)lccccacc)acca
TYR    (ggtggggltt(cccglagcggccaalaggg)a(gcagalctgtaaaltctgc)          cgtcacagactt c(gaagglttcgaatlccttc)lccccacc)acca
```

Figure 13.9: *tRNA from* E. coli

Problems

Problem 13.1 Find $S(n)$ for $n = 1, 2, \ldots, 8$.

Problem 13.2 Generalize Theorem 13.1 to count the number of structures on a sequence $\mathbf{a} = a_1 a_2 \ldots a_n$ with a pairing function $\rho(a, b) \in \{0, 1\}$ as in Section 13.2.

Problem 13.3 A stem is a paired region with a bases on one side and b bases on the other side with the first and last bases in each side paired. Show there are $\binom{a+b-4}{a-2}$ stems.

Problem 13.4 A hairpin is required to have an end loop of at least m bases, where $m \geq 1$ is fixed. Find $L(n)$, the number of hairpins for $[1, n]$.

Problem 13.5 End loops are required to have at least m bases, where $m \geq 1$ is fixed. Generalize Theorem 13.1 to count secondary structures with this restriction.

Problem 13.6 Keeping the requirement of no crossing arcs, what is the recursion rule and boundary conditions for $m = 0$ in the generalization of Theorem 13.1?

RNA Secondary Structure

Problem 13.7 Generalize Lemma 13.1 and Theorem 13.2 for the case $m = 0$.

Problem 13.8 Generalize Lemma 13.1 and Theorem 13.2 for the case $m = 2$.

Problem 13.9 Let $\mathbf{A} = A_1 A_2 \ldots A_n$ be iid with alphabet $\mathcal{A} = \{\text{A,U,G,C}\}$ and $p = \mathbb{P}(\rho(A_i, A_j)) = 1$, $i \neq j$. Let $R(n)$ be the number of secondary structures when we no longer count all possible structures (with $\rho \equiv 1$) but require $\rho(A_i, A_j) = 1$ if they are paired in a structure. Show that $\mathbb{E}(R(n)) \sim \beta n^{-3/2} \alpha^n$, where $\alpha = \left(1 + \sqrt{1 + 4\sqrt{p}}\right)^2 / 4$ and $\beta = \alpha(1 + 4\sqrt{p})^{1/4}/2\sqrt{\pi p^{3/4}}$.

Problem 13.10 Prove Theorem 13.4.

Problem 13.11 For $\mathbf{a} = a_1 a_2 \ldots a_n$, define the vector $\mathbf{v}^T = (v_1, v_2, \ldots, v_n)$ by

$$v_i = \begin{cases} +1 & \text{if } a_i = \text{A}, \\ -1 & \text{if } a_i = \text{T}, \\ +i & \text{if } a_i = \text{G}, \\ -i & \text{if } a_i = \text{C}. \end{cases}$$

The $n \times n$ pairing matrix M is defined by $m_{i,j} = -1 = m_{j,i}$ if $i \neq j$ are paired and $m_{i,j} = 0$ otherwise. Show that $M\mathbf{v} = \mathbf{v}$ if and only if M corresponds to a secondary structure for $\mathbf{a}$.

Chapter 14

Trees and Sequences

Throughout this book, evolutionary processes have been described informally. At the molecular level, evolution can proceed by insertions, deletions, substitutions, inversions, and transpositions. In this chapter, we take up the question of inferring the evolutionary relationships among a set of contemporary organisms by examining their DNA or protein sequences. In the past, characters from morphology (e.g., forearm length, presence of wings, etc.) or biochemistry (e.g., amino acid synthesis pathways, etc.) have been used to infer ancestral relationships. Today, using data from molecular sequence comparisons for these problems is increasingly popular, due in part to the ease of obtaining sequence data.

The idea of using trees to represent evolution dates to Darwin. If there is an ancestral sequence, labeled in Figure 14.1 by *root*, then each branching of the tree represents the time of a divergence. The set $\{1,2,...,8\}$ denotes an arbitrary labeling of the modern sequences. In the unrooted representation, we can include the root (there labeled 9) or just omit it. In much of the following we will look at unrooted trees.

There are two distinct features of rooted or unrooted trees. First is the branching structure or topology of the tree. Second is the branch lengths of the tree, which often indicate the distance in time or evolutionary events between the vertices. In the first section, we will study the topology of trees. Then we study three approaches to evolutionary trees: distance, parsimony and maximum likelihood.

14.1 Trees

A *tree* is a cycle-free connected graph. A *binary tree* has all vertices of degree one (the terminal or labeled vertices) or of degree three. We will assume trees have all vertices of degree 1 or $d \geq 3$. A tree is *labeled* if its terminal vertices are labeled. Some easy combinatorics for these trees is in the next proposition.

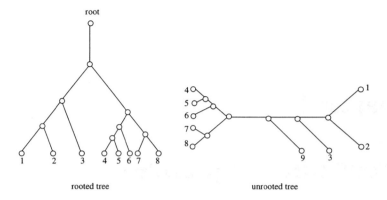

Figure 14.1: *Evolutionary relationships as a rooted or unrooted tree (with the "root" labeled as 9)*

Proposition 14.1 *Let T be a labeled binary tree with $n \geq 3$ terminal vertices. Then T has $n-2$ internal vertices and $n-3$ internal edges. There are $\prod_{j=3}^{n}(2j-5) = 1 \cdot 3 \cdot 5 \cdots (2n-5)$ different trees.*

Proof. For $n = 3$, there is a unique tree topology with $n - 2 = 3 - 2 = 1$ internal vertex and $n - 3 = 3 - 3 = 0$ internal edges.

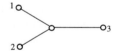

Proceeding by induction, adding a new terminal vertex adds one new internal vertex and one new internal edge. This shows that tree T with $n \geq 3$ terminal vertices has $n - 2$ internal vertices and $n - 3$ internal edges. The number of different trees follows in the same way. Let $g(n)$ be the number of distinct trees for n terminal vertices. Now $g(3) = 1$ and $g(4) = g(3) \cdot 3 = 1 \cdot 3$ because there are 3 edges for $n = 3$ terminal vertices and we can add the next terminal vertex to any of these 3 edges. In general a tree with $n - 1$ terminal vertices has a total of $(n-1) + (n-1) - 3 = 2n - 5$ edges, so $g(n) = g(n-1)(2n-5)$. ∎

Note that a tree is a graph with some special vertices labeled. Two trees $T = (V, E)$ and $T' = (V', E')$ on $\{1, 2, \ldots, n\}$ are equivalent if there is a graph isomorphism $\psi : V \to V'$ of the trees that is compatible with the labels; that is, if φ is the labeling for T and φ' for T', then $\psi(\varphi(i)) = \varphi'(i)$ for $1 \leq i \leq n$.

14.1.1 Splits

For a labeled tree T with labels $\mathcal{L}$, define the set $\mathcal{S}$ of *splits* as follows. For each edge e of T, let the removal of e result in two tree structures T_1 and T_2 defined on L and L^c with $L \cup L^c = \mathcal{L}$ with both L and L^c nonempty. Note that if T is a binary tree, T_1 (and T_2) is not a binary tree. This is due to the vertex $v \in e$: If $\deg(v) = d$ in T, then $\deg(v) = d - 1$ in T_1. The pair (L, L^c) is a split of (T, φ). Each tree defines its set of splits. It is intuitive that the set of splits is sufficient to recover T. Systems of splits are characterized in the next theorem. Note that we include the splits $(\{i\}, \{i\}^c)$ in our set $\mathcal{S}$ for each $i \in \mathcal{L}$.

Theorem 14.1 *$\mathcal{S}$ is the set of splits for a tree T with label set $\mathcal{L}$ if and only if for any two members (L_1, L_1^c) and (L_2, L_2^c) of $\mathcal{S}$, exactly one of the following four sets is empty: $L_1 \cap L_2$, $L_1 \cap L_2^c$, $L_1^c \cap L_2$, and $L_1^c \cap L_2^c$.*

Proof. If $\mathcal{S}$ is constructed from a tree T with arbitrary edges e_1 and e_2 ($e_1 \neq e_2$), then (L_i, L_i^c) is obtained by removing edge e_i, $i = 1, 2$. The edge e_1 partitions the tree T into subtrees T_1 and T_2. Without loss of generality, let $e_2 \in T_2$. Let the label set L_i be a subset of the labels of T_i, $i = 1, 2$. Then $L_1 \cap L_2 = \emptyset$ and, therefore, $L_2 \subset L_1^c$ and it follows easily that $L_1 \cap L_2^c = L_1 \neq \emptyset$ and $L_1^c \cap L_2 = L_2 \neq \emptyset$. In addition, $L_1^c \cap L_2^c \neq \emptyset$. To see this, let $v \in e_1$ be in T_2. Recall that in T_2, v has degree $d(v) - 1 \geq 2$. The edge e_2 is a descendent of one of these edges. Let M be the nonempty set of labels descendent to the other edge. By construction, $M \subset L_2^c$ and $M \subset L_1^c$ so that $L_1^c \cap L_2^c \neq \emptyset$.

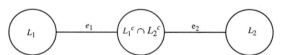

Assume now that the set of splits $\mathcal{S}$ of $\mathcal{L}$ satisfies the condition of the theorem. We will inductively derive a tree T with split set $\mathcal{S}$. The induction hypothesis is that a set $\mathcal{S}_k$ of k splits can be represented by a tree structure of $k + 1$ vertices (each vertex is labeled by a set and the union is the set $\mathcal{L}$) and k edges, where each edge partitions the label set into one of the splits in $\mathcal{S}_k$. Begin with a single vertex consisting of all labels $\mathcal{L}$:

Then, the first split (L_1, L_1^c) separates $\mathcal{L}$ into two vertices.

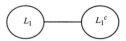

Now assume that k splits $\mathcal{S}_k$ have been used to generate a tree structure T_k, where each edge corresponds to a split. Note that for $k \geq 1$, T_k must have at least two terminal vertices corresponding to splits (L, L^c) and (M, M^c).

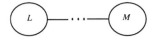

The split (L_{k+1}, L_{k+1}^c) cannot partition both L and M; that is, assume $L_{k+1} \cap L \neq \emptyset$ and $L_{k+1}^c \cap L \neq \emptyset$, and that $L_{k+1} \cap M \neq \emptyset$ and $L_{k+1}^c \cap M \neq \emptyset$. As $M \subset L^c$, we have $L_{k+1} \cap L^c \neq \emptyset$ and $L_{k+1}^c \cap L^c \neq \emptyset$, obtaining a contradiction. Therefore, we can apply this induction hypothesis to the subtree T_k' and split (L_{k+1}, L_{k+1}^c) with L (say) removed. ∎

Next we illustrate the algorithm implicit in the proof of Theorem 14.1. It is not necessary to include splits $(\{i\}, \{i\}^c)$, and several trivial steps are eliminated by that omission. Denote $\mathcal{S}^* = \mathcal{S} \sim \cup_i(\{i\}, \{i\}^c)$. Let $\mathcal{L} = \{1, 2, \ldots, 9\}$ and

$$\mathcal{S}^* = \{(\{1,2\}, \{1,2\}^c), (\{1,2,3\}, \{1,2,3\}^c), (\{1,2,3,9\}, \{1,2,3,9\}^c),$$
$$(\{4,5\}, \{4,5\}^c), (\{4,5,6\}, \{4,5,6\}^c), (\{7,8\}, \{7,8\}^c)\}.$$

The algorithm proceeds as follows using splits in the order given in $\mathcal{S}^*$:

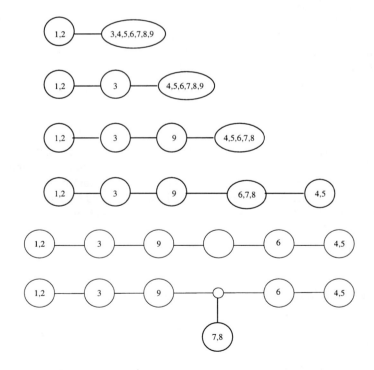

Now, a final step separates all single and double vertices [by using $(\{i\}, \{i\}^c)$ for $i = 1, \ldots, 9$]:

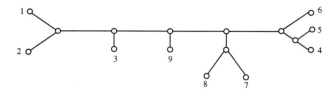

obtaining the unrooted tree of Figure 14.1.

The above algorithm involves looking at each edge that corresponds to (L_1, L_1^c) say. Then, with a new split L_2, L_2^c, we must test the condition of Theorem 14.1. If there are n labels, this can be done in time $O(n)$. Therefore, constructing a tree from m splits takes time proportional to

$$\sum_{k=1}^{m} kn = O(m^2 n).$$

Below, we improve this to time $O(mn)$.

First, we discuss splits determined by binary characters. Our set of labels often correspond to a set of species. Suppose we associate k binary characters to each species; that is, species α has the character j ($c_{\alpha j} = 1$) or does not have the character j ($c_{\alpha j} = 0$). Define the label split induced by character j by $L_j = \{\alpha \in \mathcal{L} : c_{\alpha j} = 1\}$ and $L_j^c = \{\alpha \in \mathcal{L} : c_{\alpha j} = 0\}$. Our theorem says that at least one of $L_i \cap L_j$, $L_i \cap L_j^c$, $L_i^c \cap L_j$, and $L_i^c \cap L_j^c$ must be empty. Put another way, $\cup_{\alpha \in \mathcal{L}} \{c_{\alpha i} C_{\alpha j}\}$ must be a proper subset of $\{00, 01, 10, 11\}$. If, for example, $c_{\alpha i} c_{\alpha j} = 00, c_{\beta i} c_{\beta j} = 01, c_{\gamma i} c_{\gamma j} = 10$, and $c_{\delta i} c_{\delta j} = 11$, then $L_i \supset \{\gamma, \delta\}, L_i^c \supset \{\alpha, \beta\}, L_j \supset \{\beta, \delta\}$, and $L_j^c \supset \{\alpha, \gamma\}$, so that all four intersections are nonempty.

This correspondence among indicators, characters, and splits leads to an improved algorithm for constructing trees from splits. A split can be viewed as assigning a binary $\{0, 1\}$ character to each member of $\mathcal{L}$. Obviously, L in (L, L^c) can be identified with 0 or 1. We will view $(0, 0, \ldots, 0)$ as a new member of $\mathcal{L}$ that defines a root. Then, for each (L, L^c), we assign 1 to the smaller set, $\min\{|L|, |L^c|\}$. Each split in S^* above corresponds to an edge e that we summarize in the table below.

	Edges					
Labels	e_1	e_2	e_3	e_4	e_5	e_6
1	1	1	1	0	0	0
2	1	1	1	0	0	0
3	0	1	1	0	0	0
4	0	0	0	1	1	0
5	0	0	0	1	1	0
6	0	0	0	0	1	0
7	0	0	0	0	0	1
8	0	0	0	0	0	1
9	0	0	1	0	0	0
Sum	2	3	4	2	3	2

Then we sort the edges in order of sum of 1's:

	Edges					
Labels	e_3	e_2	e_5	e_1	e_4	e_6
1	1	1	0	1	0	0
2	1	1	0	1	0	0
3	1	1	0	0	0	0
4	0	0	1	0	1	0
5	0	0	1	0	1	0
6	0	0	1	0	0	0
7	0	0	0	0	0	1
8	0	0	0	0	0	1
9	1	0	0	0	0	0
Sums	4	3	3	2	2	2

This order is not unique. For example, columns e_2 and e_5 can be interchanged because they have the same sum. The logic behind this rearrangement is that label 1 (for example) appears after edges e_3, e_2, and e_1, and because e_3 has the largest sum it must be first, e_2 next, and e_1 last. This gives the tree with root at 0:

⓪──e_3──◯──e_2──◯──e_1──①

Because label 2 has identical row to label 1, 2 joins 1 on this tree:

⓪──e_3──◯──e_2──◯──e_1──①,②

Now, label 3 should appear at the end of edge e_2:

⓪──e_3──◯──e_2──③──e_1──①,②

Next, label 4 is after edges e_5 and e_4, and proceeding through the labels we obtain the rooted tree:

Trees and Sequences

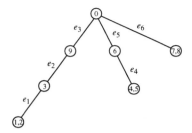

Using the nine edges from $(i, \{i\}^c)$, $i = 1$ to 9, and unrooting the tree by removing 0 gives the original unrooted tree of Figure 14.1. The running time of this algorithm is $O(nm)$.

14.1.2 Metrics on Trees

The set $\mathcal{T}_n$ of labeled trees on $\mathcal{L} = \{1, 2, \ldots, n\}$ is a natural object of study. Below, we will pose optimization problems for certain real-valued functions on $\mathcal{T}_n$ that will usually be NP-hard. Thus, it is useful to have a metric on $\mathcal{T}_n$ so that neighborhoods can be defined and heuristic search methods such as the gradient method can be based on the neighborhoods.

The Splits Metric

Our first metric is easy to motivate. One operation α deletes a split in the tree, whereas its inverse α^{-1} inserts a split in the tree. We will consider trees with interior vertices that can be labeled. Below, the edge e associated with $S = (\{1, 2, 3\}, \{4, 5\})$ is deleted (α) and inserted (α^{-1}).

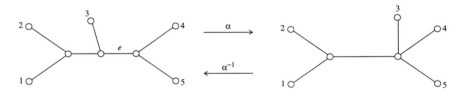

The metric $\rho(T_1, T_2)$ is easily defined:

$$\rho(T_1, T_2) = \min\{k : \text{There exist } \alpha_1, \ldots, \alpha_k \\ \text{such that } \alpha_k \circ \alpha_{k-1} \circ \cdots \circ \alpha_1(T_1) = T_2\}.$$

In this equation, α_i can be either an insertion or deletion. It is easy to show that ρ is a metric on $\mathcal{T}$. Perhaps, surprisingly, ρ can be computed without finding a sequence $\alpha_1 \cdots \alpha_k$.

Theorem 14.2 *If S_1 and S_2 are the sets of splits for trees T_1 and T_2, respectively, where $T_1, T_2 \in \mathcal{T}$, then*

$$\rho(T_1, T_2) = |S_1 \sim (S_1 \cap S_2)| + |S_2 \sim (S_1 \cap S_2)|$$
$$= |S_1| + |S_2| - 2|S_1 \cap S_2|.$$

Proof. If $T_1 = T_2$, then $S_1 = S_2$ and $\rho(T_1, T_2) = 0$. If $T_1 \neq T_2$, then, with $|S_1 \sim (S_1 \cap S_2)|$ α-transformations followed by $|S_2 \sim (S_1 \cap S_2)|$ α^{-1}-transformations, T_1 can be transformed to T_2. Therefore,

$$\rho(T_1, T_2) \leq |S_1 \sim (S_1 \cap S_2)| + |S_2 \sim (S_1 \cap S_2)|.$$

But, as a split cannot be deleted or inserted without an α or α^{-1} transformation, this is an equality. ∎

It can be shown that when $n \geq 3$, the diameter of the metric on $\mathcal{T}_n$ is $2n - 6$. See Problem 14.4. Although the metric is very easy to compute, many pairs of trees are the same distance apart and it has not proved too useful for guiding searches for optimal trees.

The Nearest Neighbor Interchange Metric

The next metric is defined from the split $(A_1 \cup A_2, B_1 \cup B_2)$, where $A_1 \cap A_2 = B_1 \cap B_2 = \emptyset$ and (A_i, A_i^c) and (B_i, B_i^c) are all splits. The *nearest neighbor interchange* (nni) metric is defined from the transformation β where tree T with split $(A_1 \cup A_2, B_1 \cup B_2)$ is replaced by one of $(A_1 \cup B_1, A_2 \cup B_2)$ and $(A_1 \cup B_2, A_2 \cup B_1)$. The three trees shown next are all 1 nni apart:

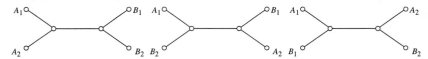

One way to picture a nni transformation β is to move the branch associated with A_2 (say) and move it to the branch associated with B_1, thereby interchanging A_2 and B_2 in the tree. The nni metric is defined as the minimum number of transformations to carry one tree to another:

$$\rho(T_1, T_2) = \min\{k : \beta_1 \cdots \beta_k \text{ are nni's and } \beta_k \circ \beta_{k-1} \circ \cdots \circ \beta_1(T_1) = T_2\}.$$

Again, ρ can be shown to be a metric on $\mathcal{T}$. Note that a nni can be accomplished by deleting and then inserting an edge. Therefore, the nni distance between two trees is bounded by twice the split distance.

A labeled binary tree in $\mathcal{T}$ has $n - 3$ interior edges so that there are a total of $2n - 3$ trees in $\mathcal{T}$ that are a distance of 1 nni. This makes a convenient search strategy. One drawback to this metric is the difficulty of computing $\rho(T_1, T_2)$. To date there is no efficient algorithm. The diameter, however, is known.

There is nothing restricting our choice of adjacent branches to attach A_1, for example, and search strategies have employed these larger neighborhoods.

14.2 Distance

It is natural to have distances between the objects in our set. For example, we can have a set of sequences and then compute all pairwise distances between them. These distances could be found using the dynamic programming algorithms from Chapter 9. For now, we will just take it that we have a matrix of distances $D = (d(i, j))$ for all objects $i, j \in \mathcal{L}$ where

$$d(i, j) = d(j, i) \geq 0, \text{ for all } i, j,$$

and

$$d(i, j) \leq d(i, k) + d(k, j), \text{ for all } i, j, k, l.$$

For convenience, we will take $d(i, j) = 0$ if and only if $i = j$, although the results can be derived without this assumption. D defines a metric on $\mathcal{L} = \{1, 2, \ldots, n\}$.

14.2.1 Additive Trees

For an introduction to trees and distances, we will characterize those D whose distances perfectly fit some tree. Take a tree T defined by splits $\mathcal{S} = \{S\}$. Then, for each split S, define

$$\delta_S(i, j) = \begin{cases} 1 & \text{if } S \text{ separates } i \text{ from } j, \\ 0 & \text{otherwise.} \end{cases}$$

For a set of weights $\alpha_S > 0$, define the *additive tree metric* Δ by

$$\Delta(i, j) = \sum_{S \in T} \alpha_S \delta_S.$$

This simply means that we can find the distance between i and j by adding edge weights on the path between them. We will now characterize those d that can be represented in this fashion.

The key ingredient is how to take d and construct the weight at each split $S = (L, L^c)$. Define

$$\mu_S = \frac{1}{2} \min\{d(i, k) + d(j, l) - d(i, j) - d(k, l) : i, j \in L, k, l \in L^c\}.$$

The first lemma shows that whenever $\mu_{S_1} > 0$ and $\mu_{S_2} > 0$, S_1 and S_2 satisfy the conditions of Theorem 14.1.

Lemma 14.1 *If $S_1 = (L_1, L_1^c)$ and $S_2 = (L_2, L_2^c)$ are splits with $\mu_{S_1} > 0$ and $\mu_{S_2} > 0$, then at least one of the following four sets is empty: $L_1 \cap L_2, L_1^c \cap L_1, L_2^c \cap L_1,$ and $L_1^c \cap L_2^c$.*

Proof. If all four sets are nonempty, then there exists i, j, k, l so that $i, j \in L_1; k, l \in L_1^c; i, k \in L_2; j, l \in L_2^c$. From $\mu_{S_1} > 0$,

$$d(i,k) + d(j,l) - d(i,j) - d(k,l) > 0,$$

but, from $\mu_{S_2} > 0$, it must also be negative. ∎

Given d we can define $T_d = \{S : \mu_S > 0\}$ and by Lemma 14.1 this set of splits is a tree. Define the additive tree metric

$$\Delta_d(i,j) = \sum_{S \in T_d} \mu_S \delta_S = \sum_S \mu_S \delta_S.$$

Proposition 14.2 $\Delta_d \leq d$.

Proof. For each pair $i, j \in \mathcal{L}$, there is a path between i and j, say of splits $S_1 S_2 \cdots S_p$ where $i \in L_k$ and $j \in L_k^c, 1 \leq k \leq p$. These are the splits that separate i from j. At each vertex between splits S_{k-1} and S_k, we choose an l_k such that $l_k \in L_k$ and $l_k \notin L_{k-1}$; that is,

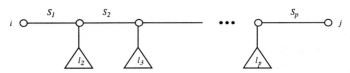

By construction of μ_S,

$$\mu_{S_1} \leq \frac{1}{2}\{d(i,j) + d(i,l_2) - d(i,i) - d(j,l_2)\},$$

$$\mu_{S_2} \leq \frac{1}{2}\{d(i,l_3) + d(j,l_2) - d(i,l_2) - d(j,l_3)\},$$

$$\mu_{S_3} \leq \frac{1}{2}\{d(i,l_4) + d(j,l_3) - d(i,l_3) - d(j,l_4)\},$$

$$\vdots$$

$$\mu_{S_k} \leq \frac{1}{2}\{d(i,j) + d(j,l_p) - d(i,l_p) - d(j,j)\}.$$

Summing yields

$$\sum_{k=1}^p \mu_{S_k} \leq d(i,j).$$

Because we took the path from i to j, then

$$\Delta_d(i,j) = \sum_S \mu_S \delta_S = \sum_{k=1}^p \mu_{S_k}.$$

∎

Next, we prove that an additive tree metric determines a unique tree.

Trees and Sequences

Theorem 14.3 *Suppose that for trees T_1 and T_2*

$$\sum_{S \in T_1} \alpha_S \delta_S = \sum_{S \in T_2} \beta_S \delta_S \text{ where } \alpha_S > 0, \beta_S > 0.$$

Then, $T_1 = T_2$ and $\alpha_S = \beta_S$ for all $S \in T_1 = T_2$.

Proof. Define a metric d by

$$d = \sum_{S \in T_1} \alpha_S \delta_S.$$

Then, if some split S in T_1 separates i, j from k, l,

$$\frac{1}{2}\{d(i,k) + d(j,l) - d(i,j) - d(k,l)\}$$

equals the sum of all α_{S^*} where $S^* \in T_1$ separates i, j from k, l. Because S separates i, j from k, l, the quantity is greater than or equal to α_S. Therefore, $\mu_S \geq \alpha_S$. But $\Delta_d > d$ is a contradiction, so equality must obtain. ∎

Now we give the condition that characterizes additive trees. We say that d satisfies the *four point condition* if for all $i, j, k, l \in \mathcal{L}$, of the three sums

$$d(i,j) + d(k,l),$$
$$d(i,k) + d(j,l),$$
$$d(i,l) + d(j,k),$$

two are equal and not less than the third. For example, this might result in

$$d(i,j) + d(k,l) \leq d(i,k) + d(j,l) = d(i,l) + d(j,k).$$

Theorem 14.4 $\Delta_d = d$ *if and only if the four point condition holds.*

Proof. Assume $\Delta_d = d$. Then if $i, j \in L$ and $k, l \in L^c$ for some $S = (L, L^c)$, we have the following tree where the weights λ are the sum of the edge weights joining appropriate vertices.

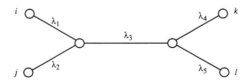

Clearly,

$$d(i,j) + d(k,l) = (\lambda_1 + \lambda_2) + (\lambda_4 + \lambda_5)$$

and

$$d(i,k) + d(j,l) = (\lambda_1 + \lambda_3 + \lambda_4) + (\lambda_2 + \lambda_3 + \lambda_5)$$
$$= (\lambda_1 + \lambda_3 + \lambda_5) + (\lambda_2 + \lambda_3 + \lambda_4)$$
$$= d(i,j) + d(j,k).$$

Now we assume the four point condition holds. For a pair $i, j \in \mathcal{L}$, define

$$f(x) = d(i,x) - d(j,x).$$

Because $d(i,x) + d(i,j) \geq d(j,x)$, $f(x) \geq -d(i,j)$, and because $-d(i,j) - d(j,x) \leq -d(i,x)$, $f(x) \leq d(i,j)$. Note that $f(i) = -d(i,j)$, and $f(j) = d(i,j)$ so the range of $f(x)$ is $[-d(i,j), d(i,j)]$. Let $\alpha < \alpha'$ be real numbers in this closed interval such that no $f(x) \in (\alpha, \alpha')$. Then the nonempty sets

$$\{x : f(x) \leq \alpha\} \quad \text{and} \quad \{x : \alpha' \leq f(x)\}$$

determine a split S. For this split, we will show $\mu_S \geq \frac{1}{2}(\alpha' - \alpha)$.

By definition of μ_S, we have $u, v, w, z \in \mathcal{L}$ such that

$$\mu_S = \frac{1}{2}\{d(u,w) + d(v,z) - d(u,v) - d(w,z)\},$$

where u, v and w, z are separated by S. Let $f(u) \leq f(v) \leq \alpha < \alpha' \leq f(w) \leq f(z)$. Now, $f(u) < f(w)$ means that $d(i,u) - d(j,u) < d(i,w) - d(j,w)$ and $d(i,u) + d(j,w) < d(i,w) + d(j,u) = d(i,j) + d(u,w)$, applying the four point condition. This means

$$d(u,w) = d(i,w) + d(j,u) - d(i,j).$$

Similarly, $f(v) < f(z)$ implies

$$d(v,z) = d(i,z) + d(j,v) - d(i,j).$$

Using $f(u) \leq f(v)$ and $f(w) \leq f(x)$, we can obtain

$$d(u,v) \leq d(u,j) + d(v,i) - d(i,j)$$

and

$$d(w,z) \leq d(w,j) + d(z,i) - d(i,j).$$

Recall that

$$\mu_S = \frac{1}{2}\{d(u,w) + d(v,z) - d(u,v) - d(w,z)\}$$
$$\geq \frac{1}{2}\{d(i,w) + d(j,v) - d(i,v) - d(j,w)\}$$
$$= (f(w) - f(v)) \geq \frac{1}{2}(\alpha' - \alpha).$$

Trees and Sequences

To complete the proof, order the values of $f(x)$ in order $\alpha_1 \leq \alpha_2 \leq \cdots \leq \alpha_p$. Above, we showed $f(x) \in [-d(i,j), d(i,j)]$, so that $\alpha_1 = -d(i,j)$ and $\alpha_p = d(i,j)$. For each pair α_k, α_{k+1}, we obtain a split S_k and $\mu_{S_k} \geq \frac{1}{2}(\alpha_{k+1} - \alpha_k)$. Then

$$\Delta_d(i,j) \geq \sum_{k=1}^{p} \mu_{S_k} \geq \frac{1}{2}\sum_{k=1}^{p}(\alpha_{k+1} - \alpha_k)$$
$$= d(i,j).$$

∎

Now we show how to construct the unique additive tree, if one exists, in time $O(|\mathcal{L}|^2)$. First, start with three vertices i, j and k. Then

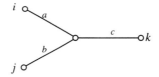

form a unique tree with edge lengths determined by

$$d(i,j) = a + b, d(i,k) = a + c \text{ and } d(j,k) = b + c,$$

so that

$$\begin{aligned} a &= (d(i,j) + d(i,k) - d(j,k))/2, \\ b &= (d(i,j) + d(j,k) - d(i,k))/2, \\ c &= (d(i,k) + d(j,k) - d(i,j))/2. \end{aligned} \quad (14.1)$$

Now, assume the tree has been uniquely determined for a subset C_k of $k \geq 2$ members of $\mathcal{L}$. Take $l \in C_k^c$ and $i, j \in C_k$ and find the unique tree for i, j, l. If the new vertex does not coincide with a vertex already on the path between i and j, add l. If the vertex does coincide with a vertex already on the path between i and j, replace j (say) with a member of C_k that is appended to that vertex. The worst case running time of this algorithm is $O(\sum_{k=1}^{n} k) = O(n^2)$. If there is an additive tree, the algorithm will produce it. If there is not an additive tree, this can be determined by checking the tree distances with D. Therefore, we can test D for additivity and produce the unique additive tree if there is one in time $O(n^2)$.

14.2.2 Ultrametric Trees

There is an even more restrictive condition on $D = (d(i,j))$ than the four point condition called the *ultrametric condition*. This condition states that, for all

$i, j, k \in \mathcal{L}$, of the three distances

$$d(i,j), d(i,k), d(j,k),$$

two are equal and not less that the third. This might, for example, be obtained by

$$d(i,j) \le d(i,k) = d(j,k).$$

First, we relate this metric to additive trees; the ultrametric condition looks a little like the four point condition. The proof is left as an exercise.

Proposition 14.3 *If $D = (d(i,j))$ is an ultrametric, it satisfies the four point condition.*

These are very special additive trees. If $d(i,k) \le d(j,i) = d(j,k)$, then the tree can be constructed according to Equation (14.1). It has edge lengths a and b where

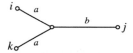

Moreover, because $2a \le a + b$, $a \le b$. This simple observation determines essential properties of the tree. Let $d(i,j) = \max\{d(k,l) : k, l \in \mathcal{L}\}$. Then all k will join the path between i and j at distance $\le d(i,j)/2$ from i or j. We partition $\mathcal{L}$ by $C_i = \{k : d(i,k) < d(i,j)\}$ and $C_j = \{k : d(i,k) = d(i,j)\}$:

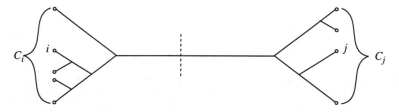

This means that essentially we have a rooted tree with uniform rates of evolution:

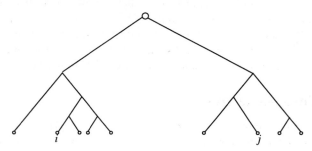

Trees and Sequences

All paths from labels are equidistant to the "root" placed at distance $d(i,j)/2$, halfway between i and j. This is the most ideal of all biologically motivated evolutionary trees: Evolution proceeds at a constant rate on all branches of the tree. If branching happens at random times, say according to independent exponential distributions, then there is a distributional theory available by applying a continuous time branching process known as the pure birth process.

In this model, let $X(t) \geq 0$ denote the number of branchings that have occurred in time $[0,t]$ The process satisfies ($h \to 0+$) for each k:

(a) $\mathbb{P}(X(t+h) - X(t) = 1 | X(t) = k) = \beta_k h + o(h)$,

(b) $\mathbb{P}(X(t+h) - X(t) = 0 | X(t) = k) = 1 - \beta_k h + o(h)$,

(c) $X(0) = 0$.

It then follows that for $P_n(t) = \mathbb{P}(X(t) = n | X(n) = 0)$,

$$P_0'(t) = \beta_0 P_0(t),$$

$$P_n'(t) = -\beta_n P_n(t) + \beta_{n-1} P_{n-1}(t), \quad n \geq 1,$$

with $P_0(0) = 1$ and $P_n(0) = 0$, $n > 0$. The time T_k between the k-th and $(k+1)$-st branching has an exponential distribution with mean $1/\beta_k$. (An additional assumption of $\sum_{k=1}^{\infty} 1/\beta_k = \infty$ is required so that the process is defined for all $t \geq 0$.)

The Yule process is a pure birth process with $\beta_k = (1/\lambda) k$, $\lambda > 0$. Here, the number of branchings is proportional to the "population size" and the waiting time between the k-th and $(k+1)$-st branching has mean λ/k. This makes the tree very thick with branching as time increases. Trees more pleasing to the eye arise where $\beta_k = 1/\lambda$ and the mean time between the k-th and $(k+1)$-st branching is λ.

14.2.3 Nonadditive Distances

The techniques of cluster analysis have long been applied to distance data. The idea of the first class of algorithms is to repeatedly merge or cluster pairs of labels. The algorithms are called pair group methods (PGM).

Algorithm 14.1 (PGM)
```
input:   D = (d(i,j)) for L = {1, 2, ..., n}
  1. find closest i, j
     d(i,j) = min{d(k,l) : k, l ∈ L}
  2. cluster {i, j} = c.
     L ← L - {i, j}
     L ← L ∪ {c}
     if |L| = 2, stop
```

3. calculate $d(c, k), k \in \mathcal{L}$
4. go to 1

The output of PGM is a set of splits that determine a tree topology. Critical details remain to be specified in step 3. Calculate $d(c, k)$, where $c = \{i, j\}$. One obvious method is

$$d(c, k) = \frac{d(i, k) + d(j, k)}{2}$$

and this determines the algorithm UPGMA (unweighted pair group method using arithmetic averages). When i and j and k might all be clusters themselves, weighted PGMA (WPGMA) uses

$$d(c, k) = \frac{\|i\| d(i, k) + \|j\| d(j, k)}{\|i\| + \|j\|},$$

where $\|i\|$ and $\|j\|$ denote the respective cluster sizes. The size of a cluster k is irrelevant to the calculation.

Branch lengths can be calculated by adding a step 3* to the algorithm. When i and j are clustered, a node representing the cluster can be placed midway between them. Another method is to simply determine the tree topology, and after that, estimate the branch lengths best fitting the data by least squares or some other method.

Yet another variant known as neighbor joining (NJ) is derived by modifying steps 3 and 3*. The node between 1 and 2 can be placed at distance d_1 from 1, where

$$d_1 = \left(d(1, 2) + \frac{1}{n-2} \sum_{k=3}^{n} d(1, k) - \frac{1}{n-2} \sum_{k=2}^{n} d(2, k) \right) \Big/ 2$$

and

$$d_2 = \left(d(1, 2) + \frac{1}{n-2} \sum_{k=3}^{n} d(2, k) - \frac{1}{n-2} \sum_{k=2}^{n} d(2, k) \right) \Big/ 2.$$

This is consistent with ideas from additive trees and, in fact, produces the additive tree if $D = (d_{ij})$ determines one.

Other methods involve searches over the set $\mathcal{T}$ of all labeled trees on $\mathcal{L} = \{1, 2, \ldots, n\}$. The searches are necessarily heuristic. For a tree $T \in \mathcal{T}$ and an assignment of edge lengths $\lambda = (\lambda_1, \lambda_1, \ldots, \lambda_{2n-3})$ to T, there is a function $f(T, \lambda; D)$ that we want to minimize

$$\min\{f(T, \lambda; D) : T \in \mathcal{T}, \lambda \geq 0\}. \tag{14.2}$$

Note that we add the requirement of $\lambda_i \geq 0$ for all i. Here are three functions of edge weights where we take $D = (d(i, j))$ as fixed. The notation $i \in (k - l)$ means edge i on the path between k and l.

Trees and Sequences

$$f_1(T, \lambda) = \sum_{i=1}^{2n-3} \lambda_i;$$

$$f_2(T, \lambda) = \sum_{k,l} \left(\sum_{i \in (k-l)} \lambda_i - d(k,l) \right) \Big/ d^\alpha(k,l), \quad (\alpha \geq 0),$$

and

$$f_3(T, \lambda) = \sum_{k,l} \left(\sum_{i \in (k-l)} \lambda_i - d(k,l) \right)^2 \Big/ d^2(k,l).$$

For each of these objective functions we add the linear constraints

$$\lambda_i \geq 0, \quad i = 1, 2, \ldots, 2n-3, \tag{14.3}$$

and

$$\sum_{i \in (k-l)} \lambda_i \geq d(k,l). \tag{14.4}$$

The last constraint (14.4) is to require at least as much evolutionary distance in the tree as between the contemporary species. This, by Proposition 14.2, rules out Δ_d in all cases that d is not already additive.

Certainly, all instances of f_1 and f_2 can be solved using linear programming, whereas f_3 with constraint (14.3) is solved by constrained least squares. The technique to solve these global optimum problems for all $T \in \mathcal{T}$ is to start at a tree T_0 and compute the minimum of the objective function. Then, using the members of a neighborhood of T_0, gradient search can be employed to find a local optimum. Each time a new tree is considered, the optimal $f_i(T)$ must be found. Many variants of these ideas exist, including using simulated annealing to combine the efficiencies of Monte Carlo and gradient searches.

14.3 Parsimony

So far, our discussion of trees has not directly involved sequences. Distances may be derived from sequence comparisons, but only the distances are used to infer a tree. In this section, a set of aligned sequences, one sequence for each label, is assumed to be given. A tree T is a good explanation of the data if at each position of the aligned sequences, the pattern of change is consistent with the tree. For example, here is an example with $n = 4$ sequences:

	Position	Position
	i	j
1	T	A
2	T	T
3	T	T
4	A	A

We place the letters of each position at the terminal vertices of a tree. For a rooted tree T, we consider for example position i:

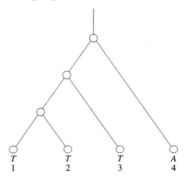

The vertices u "below" each vertex v, where $\{u, v\} \in E$ are called the *children of v*; v is called the *parent of u*. The above tree looks consistent with position i because it is easy to assign letters to the other vertices in a way that, except the root, requires no evolutionary changes. The root

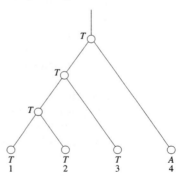

could be assigned A or T, reflecting the assignment of the children of the root. The assignment above has one edge $\{u, v\}$, where the letters at the vertices u and v differ. The problem of parsimony is to find the assignment of letters at the vertices of the tree to minimize the number of changes.

First, we give a somewhat more formal presentation of the problem. The rooted tree T is fixed for the problem. Let

$$\delta(v) = \{w \in V : w \text{ is a child of } v\}$$

Trees and Sequences

and

$$T(v) = \text{the subtree rooted at } v.$$

A *fit* f *to* T is an assignment of letters $f(v) \in \mathcal{A}$ to the vertices $v \in V$. The *cost of the fit* f, $\|f\| = \|f(T)\|$, is the number of changes in the tree:

$$\|f(T)\| = \sum_{v \in T} \sum_{w \in \delta(v)} \mathbb{I}\{f(v) \neq f(w)\}.$$

Also, let $L(T) = \min\{\|f(T)\| : f \text{ is a fit to } T\}$. An f satisfying $L(T) = \|f\|$ is a *minimum mutation fit*.

First, we give an algorithm to find $L(T)$. The algorithm begins with the letters at the tips of the trees and recursively assigns sets of letters and values of $L(T(v))$ at each vertex v.

Algorithm 14.2 (Parsimony Score)

```
Forward recursion

    Given F(v) ⊂ A and L(v) = 0 at v = terminal vertices
    For each v until root
```

$$n_v(a) = |\{w : w \in \delta(v) \text{ and } a \in F(w)\}|$$
$$m = \max\{n_v(a) : a \in \mathcal{A}\}$$
$$F(v) = \{a : n_v(a) = m\}$$
$$L(T(v)) = \left\{\sum_{w \in \delta(v)} (L(T(w)) + 1)\right\} - m$$

This easily applies to the above example to give $\{A, T\}$ and $L = 1$ at the root. For a more revealing case, consider position j:

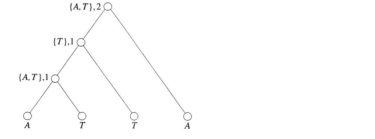

This seems to give two trees by simple backtracking:

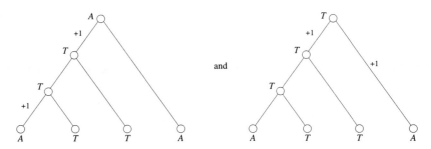

However, there is an unexpected third fit with cost 2:

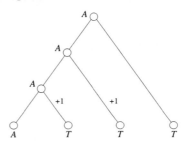

Clearly, we cannot easily find all solutions using the sets F, at least not in an obvious way. This will require defining another set $S(v)$ at each vertex. We will call $F(v)$ the *first set* and $S(v)$ the *second set*. The values of $f(v)$ that are part of an optimal assignment will also be defined.

$$O(v) = \{a : \text{there exists a minimum mutation to } T \text{ with } f(v) = a\},$$
$$F(v) = \{a : \text{a fit with } f(v) = a \text{ has cost } L(T(v))\},$$
$$S(v) = \{a : \text{a fit with } f(v) = a \text{ has cost } + 1\}.$$

Now we can state the full algorithm, which is proved in the theorems that follow.

Algorithm 14.3 (Parsimony)
```
Forward recursion
     Given F(v) ⊂ A and L(v) = 0 at v = terminal vertices
     For each v until root
```

$$n_v(a) = |\{w : w \in \delta(v) \text{ and } a \in F(v)\}|$$
$$m_v = \max\{n_v(a) : a \in \mathcal{A}\}$$
$$F(v) = \{a : n_v(a) = m_v\}$$
$$S(v) = \{a : n_v(a) = m_v - 1\}$$
$$L(T(v)) = \left\{\sum_{w \in \delta(v)} (L(T(w)) + 1)\right\} - m_v$$

Trees and Sequences

```
Backward recursion
    for w ∈ δ(v) and f(v) = a
```

$$
\begin{aligned}
&\text{if } a \in F(w), &&\text{then } f(w) = a \\
&\qquad a \in S(w), &&\qquad f(w) = \{a\} \cup F(w) \\
&\qquad a \notin F(w) \cup S(w), &&\qquad f(w) = F(w)
\end{aligned}
$$

Theorem 14.5 *For all $v \in V(T)$, $O(v) \subset F(v) \cup S(v)$.*

Proof. If $v = \text{root}$, then $T(v) = T$ and, therefore, $O(v) = F(v)$.

For $v \neq \text{root}$, let u be the parent of v. Then the cost of the fit f can be decomposed into three parts:

$$\|f\| = \|f(T(v))\| + \mathbb{I}\{f(v) \neq f(u)\} + \sum_{\substack{x,y \notin \delta(v) \\ x \in \pi(y)}} \mathbb{I}\{f(x) \neq f(y)\}.$$

Therefore, if $f(v) \neq f(u)$, $\|f(T)\|$ is optimal only if $\|f(T(v))\|$ is optimal. Otherwise, $\|f(T(v))\|$ could be decreased without effecting the other sums. We have $f(v) \in F(v)$. The other case is $f(v) = f(u)$. Now the fit f restricted to $T(v)$ is bounded by $L(T(v)) + 1$, which can be achieved by assigning the optimal fit to $T(v)$. Therefore, $v \in F(v) \cup S(v)$. ∎

Note that the definitions of $F(v)$, $S(v)$, and $L(T(v))$ are changed in their assignments in the algorithm. The next theorem proves that both forms are equivalent.

Theorem 14.6 *For all $v \in V(T)$, define $n_v(a)$ and m_v as above. Then*

$$
\begin{aligned}
&(i) \ F(v) = \{a : n_v(a) = m_v\}, \\
&(ii) \ S(v) = \{a : n_v(a) = m_v - 1\},
\end{aligned}
$$

and

$$(iii) \ L(T(v)) = \left\{ \sum_{w \in \delta(v)} (L(T(w)) + 1) \right\} - m_v.$$

Proof. First, note that

$$\|f(T(v))\| = \sum_{w \in \delta(v)} \left(\|f(T(w))\| + \mathbb{I}\{f(v) \neq f(w)\} \right).$$

Consider the optimal fit to $T(w)$, assuming $f(v) = a$. If $a \in F(w)$, it follows that $f(w) = a$ and $\|f(T(w))\| + \mathbb{I}\{f(v) \neq f(w)\} = L(T(w))$. If $a \notin F(w)$, then this cost cannot be less than $L(T(w)) + 1$, which is achieved by $f(w) = a$ when

$a \in S(w)$ or by $f(w) = b$ for any $b \in F(w)$. In general, if $f(v) = a$, summing over all $w \in \delta(v)$,

$$\|f(T(v))\| = \sum_{w \in \delta(v)} (\|f(T(w))\| + 1) - n_v(a).$$

If $\|f(T(v))\| = L(T(v))$, then (i), (ii), and $n_v(a) = m_v$ follow. ∎

Now we prove the backward recursion is correct.

Theorem 14.7 *For all $v \in V(T)$, $f(v) = a \in O(v)$, and $w \in \delta(v)$, define*

$$f_a(w) = \{b : b \in O(w) \text{ and } a \in O(v)\}.$$

Then

$$\begin{aligned} \text{if } a \in F(w), \quad & \text{then } f(w) = a, \\ a \in S(w), \quad & f(w) = \{a\} \cup F(w), \\ a \notin F(w) \cup S(w), \quad & f(w) = F(w). \end{aligned}$$

Proof. Theorem 14.5 tells us that the subtree consisting of v and $\delta(v)$ has T-optimal cost of $L(T(w))$ or $L(T(w)) + 1$.

In the case $a \in F(w)$, $f(w) = a$ has subtree cost $L(T(w)) + \mathbb{I}(f(w) \neq f(v)) = L(T(w))$. Any other assignment at w increases this cost.

In the case $a \in S(w)$, cost $L(T(w))$ cannot be achieved. The cost $L(T(w))+1$ is obtained for $f(w) = a$ or $f(w) \in F(w)$. Any other assignment at w increases this cost.

Finally, assume $a \notin F(w) \cup S(w)$. Then $f(w) \in S(w) \cup (S(w) \cup F(w))^c$ implies cost at least $L(T(w)) + 2$. When $f(w) \in S(w)$, $f(v) \neq f(u)$ and the cost is bounded by $1 + L(T(w)) + 1$. When $f(w) \in (S(w) \cup F(w))^c$, the cost is at least $L(T(w)) + 2$ by definition. However, $f(w) \in F(w)$ has cost $L(T(w)) + \mathbb{I}(f(w) \neq a) = L(T(w)) + 1$, which must be optimal. ∎

It is natural to consider *weighted parsimony*, where $\lambda(a, b)$ replaces $\mathbb{I}(a \neq b)$ at each edge of T. Define $\|f_v(a)\|$ to be the minimum cost of all fits to $T(v)$, where $f(v) = a$. The new cost is, of course,

$$\|f(T)\| = \sum_{v \in T} \sum_{w \in \delta(v)} \lambda(f(v), f(w)).$$

To begin the recursion, $\|f_v(a)\| = 0$ at a terminal vertex v if a is assigned to v and $\|f_v(a)\| = \infty$ otherwise. Then, recursively calculate for all $a \in \mathcal{A}$ and $v \in V$,

$$\|f_v(a)\| = \sum_{w \in \delta(v)} \min\{f_w(b) + \lambda(a, b) : b \in \mathcal{A}\}$$

and
$$M_w(a) = \{b : f_w(b) + \lambda(a,b) \text{ is minimal}\}.$$

The optimal cost of fitting T is

$$\|f(T)\| = \min\{\|f_{\text{root}}(a)\| : a \in \mathcal{A}\},$$

and the assignment can be obtained by backtracking using $M_w(a)$.

The statistical properties of parsimony scores are, of course, of interest. The Azuma-Hoeffding lemma 11.1 gives a large deviation bound on the deviations of the score for fitting a single position. The proof is similar to that for the large deviations global alignment result.

Proposition 14.4 *Let $A_1, A_2, \ldots$ be iid letters assigned to the terminal vertices of a tree T. Then, if we define $S_n = \|f(T)\|$,*

$$\mathbb{P}(S_n - \mathbb{E}(S_n) \geq \gamma n) \leq e^{-\gamma^2 n}.$$

Of course, parsimony scores are derived from fitting a number of positions and adding the costs. If the positions are iid, the Central Limit Theorem applies.

Proposition 14.5 *Suppose N iid positions are fit to a tree T. If the fit of each position (i) has mean μ and standard deviation σ, then*

$$S_N = \sum_{i=1}^{N} \|f_i(T)\| \stackrel{d}{\Rightarrow} \mathcal{N}(N\mu, N\sigma^2).$$

14.4 Maximum Likelihood Trees

In this section, we will model sequence evolution as a stochastic process. When we discussed distance and parsimony methods, assumptions about the evolutionary process were not made explicit. Instead, we appealed to intuition. Distance along the branches of the tree should be at least as great as distance between contemporary species. Parsimony should work well when there are only a few changes in the tree at each position. Although appealing to an intuitive sense of the evolutionary process, these sentences are difficult to pin down. In this section, we give a set of specific, stochastic models. Then we can study the expected number of changes between any two species. Maximum likelihood is used to choose a tree.

14.4.1 Continuous Time Markov Chains

The state space of our Markov chain is $\mathcal{A}$, and to be definite, we use $\mathcal{A} = \{A,G,C,T\}$. The random variable of interest is $X(t) \in \mathcal{A}$, defined for $t \geq 0$. This random variable will describe the substitution process of a sequence position. No indels are allowed in our model. The Markov assumption asserts that

$$P_{ij}(t) = \mathbb{P}(X(t+s) = j | X(s) = i) \tag{14.5}$$

is independent of $s \geq 0$ for $t > 0$. From this property, it follows that

(i) $P_{ij}(t) \geq 0$, $i, j \in \mathcal{A}$;

(ii) $\sum_{j \in \mathcal{A}} P_{ij}(t) = 1$, $i \in \mathcal{A}$;

(iii) $P_{ik}(t+s) = \sum_{j \in \mathcal{A}} P_{ij}(t) P_{jk}(s)$, $t, s \geq 0$, $i, k \in \mathcal{A}$.

The following assumption is added:

(iv) $\lim_{t \to 0+} P_{ij}(t) = \mathbb{I}(i = j)$.

This last assumption leads us to define $P_{ii}(0) = 1$, and $P_{ij}(0) = 0$ if $i \neq j$.

This is most easily presented in matrix form, $\mathbf{P}(t) = (P_{ij}(t))$. The Chapman-Kolmogorov equations (iii) can all be written as $\mathbf{P}(t+s) = \mathbf{P}(t)\mathbf{P}(s)$, and setting $\mathbf{P}(0) = \mathbf{I}$, the identity matrix, (iv) states that $\mathbf{P}(t)$ is right continuous at 0. Derivatives at 0 can also be shown to exist and we define q_i and q_{ij} by

$$\lim_{h \to 0+} \frac{1 - P_{ii}(h)}{h} = q_i$$

and

$$\lim_{h \to 0+} \frac{P_{ij}(h)}{h} = q_{ij} \quad i \neq j.$$

It easily follows that $q_i = \sum_{j \neq i} q_{ij}$. Now it is easy to present these derivatives in matrix form. With $Q = (q_{ij})$ defined as

$$Q = \begin{pmatrix} -q_1 & q_{12} & q_{13} & q_{14} \\ q_{21} & -q_2 & q_{23} & q_{24} \\ q_{31} & q_{32} & -q_3 & q_{34} \\ q_{41} & q_{42} & q_{43} & -q_4 \end{pmatrix},$$

$\sum_j q_{ij} = 0$ and we have

$$\lim_{h \to 0+} \frac{\mathbf{P}(h) - \mathbf{I}}{h} = Q. \tag{14.6}$$

From this formula and $\mathbf{P}(t+s) = \mathbf{P}(t)\mathbf{P}(s)$, it follows that

$$\mathbf{P}'(t) = \mathbf{P}(t)Q = Q\mathbf{P}(t), \tag{14.7}$$

where $\mathbf{P}'(t) = (P'_{ij}(t))$. It is possible to solve (14.7) to obtain

$$\mathbf{P}(t) = e^{Qt} = \mathbf{I} + \sum_{n=1}^{\infty} \frac{Q^n t^n}{n!}. \tag{14.8}$$

We will want to compute $\mathbf{P}(t) = e^{Qt}$, for a given Q. Although Equation (14.8) is easy to calculate for reals, it is by no means easy numerically in general, due to the fact that Q has both positive and negative elements. Let $\lambda = \max q_i$. Then define Q^* by

$$Q = \lambda \begin{pmatrix} -q_1/\lambda & q_{12}/\lambda & q_{13}/\lambda & q_{14}/\lambda \\ q_{21}/\lambda & -q_2/\lambda & q_{23}/\lambda & q_{24}/\lambda \\ q_{31}/\lambda & q_{32}/\lambda & -q_3/\lambda & q_{34}/\lambda \\ q_{41}/\lambda & q_{42}/\lambda & q_{43}/\lambda & -q_4/\lambda \end{pmatrix}$$

$$= \lambda(Q^* - \mathbf{I}),$$

and as $q_i = \sum_{j \neq i} q_{ij}$, Q^* is a stochastic matrix with row sums equal to 1. Therefore,

$$\begin{aligned} e^Q &= e^{\lambda(Q^* - \mathbf{I})} \\ &= e^{\lambda Q^*} e^{-\lambda \mathbf{I}} = e^{\lambda Q^*} e^{-\lambda} \mathbf{I} \\ &= e^{-\lambda} \sum_{n \geq 0} \frac{\lambda^n}{n!} (Q^*)^n. \end{aligned}$$

This sum, because $(Q^*)^n$ is stochastic for all n, has good convergence properties.

A stationary distribution $\pi = (\pi_1 \cdots \pi_4)$ usually means that $\pi \mathbf{P}(t) = \pi$ holds for all t. Note that $\pi \mathbf{I} = \pi$ for all π and this implies from Equation (14.6) that

$$\pi Q = \mathbf{0}. \tag{14.9}$$

A continuous time Markov chain is said to be *stationary* if Equation (14.9) holds and $\pi \mathbf{P}(0) = \pi$. Certainly, Equation (14.9) implies $\pi \mathbf{P}(t) = \pi$ for all t.

14.4.2 Estimating the Rate of Change

It is very natural to wish to estimate quantities related to the rate matrix Q. Note that because we assume $\mathbf{P}(t) = e^{Qt}$ that t and Q are confounded, $Qt = (\alpha Q)(t/\alpha)$ for any $\alpha \neq 0$. This is very sensible, as twice the rate at half the time has the same results.

At time $t = 0$, two species separate and evolve independently according to a continuous Markov chain $\mathbf{P}(t)$ with stationary distribtion π. The letters at time t will be $X(t)$ and $Y(t) \in \mathcal{A}$:

When the initial distribution at $t = 0$ is π, then the rate R of change is easy to calculate:

$$R = \lim_{h \to 0+} \frac{\mathbb{P}(X(t+h) \neq X(t))}{h}$$
$$= \lim_{h \to 0+} \sum_i \frac{\mathbb{P}(X(t+h) \neq i | X(t) = i)}{h} \mathbb{P}(X(t) = i)$$
$$= \sum_i -q_{ii}\pi_i. \tag{14.10}$$

To provide an estimate of R, we need an explicit model. One of the more general models has

$$Q = \begin{pmatrix} -(\gamma - \gamma_1) & \gamma_2 & \gamma_3 & \gamma_4 \\ \gamma_1 & -(\gamma - \gamma_2) & \gamma_3 & \gamma_4 \\ \gamma_1 & \gamma_2 & -(\gamma - \gamma_3) & \gamma_4 \\ \gamma_1 & \gamma_2 & \gamma_3 & -(\gamma - \gamma_4) \end{pmatrix}, \tag{14.11}$$

where $\gamma = \gamma_1 + \gamma_2 + \gamma_3 + \gamma_4$.

Proposition 14.6 *Assuming the Markov chain with rate Q of Equation (14.11),*

(i) $\pi Q = 0$ *implies* $\pi = (\gamma_1/\gamma, \gamma_2/\gamma, \gamma_3/\gamma, \gamma_4/\gamma)$

and

(ii) $R = \gamma(1 - \sum_i \pi_i^2)$.

Proof. (i) Solve $\pi Q = 0$. (ii):

$$-\sum_i \pi_i q_{ii} = \sum_i (\gamma - \gamma_i)\gamma_i/\gamma$$
$$= \gamma \sum_i (1 - \pi_i)\pi_i$$
$$= \gamma(1 - \sum_i \pi_i^2). \qquad \blacksquare$$

We set $F = \sum_i \pi_i^2$. The above proposition tells us that d, the expected number of substitutions between $X(t)$ and $Y(t)$, is

$$d = 2t\gamma(1 - F),$$

Trees and Sequences

as the time separating $X(t)$ and $Y(t)$ is $2t$. We show the quantity $t\gamma$ can be estimated from sequence data. A Markov chain is *reversible* if the probability of going from i to j in time t is the same of going from j to i in time t. From the equations

$$\pi_i P_{ij}(t) = \pi_j P_{ji}(t), \quad \text{all } i, j, t,$$

it follows that $\pi_i q_{ij} = \pi_j q_{ji}$, all i, j.

Lemma 14.2 *Assume that the rate Q of Equation (14.11) is reversible. Then*

$$\mathbb{P}(X(t) = a, Y(t) = b) = \pi_a P_{ab}(2t)$$
$$= \begin{cases} \pi_a(1 - e^{-2\gamma t})\pi_b, & a \neq b, \\ \pi_a(e^{-2\gamma t} + (1 - e^{-2\gamma t})\pi_a), & a = b. \end{cases}$$

Proof.

$$\mathbb{P}(X(t) = a, Y(t) = b) = \sum_c \pi_c P_{ca}(t) P_{cb}(t)$$
$$= \sum_c \pi_a P_{ac}(t) P_{cb}(t)$$
$$= \pi_a \sum_c P_{ac}(t) P_{cb}(t) = \pi_a P_{ab}(2t).$$

To compute the probability, we use $\mathbf{P}(t) = e^{Qt}$. First, note that $Q = -\gamma(\mathbf{I} - P)$, where

$$P = \begin{pmatrix} \pi_1 & \pi_2 & \pi_3 & \pi_4 \\ \pi_1 & \pi_2 & \pi_3 & \pi_4 \\ \pi_1 & \pi_2 & \pi_3 & \pi_4 \\ \pi_1 & \pi_2 & \pi_3 & \pi_4 \end{pmatrix}.$$

Then

$$e^{Qt} = e^{-\gamma t(\mathbf{I}-P)} = e^{-\gamma t} e^{\gamma t P}$$
$$= e^{-\gamma t} \sum_{m=0}^{\infty} \frac{(\gamma t)^m}{m!} P^m$$
$$= e^{-\gamma t} \mathbf{I} + \left(1 - e^{-\gamma t}\right) P,$$

because $P^m = P$ for $m \geq 1$. The formula follows from this matrix equation. ∎

This lemma shows that under these assumptions the location of the root does not change $\mathbb{P}(X(t) = a, Y(t) = b)$. For example, $\mathbb{P}(X(t/2) = a, Y(3t/2) = b) = \pi_a P_{ab}(2t)$. Now we compute the probability that $X(t) \neq Y(t)$.

$$\mathbb{P}(X(t) \neq Y(t)) = \sum_{a \neq b} \pi_a P_{ab}(2t)$$
$$= \sum_{a \neq b} \pi_a \pi_b \left(1 - e^{-2\gamma t}\right)$$
$$= (1 - F)\left(1 - e^{-2\gamma t}\right)$$
$$= B\left(1 - e^{-d/B}\right),$$

where $B = 1 - F$. A famous estimate follows: If D is the random fraction of positions at which $X(t) \neq Y(t)$, we replace $\mathbb{P}(X(t) \neq Y(t))$ by its maximum likelihood estimator D:

$$D = B\left(1 - e^{-d/B}\right)$$

or

$$d = -B \log\left(1 - D/B\right).$$

Note that when D is small, $d \approx -B(-D/B) = D$, supporting the intuition that when the number of changes is small, the maximum likelihood estimate of d is D.

14.4.3 Likelihood and Trees

In Section 14.4.2, we showed how to calculate the probability that $X(t) = a$ and $Y(t) = b$ when the two iid continuous reversible Markov chains had each run for time t, beginning in equilibrium: $\mathbb{P}(X(t) = a \text{ and } Y(t) = b) = \pi_a P_{ab}(2t)$. In fact, if X had run for time t_1 and Y for time t_2, then $\mathbb{P}(X(t_1) = a \text{ and } Y(t_2) = b) = \pi_a P_{ab}(t_1 + t_2)$, so that the "common ancestor" can be placed at any point on the path between X and Y without changing the probability.

We need to generalize this notation to compute the likelihood of a tree with given topology (root $= r$), branch lengths (times), and Q. Just as in the proof of Lemma 14.2, we summed over all assignments of letters to the root vertex; here, we must sum over all fits f to T where $f(v) \in \mathcal{A}$ for $v \in V(T)$. The fit at the tips is determined by the data $\mathbf{a} = (a_1 \cdots a_m)$, where we have m aligned sequences at a position. Then, at each edge $(v, w), w \in \delta(v)$, there is a term $P_{f(v),f(w)}(t(v,w))$, where $t(v,w)$ is the length of edge (v,w). Therefore, the probability of seeing $\mathbf{a}$ at the tips of T is

$$\mathbb{P}(\mathbf{a}) = \sum_f \pi_{f(r)} \prod_{v \in V} \prod_{w \in \delta(v)} P_{f(v),f(w)}(t(v,w)). \qquad (14.12)$$

In case the root has two descendents with letters a and b and branch lengths each t, Equation (14.12) just becomes

$$\mathbb{P}(\mathbf{a}) = \sum_{c \in \mathcal{A}} \pi_c P_{ca}(t) P_{cb}(t).$$

Trees and Sequences

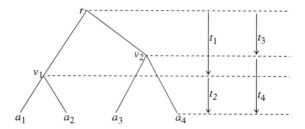

Figure 14.2: *A tree for* $\mathbf{a} = (a_1, a_2, a_3, a_4)$

Recall that $\mathbb{L} = \pi_a P_{ab}(2t) = \pi_b P_{ba}(2t)$ in this case. The fit to the tree in Figure 14.2 assigns letters to r, v_1, and v_2. Equation (14.12) becomes

$$\mathbb{P}(\mathbf{a}) = \sum_{(c,d,e) \in \mathcal{A}^3} \pi_c \left\{ P_{c,d}(t_1) \left(P_{d,a_1}(t_2) P_{d,a_2}(t_2) \right) \right\}$$
$$\times \left\{ P_{c,e}(t_3) \left(P_{e,a_3}(t_4) P_{e,a_4}(t_4) \right) \right\},$$

where $t_1 + t_2 = t_3 + t_4 = t$, $f(r) = c$, $f(v_1) = d$, and $f(v_2) = e$. Because the sum in (14.12) has approximately $|\mathcal{A}|^{|V|}$ terms, usually it is not practical to compute in that form. Instead, there is a method that greatly reduces the computation.

The following algorithm is sometimes called the peeling algorithm. In this context, $T(v)$ is the event that the tree rooted at v has the given letters at the tips. Then

$$A(a, v) = \mathbb{P}(T(v) | f(v) = a)$$

is the probability that with a assigned to v, we see the given letters at the tips of $T(v)$. The logic is similar to that already employed.

$$A(a, v) = \prod_{w \in \delta(v)} \sum_{b \in \mathcal{A}} P_{ab}(t(v, w)) A(b, w)$$
$$= \prod_{w \in \delta(v)} (\mathbf{P}(t(v, w)) \mathbf{A}(w))_a,$$

where $\mathbf{A}(w)$ is a $|\mathcal{A}| \times 1$ column vector. The Schur product $(x_1, x_2, \ldots) \otimes (y_1, y_2, \ldots) = (x_1 y_1, x_2 y_2, \ldots)$ allows us to write these equations in a compact form:

$$\mathbf{A}(v) = \prod_{w \in \delta(v)} \otimes \mathbf{P}(t(v, w)) \mathbf{A}(w).$$

Now we can give the algorithm.

Algorithm 14.4 (Peeling)   until $v = $ root,

$$\mathbf{A}(v) \leftarrow \prod_{w \in \delta(v)} \otimes \mathbf{P}(t(v,w)) A(w)$$

$\mathbb{L} = \pi \otimes \mathbf{A}^t$ (root)

Some Statistical Issues

So far the discussion has been about models and their properties. Difficulties arise when we turn to developing statistical tests for a given model M. Formally we will parameterize M by $\theta \in \Theta$. First, we formally present a null hypothesis. The data are a set of m sequences,

H_0: (1) The m sequences are related by an unknown tree $T \in \mathcal{T}_m$.
(2) The sites have evolved independently according to the model M.

If the tree is known in H_0, then the word "unknown" can be removed in part (1). $\mathcal{T}_m$ denotes the set of labeled trees. Generally, of course, the tree is unknown. We now find the maximum likelihood over all members of H_0:

$$\widehat{\mathbb{L}_0} = \max_{H_0} \mathbb{L}_0 = \max\{\mathbb{L}_0 : \theta \in \Theta, T \in \mathcal{T}_m\}.$$

To assess the likelihood $\widehat{\mathbb{L}_0}$, we need an alternative hypothesis H_1. We make the most general alternate hypothesis possible. The sequences at a given site have one of 4^m possible patterns, which we will denote as $\gamma \in \Gamma$. The most general model for H_1 is that, at each site, pattern γ occurs with probability $p_\gamma \geq 0, \sum_{\gamma \in \Gamma} p_\gamma = 1$. Therefore, the likelihood is given by a multinomial distribution:

$$\mathbb{L}_1 = \prod_{i=1}^{N} p_{\gamma_i}$$
$$= \prod_{\gamma \in \Gamma} (p_\gamma)^{n(\gamma)},$$

where $n(\gamma) = $ the number of sites with pattern γ and there are a total of N sites. The maximum likelihood solution has $\widehat{p}_\gamma = \frac{n(\gamma)}{N}$ and

$$\widehat{\mathbb{L}_1} = \prod_{\gamma \in \Gamma} \left(\frac{n(\gamma)}{N}\right)^{n(\gamma)}.$$

One advantage of H_1 is that it is the most general independent sites model and, therefore, it provides a good test of the hypothesis H_0. Obviously $H_0 \subset H_1$; the hypothesis H_0 is contained in H_1. The log likelihood ratio

$$D = \log\left(\frac{\widehat{\mathbb{L}}_1}{\widehat{\mathbb{L}}_0}\right) = \log\widehat{\mathbb{L}}_1 - \log\widehat{\mathbb{L}}_0$$

is therefore always non-negative. When this difference is small, then H_0 provides a good fit to the data relative to H_1. When the difference D is large, H_0 should be rejected.

In some cases where $H_0 \subset H_1$, there is a general result about the approximate distribution of D. The statistic $2D = 2\log(\widehat{\mathbb{L}}_1/\widehat{\mathbb{L}}_0)$ has an asymptotic χ^2 distribution with degrees of freedom equal to the number of restrictions on the parameters in H_1 required to obtain H_0. Unfortunately, this result does not apply to our situation with models on trees. The most evident reason is sample size. There are 4^m different γs and almost no data sets will have at least 5 observations for each γ that are required for the χ^2 asymptotics. For example with $m = 10$ sequences this means **at least** $5 \cdot 4^{10} = 5 \times 10^6$ sites. Our sequences, typically genes, are much smaller than this. Second, the degrees of freedom are far from obvious, being greatly complicated by the unknown tree. Almost nothing is known about parameterizing trees in situations like this. Another approach must be take! n.

Let $\widehat{\mathbb{L}}_0 = \mathbb{L}_0(\widehat{\theta}, \widehat{T})$ be maximized at $\theta = \widehat{\theta}$ and $T = \widehat{T}$. Then we will use the values $(\widehat{\theta}, \widehat{T})$ to specify a model and to generate m sequences of specified length.

Algorithm 14.5 (Model Test)
```
input:    m sequences a₁ ··· aₘ; N
output:   p-value
   1. estimate H₀ parameters θ̂,T̂
```

$$\widehat{\mathbb{L}}_0 = \mathbb{L}_0(\widehat{\theta}, \widehat{T}) = \max\{\mathbb{L}(\theta, T) : \theta \in \Theta, T \in \mathcal{T}_m\}$$
$$D = \log\widehat{\mathbb{L}}_1 - \log\widehat{\mathbb{L}}_0$$

```
   2. for M repetitions (i = 1 to M)
          generate m sequences of length N from θ̂,T̂
          compute Dᵢ = log L̂₁ - log L̂₀
   3. use the histogram of Dᵢ, i = 1 to M, to estimate the
      p-value of D
```

We emphasize that in step 2 the m sequences are generated from $\widehat{\theta}$ and $\widehat{T}$. Then, $\widehat{\mathbb{L}}_1$ is computed by taking maximum over Θ and $\mathcal{T}_m$.

This plausible procedure is not the final word. Real DNA sequences do not have independent sites nor is the rate of evolution equal at all sites. This is an area of active current research.

Problems

Problem 14.1 Show for all splits $S = (L_S, L_S^c)$ separating i and j that if we take $i \in L_S$, L_S can be ordered by inclusion.

Problem 14.2 Prove that the split distance defines a finite metric on $\mathcal{T}$.

Problem 14.3 Find all eight trees on $\{1, 2, 3\}$ when interior and terminal vertices can be labeled with subsets of $\{1, 2, 3\}$. Show that $\max\{\rho(T_1, T_2) : T_1 \text{ and } T_2 \text{ are trees on } \{1, 2, 3\}\} = 3$, where ρ is the splits metric.

Problem 14.4 Find two trees $T, S \in \mathcal{T}_n$ such that the splits metric is $\rho(T, S) = 2n - 6$.

Problem 14.5 Prove that the nni distance is a finite metric on $\mathcal{T}$.

Problem 14.6 Find the split distance and nni distance between the trees

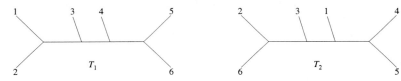

Problem 14.7 Prove Proposition 14.3.

Problem 14.8 Assume $w \in \pi(v)$. Prove: (i) If $F(w) \supset O(v)$, then $O(w) = O(v)$, and (ii) otherwise, $O(w) = F(w) \cup (O(v) \cap S(w))$.

Problem 14.9 Characterize $(F(v), S(v)) : v \in V(T)$ where there is a unique assignment.

Problem 14.10 The quantities $\|f_v(a)\|$ and $M_v(a)$ have been defined for general parsimony. Relate these quantities to the quantities $L(v), F(v)$, and $S(v)$ in case $\lambda(a, b) = \mathbb{I}(a \neq b)$.

Problem 14.11 Prove Proposition 14.4.

Problem 14.12 Let T_n the symmetric binary tree with 2^n terminal vertices (tips) and n edges between each tip and the root. Place iid random letters at the tips according to $\mathbb{P}(\{a\}) = \alpha = 1 - \mathbb{P}(\{b\})$. The first set at each vertex v of height h (where the root has height n) is $F_h(v)$. Define $p_s(h) = \mathbb{P}(F_h(v) = S)$ and $D(h) = p_{\{a\}}(h) - p_{\{b\}}(h)$. Show that

(i) $D(h) = D(h-1)\{1 + p_{\{a,b\}}(h-1)\}$;

(ii) if $p_S(h) \to p_S$ for $S = \{a\}, \{b\}, \{a, b\}$, show $p_{\{a\}} = p_{\{b\}} = p_{\{a,b\}} = 1/3$ if $\alpha = 1/2$ and $p_{\{a\}} = 1, p_{\{b\}} = p_{\{a,b\}} = 0$ if $\alpha > 1/2$.

Chapter 15

Sources and Perspectives

In this chapter, I give some background references and citations to key original material, especially in the cases where proofs are omitted in the text. Also, I will discuss some of the interesting current work in the area. In this way I hope to allow an interested reader to trace to background of the ideas and to find where the action is today. A complete survey of the literature would double the size of this book.

15.1 Molecular Biology

Two historical papers are well worth reading: Pauling et al. (1951) describe the helical structure of protein, and Watson and Crick (1953) describe the double stranded helical structure of DNA. The textbooks by Watson et al. and by Lewin must be under constant revision, and they are frequently revised. If you read the latest edition of one of these, you will have a general knowledge of molecular biology. The field, as I have emphasized, is rapidly changing. To keep in touch with those developments, *Science* and *Nature* are recommended. They often have scientific news as well as discussions and reviews of current developments for the non-specialist. Many other journals feature molecular biology, but it is difficult for a non-biologist to follow all of them. Many interesting articles dealing with mapping and sequencing appear in *Genomics*.

15.2 Physical Maps and Clone Libraries

Restriction maps are discussed by Nathans and Smith (1975). Benzer (1959) established the fact that the DNA making up a bacterial gene was linear. His data and analysis of the overlap of gene fragments motivated the development of interval graphs. The relationship between restriction maps and interval graphs in

Chapter 2 is a return to the ideas of Benzer. The material in Chapter 2 is taken from Waterman and Griggs (1986).

While developing an algorithm to solve DDP, Goldstein and Waterman (1987) discovered the phenomenon of multiple solutions (Chapter 3). They prove Theorems 3.2 and 3.3. Schmitt and Waterman (1991) try to get at the basic combinatorial nature of multiple solutions, and present the idea of cassette equivalence classes that were later characterized by Pevzner (1994). Pevzner's surprising solution is given here as Theorem 3.9. Subadditive methods give no information about the expected number of cassette equivalence classes. Daniela Martin recently made progress on this problem.

The proof that DDP is NP-complete appears in Goldstein and Waterman (1987). The standard reference for NP-complete problem is Garey and Johnson (1979). The simulated annealing algorithm gives a nice motivation for presenting Markov chains. The heuristic for Theorem 4.4 is much clearer than I have seen elsewhere; Larry Goldstein showed it to me. A number of DDP algorithms have been proposed, but to my knowledge, none of them is completely satisfactory when applied to real data.

DDP is far from the only restriction mapping problem that has been studied. For example, the data might be obtained from a partial digest which gives lengths between all pairs of restriction sites. Some very interesting results have been obtained, some based on a paper of Rosenblatt and Seymour (1982) which solves a related combinatorial problem by an algebraic method. See Skiena, Smith, and Lemke (1990) and Skiena and Sundaram (1994).

The use of length data to make restriction maps has been around 20 years. Recently, Schwartz et al. (1993) have been able to observe cuts made in single DNAs by restriction enzymes and to estimate the length of the fragments, thus reducing the problem to one of observation rather than inference. It is not clear how these techniques will develop, but I am optimistic that they will be successful. Of course, new computational problems are associated with new techniques.

The representation of genomic DNA in clone libraries was taken up by Brian Seed in two papers in 1982. Also, see Clarke and Carbon (1976). Chapter 5 represents my attempts to understand his papers. Theorem 5.2 is from work of Port and myself and was motivated by Tang and Waterman (1990).

Physical mapping of genomes began with two papers Coulson et al. (1986) and Olson et al. (1986) and was followed by the paper of Kohara et al. (1987) cited in Chapter 6. Figure 6.6 of the *E. coli* genome was provided by Ken Rudd. [See Rudd (1993).] Eric Lander and I began the mathematical analysis of these experiments in our 1988 paper. Our analysis was based on a discrete, base-by-base model. When Arratia et al. (1991) gave a Poisson processes model for STS anchors, it was clear that their approach was the "correct" one for this class of problems. Sections 6.2 and 6.3 are from those papers. Now, these are experiments that in a sense both map and sequence at the same time. The ends of clones are sequenced and clone maps are made by overlapping the sequenced ends. See Edwards and Caskey (1991). There are, therefore, two types of islands:

Sources and Perspectives

overlapped clones with unsequenced regions and islands of overlapped sequence. This problem is studied in Port et al. (1995). The book of Hall (1988) gives a general treatment of coverage processes.

The clever idea of pooling clones in Section 6.3 is due to Glen Evans as a way to escape the combinatorial difficulties of oceans and islands. See Evans (1991) for a review. The material in Section 6.4 is taken from Alizadeh et al. (1992).

15.3 Sequence Assembly

The development of DNA sequencing technology is a very interesting topic. Howe and Ward (1989) provide a user's handbook for DNA sequencing. Hunkapiller et al. (1991) discuss some current issues for genome sequencing.

Testing if there exists a superstring of a given length was shown to be NP-complete by Gallant et al. (1980). The clever extension to SRP introducing the realities of errors and reversed fragments is in the thesis of Kececioglu (1991).

The first approximation algorithm for SSP was given by Blum et al. (1991) with a bound of 3 times optimal. The proof given in 7.1.2 was shown to me by Gary Benson. No example is known with a bound smaller than 2 for the greedy algorithm. The bound of 2.793 is given by Kosaraju et al. (1994). Stein and Armen (1994) have a bound of 2.75.

The greedy algorithm described in Section 7.1.3 first appears in Staden (1979). Many modifications have been made, speeding up various steps in the algorithm. See in particular the thesis of Kececioglu (1991) and Huang (1992).

The Section 7.1.4 on sequence accuracy is from Churchill and Waterman (1992).

SBH was proposed independently by several people; Drmanac et al. (1989) is a good reference. The basic computer science was laid out by Pevzner (1989). The generalized sequencing chips were proposed by Pevzner and Lipshutz (1994). Dyer et al. (1994) study the problem of unique solutions for SBH.

The idea for the hybrid algorithm of Section 7.3 came while preparing this chapter. More details and an analysis of the behavior of the algorithm in practice is given in Idury and Waterman (1995).

15.4 Sequence Comparisons

Although the material of Chapters 8, 9, and 10 overlap, it is easiest to organize the comments by chapter.

15.4.1 Databases and Rapid Sequence Analysis

There are now databases for organisms, for specific molecules (such as hemoglobin, tRNA, 16SRNA, ...), for projects (such as physical or genetic mapping of specific chromosomes), and, of course, for nucleic acid and protein sequences. For access to biological databases, see Genome Data Bank (GDB) at gopher.gdb.org. Linking some of these databases is discussed in Fasman (1994).

Martinez (1983) made the first use of suffix trees for DNA and protein sequence analysis. The expected running time of the simple algorithm 8.4 (repeats) is $O(n \log n)$ where $n =$ sequence length. He exploited the data structure to infer multiple sequence alignment. A somewhat more sophisticated algorithm runs in time $O(n)$ [Aho et al. (1983)]. Currently, there are a number of efforts to extend the application of suffix trees to find approximate matchings.

In an important paper, Dumas and Ninio (1982) describe and apply to DNA sequences the concepts of hashing (Section 8.4). This basic idea was then used as in Section 8.5 for sequence comparison by hashing [Wilbur and Lipman (1983)]. The current implementations apply the local dynamic programming algorithm (Section 9.6) restricted to regions that score high in the k-tuple analysis [Pearson and Lipman (1988)].

The ideas of Section 8.6, filtration by k-tuple matches, trace back to Karp and Rabin (1987) for exact matching. This section is taken from Pevzner and Waterman (1993). See also Baeza-Yates and Perleberg (1992) and Wu and Manber (1992).

BLAST, probably the most used fast database search technique, is not discussed in the text [Altschul et al. (1990)]. The method properly belongs in Chapter 8 by subject but uses methods that come later. In particular, a sequence is preprocessed to find all neighboring words (Section 10.6.1) that have statistically significant scores against an interval of the query sequence (Theorem 11.17). Then a technique from computer science related to suffix trees is used to simultaneously search the database for instances of the significant candidate words [Aho and Corasick (1975)].

Recently, some deep results have been obtained that allow database searches in time linear or sublinear in database size. Ukkonen, at the cost of increased storage, gave the first result of this type [Ukkonen (1985)]. Myers (1990) proposes a method that requires a prebuilt inverted index which must be built before the search. He finds long matches with less than fraction ϵ of errors. Chang and Lawler (1990) present an algorithm that is very fast for random text, basing their analysis on expected search time. This algorithm is on-line, that is, does not require preprocessing.

15.4.2 Dynamic Programming for Two Sequences

The solution to the recursion counting alignments in Theorem 9.1 is given in Laquer (1981). The problem was first posed by me and communicated to Laquer by Paul Stein. Theorem 9.3 and Corollary 9.1 are from Griggs et al. (1986).

The first dynamic programming sequence alignment algorithm was Needleman and Wunsch (1970). It was a global similarity algorithm that had gap penalty $g(k) = \alpha$ and that did not allow insertions and deletions to be adjacent. The paper of Sankoff (1972) was a significant contribution. Then, Sellers (1974) presented a global distance algorithm with single letter indel weights. The distance function $d(a, b)$ on the alphabet was general. At the same time, Wagner and Fischer (1974) gave the same algorithm to solve a string matching problem in computer science. The question of the relationship between distance and simulating became heated in some circles. Mathematicicans felt that distance was more rigorous and, therefore, the correct method. In Smith et al. (1981), the equivalence given in Theorem 9.11 is established. To accomplish this, I redefined the Needleman-Wunsch algorithm to be comparable to the distance algorithm with general gap function and removed the restrictions. This recast algorithm is what today most people know as "the Needleman-Wunsch" algorithm.

Alignment algorithms with general gap functions as in Theorem 9.5 appeared in Waterman et al. (1976). In the late 1970s, I first learned of Theorem 9.6 for the gap functions $g(k) = \alpha + \beta(k - 1)$ from a manuscript by Paul Haberli of the University of Wisconsin that was widely circulated. The same algorithm independently appears in Gotoh (1982). The concave gap problem was raised but not satisfactorily solved in Waterman (1984). See Myers and Miller (1988) and Galil and Giancarlo (1989) for the $O(n^2 \log n)$ algorithms referred to by Theorem 9.7. Position-dependent weighting is sometimes thought to be difficult, but Section 9.3.2 gives $O(n^2)$ algorithms. See, for example, Gribskov et al. (1987).

The distance version of Section 9.5 on the fitting algorithm is due to Sellers (1980). The local alignment algorithm is also known as the Smith-Waterman algorithm (1981). Declumping was developed by Waterman and Eggert (1987). The tandem repeats algorithm of Section 9.6.2 was introduced by Myers and Miller (1988), and Landau and Schmidt (1993). Hirschberg (1975) gave the elegant linear space method of Section 9.7. In a nontrivial paper, Huang and Miller (1991) extend the idea to include the local algorithm. Near-optimal alignments were introduced by Waterman (1983). Naor and Brutlag (1993) look more deeply into the combinatorial structure. The second method for near-optimal alignments (2) was first presented by Vingron and Argos (1990).

Section 9.9 on inversions is from Schöniger and Waterman (1992). At a less fine scale than sequence, comparing chromosomes by gene orders is a very interesting recent development. Assuming the order of genes is known in two chromosomes, the problem is to find the minimum number of inversions to change one gene order into the other. The first mathematical paper was Watterson et al. (1982) and then Sankoff brought the problem into prominence. See Sankoff and Goldstein (1988) and Kececioglu and Sankoff (1994). In Bafna and Pevzner (1993), the two permutations at maximum distance is established, and in Bafna and Pevzner (1994), they turn to transposition distance.

Map alignment algorithm 9.12 was first given in Waterman et al. (1984). This $O(n^4)$ (when $m = n$) algorithm was reduced to $O(n^2 \log n)$ by Myers and Huang (1992). For generalizations that include cases such as multiple matching and are motivated by experimental data, see Huang and Waterman (1992).

Fitch and Smith (1981) addressed parametric alignment. Gusfield et al. (1992)

and Waterman et al. (1992) independently developed general algorithms. Vingron and Waterman (1994) studied the implication to biological problems. Theorem 9.15 appears in a thesis by Xu (1990) and in Gusfield et al. (1992) where, in addition, Propositions 9.3 and 9.4 are found.

Problem 9.18 is taken from a paper on cryptogenes [von Haeseler et al. (1992)]. Problem 9.19 on DNA and protein alignment has been rediscovered several times.

15.4.3 Multiple Sequence Alignment

The discussion of the CF gene could have appeared in Chapter 8, but the required tools for local alignment were not developed there. Instead, we use the CF story to motivate alignment. The original analysis appeared in Riordan et al. (1989).

Whereas the naive r-sequence alignment algorithm in Section 10.2 first appeared in Waterman et al. (1976), a deeper look at the problem had already been given by Sankoff (1975) who considered the relevant problem of aligning r sequences **given** an evolutionary tree.

The volume reduction idea appears in Carillo and Lipman (1988). See also Kececioglu (1993) for a branch and bound algorithm that achieves volume reduction in the r-sequence dynamic programming problem.

Weighted-average sequences give a nice context to place several ideas in multiple alignments, especially where iterative methods are involved [Waterman and Perlwitz (1984)]. The profile method of Gribskov et al. (1987) is an example of weighted-average sequences. Statistical significance of profile scores $T(P, \mathbf{b})$ is easy to motivate but requires more analysis than might be guessed from Section 10.4.1. See Goldstein and Waterman (1994).

Hidden Markov models can be viewed as iterated weighted-average alignment. The Markov model gives a nice stochastic framework to motivate weighting, and related issues. See Krogh et al. (1994).

Consensus word analysis allows us to look for approximate matches to unknown patterns without alignment. In Galas et al. (1985), a set of *E. coli* sequences are analyzed by these methods.

15.5 Probability and Statistics

Chapters 11 and 12 are of quite distinct character. Sequence alignment scores are optimized over a large number of alignments and, thus, involve large deviations. Counts of common pattern in sequences might not involve large deviations at all.

15.5.1 Sequence Alignment

The first probability result for longest common subsequence length is Chvátal and Sankoff (1975). For more recent work on this difficult and interesting topic, see Alexander (1994) and Dančik and Paterson (1994). The Azuma-Hoeffding application to global alignment is developed in Arratia and Waterman (1994). Basic probability material such as the central limit theorem, the Borel-Cantelli lemma (Theorem 11.9), and Kingman's theorem 11.2 can be found in Durrett (1991). Theorem 11.5 is an application of Steele (1986).

The treatment of large deviations for binomials is taken from the tutorial of Arratia and Gordon (1989) where Theorem 11.8 is carefully proved.

The $\log(n)$ law for local exact matching was discovered independently by Karlin and co-workers [Karlin et al. (1983)] and by Arrata and myself at USC [Arratia and Waterman (1985)]. Karlin's first treatment gave mean and variance whereas Arratia and I first studied strong laws. Theorem 11.10 is from the basic paper of Erdös and Rényi (1970), motivating Theorems 11.11, 11.12, and 11.14. See Arratia and Waterman (1985, 1989).

Matching with scores is an important topic. Arratia et al. (1988) opens up this area with restricted scoring allowed. They prove Theorem 11.17 in that setting. The first nontrivial results on the special distributions of letters in optimal local alignments is in Arratia and Waterman (1985). Karlin and Altschul (1990) give the theorem in the generality stated here. I learned the heuristic proof from Richard Arratia.

The heuristic motivating the extreme value results in Section 11.3 was due to Louis Gordon. Theorem 11.19 is from Gordon et al. (1986) and Theorem 11.20 is from Arratia et al. (1986).

Following up a suggestion of Persi Diaconis, Arratia et al. (1989) wrote a beautiful paper that makes the work of Chen (1975) available to a general audience. Here, I give a less general but useful Theorem 11.22. This makes the $O(\frac{\log n}{n})$ bound in Equation (11.13) easy. The material in Section 11.5.2 on exact matching between sequences is the more straightforward version of a distributional Erdös-Rényi law [Arratia et al. (1990)].

Section 11.5.3 on approximate matching is taken from Goldstein and Waterman (1992).

The phase transition result in Section 11.6 is from Arratia and Waterman (1994). The motivation for the practical application Theorem 11.30 is from Karlin and Altschul (1990). The remainder of Section 11.6.2 is taken from Waterman and Vingron (1994a,b). Motivation for the Poisson clumping heuristic is contained along with many other insights in the book *Probability Approximation via the Poisson Clumping Heuristic* by David Aldous (1989).

15.5.2 Sequence Patterns

There is a large literature about patterns in sequences. Theorem 12.1 is still being rediscovered, for example. Section 12.1 and Problem 12.3 are taken from the Ph.D. thesis of Ron Lundstrum (1990) which unfortunately remains unpublished. Instead of using generating functions such as in Pevzner et al. (1989), the general Markov setting for overlapping words and central limit theorems for Markov chains can be applied. Theorem 12.3 is in Ibragimov (1975), and Theorem 12.4 is in Billingsley (1961). See Kleffe and Borodovsky (1992) for some related work.

There is another approach to central limit theorems for counts which traces back to Whittle (1955). The idea is to condition on statistics observed in $A_1, \ldots, A_n$, such as the number of letters and/or pairs of letters of each type. It is then possible to prove a conditional central limit theorem. This appears in Prum et al. (1995).

For some interesting work on word distributions in DNA sequences, see Burge et al. (1992) and Kozhukhin and Pevzner (1991). The fit of Markov models is also discussed in Pevzner (1992).

Nonoverlapping count distributions can be derived using renewal theory. Feller worked out the one pattern case in 1948, and Chapter 8 of his classic book Feller (1971) is highly recommended. Also, see Guibas and Odlyzko (1981) who were motivated by string search algorithms. Generalizing that work to mutiple patterns in the iid case is found in Breen et al. (1985). See Problems 12.8–12.10. For the Markov case, the correct probability setting is Markov renewal processes. See Biggins and Cannings (1987) for a full treatment.

A deeper analysis of Li's method (Section 12.2.2) relies on stopping times of martingales [Li (1980)]. The Markov and multiple pattern discussion is from the straightforward exposition of Biggins and Cannings (1986).

Applying Chen-Stein Poisson approximation to counting patterns is natural. Section 12.3 is an adaptation of Goldstein and Waterman (1992).

Section 12.4 on site distributions is from Churchill et al. (1990) which gave an analysis of the eight-enzyme map of *E. coli* given by Kohara et al. (1987). Later, Karlin and Macken (1991) generalized these concepts and applied them to the same data set.

Certainly, there are other interesting patterns in sequences. For example, Trifonov and Sussman (1980) use an autocorrelation function to discover periods in DNA sequences.

15.6 RNA Secondary Structure

Computational approaches to RNA secondary structure begin with work by Tinoco and colleagues (1971, 1973) who introduced a 0-1 basepairing matrix and later a matrix of free energies between potential basepairs.

Dynamic programming methods for RNA structure began with Nussinov et al. (1978) and Waterman (1978). The Nussinov method is essentially Theorem 13.6 and maximizes the number of basepairs. Waterman proposed a multiple pass method that handled energy functions in full generality. An excellent review of the literature is Zuker and Sankoff (1984). Sections 13.2.1–13.2.3 are from Waterman and Smith (1986).

As the recursions for the algorithms can also be used for combinatorics, it was natural to study the number of structures. Theorem 13.1 is from Waterman

(1978) and Theorem 13.2 is from Stein and Waterman (1978). The theorem used in the proof of Theorem 13.2 is from Bender (1974). Theorem 13.4 appears in Howell et al. (1980). Section 13.1.1 is from Schmitt and Waterman (1994). An extension of those ideas was made by applying topological techniques from Techmüller theory and the theory of train tracks in Penner and Waterman (1993). The last problem giving the $M\mathbf{v} = \mathbf{v}$ characterization of secondary structures is from Magarshak and Benham (1992), who pose the problem: Given $\mathbf{v}$, find all M that solve $(M - I)\mathbf{v} = \mathbf{0}$, an inverse eigenvalue problem.

Knotted structures are of interest as well. Here bases unpaired in the secondary structure can form pairs, possibly with a different energy function. In this way bases in loops can interact. One way of viewing this problem is as two disjoint secondary structures with free-energies X and Y. A mathematical problem is to minimize $X + Y$. This formation is due to Michael Zuker.

Clearly finding RNA secondary structure is closely related to sequence alignment, and so the parallel problem in alignment is to find disjoint common subsequences between two sequences such that the sum of the lengths is maximized. Pavel Pevzner has pointed out this problem is solved using Young tableaux. See Chastain and Tinoco (1991) for a discussion of structural elements in RNA.

15.7 Trees and Sequences

The material on splits (Section 14.1.1) is part of an interesting paper by Bandelt and Dress (1986). The algorithm for obtaining trees from splits first appeared in Meacham (1981). The $O(nm)$ algorithm is due to Gusfield (1991).

The splits metric was developed by Robinson and Foulds (1981). The nearest neighbor interchange metric is from Waterman and Smith (1978).

The beautiful results on additive trees are in a paper by Buneman (1971) contained in a volume on archaeology. The algorithm for constructing additive trees is in Waterman et al. (1977). There is a vast literature on nonadditive trees. Fitch and Margoliash (1967) and Cavalli-Sforza and Edwards (1967) are early and very influential papers. For some recent work, see Rzhetsky and Nei (1992 a,b).

The basic parsimony algorithm for sequences is due to Fitch (1971). The proof by Hartigan (1973) appeared later. Then Sankoff and Rousseau (1975) introduced weighted parsimony.

Maximum likelihood methods for trees started with Felsenstein (1981). The material in Section 14.4 is mostly taken from Tavaré (1986). Statistical tests of these models are discussed in Goldman (1993). It is popular to apply bootstrapping techniques for evolutionary confidence in evolutionary trees. Hillis and Bull (1993) present a critical empirical study.

References

Aho, A.V., and Corasick, M., (1975) Efficient string matching: An aid to bibliographic search. *Comm. ACM*, **18**, 333–340.

Aho, A.V., J.E. Hopcroft and Ullman, J.D. (1983) *Data Structures and Algorithms*, Addison-Wesley, Reading, MA.

Aldous, D. (1989) *Probability Approximations via the Poisson Clumping Heuristic*. Springer-Verlag, New York.

Alexander, K. (1994) The rate of convergence of the mean length of the longest common subsequence. *Ann. Probab.*, **4**, 1074–82.

Alizadeh, F., Karp, R.M., Newberg, L.A. and Weisser, D.K. (1992) Physical mapping of chromosomes: A combinatorial problem in molecular biology. *Symposium on Discrete Algorithms*.

Altschul, S.F., Gish, W., Miller, W., Myers, E.W. and Lipman, D. (1990) Basic local alignment search tool. *J. Mol. Biol.*, **215**, 403–410.

Armen, C., and Stein, C. (1994) A $2\frac{3}{4}$-Approximation algorithm for the shortest superstring problem. Dartmouth Technical Report PCS-TR94-214.

Arratia, R., Goldstein, L. and Gordon, L. (1989) Two moments suffice for Poisson approximation: The Chen-Stein method. *Ann. Probab.*, **17**, 9–25.

Arratia, R. and Gordon, L. (1989) Tutorial on large deviations for the binomial distribution. *Bull. Math. Biol.*, **51**, 125–131.

Arratia, R., Gordon, L. and Waterman (1986) An extreme value theory for sequence matching. *Ann. Statist.*, **14**, 971–993.

Arratia, R., Gordon, L. and Waterman, M.S. (1990) The Erdös-Rényi law in distribution, for coin tossing and sequence mathcing. *Ann. Stat.*, **18**, 539–570.

Arratia, R., Lander, E.S., Tavaré, S. and Waterman, M.S. (1991) Genomic mapping by anchoring random clones: A mathematical analysis. *Genomics*, **11**, 806–827.

Arratia, R., Morris, P., and Waterman, M.S. (1988) Stochastic scrabble: Large deviations for sequences with scores. *J. Appl. Prob.*, **25**, 106–119.

Arratia, R. and Waterman, M.S. (1985) Critical phenomena in sequence matching. *Ann. Probab.*, **13**, 1236–1249.

Arratia, R. and Waterman, M.S. (1989) The Erdös-Rényi strong law for pattern matching with a given proportion of mismatches. *Ann. Probab.*, **17**, 1152–1169.

Arratia, R. and Waterman, M.S. (1994) A phase transition for the score in matching random sequences allowing deltions *Ann. Appl. Probab.*, **4**, 200–225.

Baeza-Yates, R.A. and Perleberg, C.H. (1992) Fast and practical approximate string matching. *Lecture Notes in Computer Science, Combinatorial Pattern Matching*, **644**, 185–192.

Bafna, V. and Pevzner, P.A. (1993) Genome rearrangements and sorting by reversals. *In Proceedings of the 34th Annual IEEE Symposium on Foundations of Computer Science*, November 1993, 148–157.

Bafna, V. and Pevzner, P.A. (1995) Sorting by transpositions. *Proc. 6th ACM-SIAM Symp. on Discrete Algorithms*, 614–623.

Bandelt, H.-J. and Dress, A. (1986) Reconstructing the shape of a tree from observed dissimilarity data. *Adv. Appl. Math.*, **7**, 309–343.

Bender, E.A. (1974) Asympotitc methods in enumeration. *SIAM Rev.*, **16**(4), 485–515.

Benzer (1959) On the topology of genetic fine structure. *Proc. Natl. Acad. Sci. USA*, **45**, 1607–1620.

Biggins, J.D. and Cannings, C. (1986) Formulae for mean restriction fragment lengths and related quantities. Research Report No. 274/86, Department of Probability and Statistics, University of Sheffield.

Biggins, J.D. and Cannings, C. (1987) Markov renewal processes, counters and repeated sequences in Markov chains. *Adv. Appl. Math.*, **19**, 521–545.

Billingsley, P. (1961) *Statistical Inferences for Markov Processes*. The University of Chicago Press, Chicago.

Blum, A., Jiang, T., Li, M., Tromp, J. and Yannakakis, M. (1991) Linear approximation of shortest superstrings. *Proceedings of* the *23rd ACM Symp. on Theory of Computing*, New Orleans, USA, May 6–8,1991, ACM Press, New York, 328–336.

Breen, S., Waterman, M.S. and Zhang, N. (1985) Renewal theory for several patterns. *J. Appl. Probab.*, **22**, 228–234.

Buneman, P. (1971) The recovery of trees from measures of dissimilarity. In F.R. Hudson, D.G. Kendall and P. Tautu, eds. *Mathematics in the Archaeological and Historical Sciences.* Edinburgh Univ. Press, Edinburgh, 387–395.

Burge, C., Campbell, A.M. and Karlin, S. (1992) Over- and underrepresentation of short oligonucleotides in DNA sequences. *Proc. Natl. Acad. Sci. USA*, **89**, 1358–1362.

Carillo, H. and Lipman, D. (1988) The multiple sequence alignment problem in biology. *SIAM J. Appl. Math.*, **48**, 1073–1082.

Cavalli-Sforza, L.L. and Edwards, A.W.F. (1967) Phylogenetic analysis: Models and estimation procedures. *Am. J. Hum. Genet.*, **19**, 233–257.

Chang, W.I. and Lawler, E.L. (1990) Approximate string matching in sublinear expected time. *Proc. 31st Annual IEEE Symposium on Foundations of Computer Science*, St. Louis, MO, October 1990, 116–124.

Chastain, M. and Tinoco, I.J. (1991) Structural elements in RNA. *Progress in Nucleic Acid Res. Mol. Biol.*, **41**, 131–177.

Chen, L.H.Y. (1975) Poisson approximation for dependent trials. *Ann. Probab.*, **3**, 534–545.

Churchill, G.A., Daniels, D.L. and Waterman, M.S. (1990) The distribution of restriction enzyme sites in *Ercherichia coli*. *Nucleic Acids Res.*, **18**, 589–597.

Churchill, G.A. and Waterman, M.S. (1992) The accuracy of DNA sequences: Estimating sequence quality. *Genomics*, **89**, 89–98.

Chvátal, V. and Sankoff, D. (1975) Longest common subsequences of two random sequences. *J. Appl. Probab.*, **12**, 306–315.

Clarke, L. and Carbon, J. (1976) A colony bank containing synthetic ColE1 hybrid plasmids representative of the entire *E. coli* gene. *Cell*, **9**, 91–101.

Coulson, A., Sulston, J., Brenner, S. and Karn, J. (1986) Toward a physical map of the genome of the nematode, *Caenorhabditis elegans*. *Proc. Natl. Acad. Sci. USA*, **83**, 7821–7825.

Dančik, V. and Paterson, M. (1994) Upper bounds for the expected length of a longest common subsequence of two binary sequences. *Proc. STACS 94*, Springer-Verlag, Heidelberg, 669–678.

Dembo, A., Karlin, S. and Zeitouni, O. (1995) Multi-sequence scoring. *Ann. Probab.* In press.

Drmanac, R., Labat, I., Brukner, I. and Crkvenjakov, R. (1989) Sequencing of megabase plus DNA by hybridization: Theory of the method. *Genomics*, **4**, 114–128.

Dumas, J.P. and Ninio, J. (1982) Efficient algorithms for folding and comparing nucleic acid sequences. *Nucleic Acids Res.*, **80**, 197–206.

Durrett, R. *Probability: Theory and Examples*, Wadsworth, Inc., Belmont, CA, 1991.

Dyer, M., Frieze, A. and Suen, S. (1994) The probability of unique solutions of sequencing by hybridization *J. Comput. Biol.*, **1**, 105–110.

Edwards, A. and Caskey, C.T. (1991) Closure strategies for random DNA sequencing. METHODS: *A Companion to Methods in Enzymology*, **3**, 41–47.

Erdös, P. and Rényi, A. (1970) On a new law of large numbers. *J. Anal. Math.*, **22**, 103–111.

Evans, G.A. (1991) Combinatoric strategies for genome mapping. *BioEssays*, **13**, 39–44.

Fasman, K.H. (1994) Restructuring the genome data base: A model for a federation of biological data bases. *J. Comput. Biol.*, **1**, 165–171.

Feller, W. (1971) *An introduction to Probability Theory and its Applications*. Volume I, 3rd ed. John Wiley and Sons, Inc., Canada.

Felsenstein, J. (1981) Evolutionary trees from DNA sequences: A maximum likelihood approach. *J. Mol. Evol.*, **17**, 368–376.

Fitch, W.M. (1971) Toward defining the course of evolution: Minimum change for a specific tree topology. *Syst. Zool.*, **20**, 406–416.

Fitch, W.M. and Margoliash, E. (1967) Construction of phylogenetic trees. *Science*, **155**, 279–284.

Fitch, W.M. and Smith, T.F. (1981) Optimal sequence alignments. *Proc. Natl. Acad. Sci. USA*, **80**, 1382–1386.

Galas, D.J., Eggert, M. and Waterman, M.S. (1985) Rigorous pattern recognition methods for DNA sequences: analysis of promoter sequences from *E. coli. J. Mol. Biol.*, **186**, 117–128.

Galil, Z. and Giancarlo, R. (1989) Speeding up dynamic programming with applications to molecular biology. *Theor. Comput. Sci.*, **64**, 107–118.

Gallant, J., Maier, D. and Storer, J. (1980) On finding minimal length superstrings. *J. Comp. Systems Sci.*, **20**, 50–58.

Garey, M.R. and Johnson, D.S. (1979) *Computers and intractability: A guide to the theory of NP-completeness*. Freeman, San Francisco.

Geman, S. and Geman, D. (1984) Stochastic relaxation, Gibbs distribution, and the Bayesian restoration of images. *IEEE Trans. Pattern Anal. Mach. Intell.*, **6**, 721–741.

Goldman, N. (1993) Statistical tests of models of DNA substitution. *J. Mol. Evol.*, **36**, 182–198.

Goldstein, L. and Waterman, M.S. (1987) Mapping DNA by stochastic relaxation. *Adv. Appl. Math.*, **8**, 194–207.

Goldstein, L. and Waterman, M.S. (1992) Poisson, compound Poisson and process approximations for testing statistical significance in sequence comparisons. *Bull. Math. Biol.*, **54**(5), 785–812.

Goldstein, L. and Waterman, M.S. (1994) Approximations to profile score distributions. *J. Comput. Biol.*, **1**, 93–104.

Golumbic, M.C. (1980) *Algorithmic Graph Theory and Perfect Graphs*, Academic Press, New York.

Gordon, L., Schilling, M. and Waterman, M.S. (1986) An extreme value theory for long head runs. *Probab. Theory Rel. Fields*, **72**, 279–287.

Gotoh, O. (1982) An improved algorithm for matching biological sequences. *J. Mol. Biol.*, **162**, 705–708.

Gribskov, M., McLachlan, A.D. and Eisenberg, D. (1987) Profile analysis: Detection of distantly related proteins. *Proc. Natl. Acad. Sci. USA*, **84**, 4355–4358.

Griggs, J.R., Hanlon, P.J. and Waterman, M.S. (1986) Sequence alignments with matched sections. *SIAM J. Alg. Disc. Meth.*, **7**, 604–608.

Guibas, L.J. and Odlyzko, A.M. (1981) String overlaps, pattern matching, and nontransitive games. *J. Comb. Theory, Series A*, **30**, 183–208.

Gusfield, D. (1991) Efficient algorithms for inferring evolutionary trees. *Networks*, **21**, 19–28.

Gusfield, D., Balasubramanian, K. and Naor, D. (1992) Parametric optimization of sequence alignment. *Proc. of the Third*

Annual ACM-SIAM Symposium on Discrete Algorithms, 432–439.

Hall, P. (1988) *Introduction to the Theory of Coverage Processes*, John Wiley and Sons, New York.

Hartigan, J.A. (1973) Minimum mutation fits to a given tree. *Biometrics*, **29**, 53–65.

Hillis, D.M. and Bull, J.J. (1993) An empirical test of bootstrapping as a method for assessing confidence in phylogenetic analysis. *Syst. Biol.*, **42**, 182–192.

Hirschberg, D.S. (1975) A linear space algorithm for computing maximal common subsequences. *Commun. ACM*, **18**, 341–343.

Howe, C.M. and Ward, E.S. (1989) *Nucleic Acids Sequencing: A Practical Approach*. IRL Press, Oxford.

Howell, J.A., Smith, T.F. and Waterman, M.S. (1980) Computation of generating functions for biological molecules. *SIAM J. Appl. Math.*, **39**, 119–133.

Huang, X. (1992) A contig assembly program based on sensitive detection of fragment overlaps. *Genomics*, **14**, 18–25.

Huang, X. and Miller, W. (1991) A time-efficient, linear-space local similarity algorithm. *Adv. Appl. Math.*, **12**, 337–357.

Huang, X. and Waterman, M.S. (1992) Dynamic programming algorithms for restriction map comparison. *Comp. Appl. Biol. Sci.*, **8**, 511–520.

Hunkapiller, T., Kaiser, R.J., Koop, B.F. and Hood, L. (1991) Large-scale and automated DNA sequence determination. *Science*, **254**, 59–67.

Ibragamov, I. (1975) A note on the central limit theorem for dependent random variables. *Theory Probab. Appl.*, **20**, 135–141.

Idury, R., and Waterman, M.S. (1995) A new algorithm for shotgun sequencing. *J. Comp. Biol.* In press.

Karlin, S. and Altschul, S.F. (1990) Methods for assessing the statistical significance of molecular sequence features by using general scoring schemes. *Proc. Natl. Acad. Sci. USA*, **87**, 2264–2268.

Karlin, S., Ghandour, G., Ost, F., Tavaré, S. and Korn, L.J. (1983) New approaches for computer analysis of nucleic acid sequences. *Proc. Natl. Acad. Sci. USA*, **80**, 5660–5664.

Karlin, S. and Macken, C. (1991) Assessment of inhomogeneities in an *E. coli* physical map. *Nucleic Acids Res.*, **19**, 4241–4246.

References

Karp, R.M. and Rabin, M.O. (1987) Efficient randomized pattern-matching algorithms. *IBM J. Res. Develop.*, **31**, 249–260.

Kececioglu, J. (1991) Exact and approximate algorithms for sequence recognition problems in molecular biology. Ph.D. Thesis, Department of Computer Science, University of Arizona.

Kececioglu, J. (1993) The maximum weight trace problem in multiple sequence alignment. *Lecture Notes in Computer Science, Combinatorial Pattern Matching*, **684**, 106–119.

Kececioglu, J. and Sankoff, D. (1995) Exact and approximation algorithms for sorting by reversals, with application to genome rearrangement, *Algorithmica*, **13**, 180–210

Kleffe, J. and Borodovsky, M. (1992) First and second moment of counts of words in random texts generated by Markov chains. *CABIOS*, **8**, 433–441.

Knuth, D. E. (1969) *The Art of Computer Programming*, Vol. I-III. Addison-Wesley; Reading, London.

Kohara, Y., Akiyama, A. and Isono, K. (1987) The physical map of the *E. coli* chromosome: Application of a new strategy for rapid analysis and sorting of a large genomic library. *Cell*, **50**, 495–508.

Kosaraju, S.R., Park, J.K. and Stein, C. (1994) Long tours and short superstrings. Unpublished.

Kotig, A. (1968) Moves without forbidden transitions in a graph. *Mat. casopis*, **18**, 76–80.

Kozhukhin, C.G. and Pevzner, P.A. (1991) Genome inhomgeneity is determined mainly by WW and SS dinucleotides. *CABIOS*, **7**, 39–49.

Krogh, A., Brown, M., Mian, I.S., Sjölander, K. and Haussler, D. (1994) Hidden Markov models in computational biology: Applications to protein modeling. *J. Mol. Biol.*, **235**, 1501–1531.

Landau, G.M. and Schmidt, J.P. (1993) An algorithm for approximate tandem repeats. *Lecture Notes in Computer Science, Combinatorial Pattern Matching*, **684**, 120–133.

Lander, E.S. and Waterman, M.S. (1988) Genomic mapping by fingerprinting random clones: A mathematical analysis. *Genomics*, **2**, 231–239.

Laquer, H.T. (1981) Asymptotic limits for a two-dimensional recursion. *Stud. Appl. Math.*, **64**, 271–277.

Lewin, B. *Genes IV*, Oxford University Press, Oxford, 1990.

Li, S.-Y. (1980) A martingale approach to the study of occurrence of sequence patterns in repeated experiments. *Ann. Probab.*, **8**, 1171–1176.

Lundstrum, R. (1990) Stochastic models and statistical methods for DNA sequence data. Ph.D. Thesis, Department of Mathematics, University of Utah.

Magarshak, Y. and Benham, C.J. (1992) An algebraic representation of RNA secondary structures. *J. Biomol. Str. Dyn.*, **10**, 465-488.

Martinez, H.M. (1983) An efficient method for finding repeats in molecular sequences. *Nucleic Acids Res.*, **11**, 4629–4634.

Meacham, C.A. (1981) A probability measure for character compatibility. *Math. Biosci.*, **57**, 1–18.

Myers, E.W. (1994) A sublinear algorithm for approximate keyword searching. *Algorithmica*, **12**, 345–74.

Myers, E.W. and Huang, X. (1992) An $O(N^2 \log N)$ restriction map comparison and search algorithm. *Bull. Math. Biol.*, **54**, 599–618.

Myers, E. and Miller, W. (1988) Sequence comparison with concave weighting functions. *Bull. Math. Biol.*, **50**, 97–120.

Myers, E. and Miller, W. (1989) Approximate matching of regular expression. *Bull. Math. Biol.*, **51**, 5–37.

Naor, D. and Brutlag, D. (1993) On suboptimal alignments of biological sequences. *Lecture Notes in Computer Science, Combinatorial Pattern Matching*, **684**, 179–196.

Nathans, D. and Smith, H. O. (1975) Restriction endonucleases in the analysis and restructuring of DNA molecules. *Ann. Rev. Biochem.*, **44**, 273–293.

Needleman, S.B. and Wunsch, C.D. (1970) A general method applicable to the search for similarities in the amino acid sequences of two proteins. *J. Mol. Biol.*, **48**, 443–453.

Nussinov, R., Pieczenik, G., Griggs, J.R. and Kleitman, D.J. (1978) Algorithms for loop matchings. *SIAM J. Appl. Math.*, **35**, 68–82.

Olson, M.V., Dutchik, J.E., Graham, M.Y., Brodeur, G.M., Helms, C., Frank, M., MacCollin, M, Scheinman, R. and Frand, T. (1986) Random-clone strategy for genomic restriction mapping in yeast. *Proc. Natl. Acad. Sci. USA*, **83**, 7826–7830.

Pauling, L., Corey, R.B. and Branson, H.R. (1951) The structure of Proteins: Two hydrogen-bonded helical configurations of the polypeptide chain. *Proc. Natl. Acad. Sci. USA*, **37**, 205–211.

Pearson, W.R. and Lipman, D.J. (1988) Improved tools for biological sequence comparison. *Proc. Natl. Acad. Sci. USA*, **85**, 2444–2448.

Penner, R.C. and Waterman, M.S. (1993) Spaces of RNA secondary structure. *Adv. Math.*, **101**, 31–49.

Pevzner, P.A. (1989) l-tuple DNA sequencing: Computer analysis. *J. Biomol. Struct. Dynam.*, **7**, 63.

Pevzner, P.A. (1992) Nucleotide sequences versus Markov models. *Computers Chem.*, **16**, 103–106.

Pevzner, P.A. (1995) Physical mapping and alternating Eulerian cycles in colored graphs. *Algorithmica*, **13**, 77–105.

Pevzner, P.A., Borodovsky, M.Y. and Mironov, A.A. (1989) Linguistics of nucleotide sequences I: The significance of deviations from mean statistical characteristics and prediction of the frequencies of occurrence of words. *J. Biomol. Struct. Dynam.*, **6**, 1013–1026.

Pevzner, P.A. and Lipshutz, R.J. (1994) Towards DNA sequencing by hybridization. *19th Symposium on Mathematical Foundation in Computer Science*, **841**, 143–158.

Pevzner, P.A. and Waterman, M.S. (1993) A fast filtration for the substring matching problem. *Lecture Notes in Computer Science, Combinatorial Pattern Matching*, **684**, 197–214.

Port, E., Sun, F., Martin, D. and Waterman, M.S. (1995) Genomic mapping by end characterized random clones: A mathematical analysis. *Genomics*, **26**, 84–100.

Prum, B., Rodolphe, F. and Turckheim, E. (1995) Finding words with unexpected frequencies in DNA sequences. *Journal Royal Statist. Soc. Ser. B*, **55**.

Riordan, J.R., Rommens, J.M., Kerem, B., Alon, N., Rozmahel, R., Grzelczak, Z., Zielenski, J., Lok, S., Plavsic, N., Chou, J.-L., Drumm, M.L., Iannuzzi, M.C., Collins, F.S. and Tsui, L.-C. (1989) Identification of the Cystic Fibrosis gene: Cloning and characterization of complementary DNA. *Science*, **245**, 1066–1073.

Robinson, D.F. and Foulds, L.R. (1981) Comparison of phylogenetic trees. *Math. Biosci.*, **53**, 131–147.

Rosenblatt, J. and Seymour, P.D. (1982) The structure of homometric sets. *SIAM J. Alg. Disc. Math.*, **3**, 343–350.

Rudd, K.E. (1993) Maps, genes, sequences, and computers: An *Escherichia coli* case study. *ASM News*, **59**, 335–341.

Rzhetsky, A. and Nei, M. (1992a) A simple method for estimating and testing minimum-evolution trees. *Mol. Biol. Evol.*, **9**(5), 945–967.

Rzhetsky, A. and Nei, M. (1992b) Statistical properties of the ordinary least-squares, generalized least-squares, and minimum-evolution methods of phylogenetic inference. *J. Mol. Evol.*, **35**, 367–375.

Sankoff, D. (1972) Matching sequences under deletion-insertion constraints. *Proc. Natl. Acad. Sci. USA*, **68**, 4–6.

Sankoff, D. (1975) Minimal mutation trees of sequences. *SIAM J. Appl. Math.*, **78**, 35–42.

Sankoff, D. and Goldstein, M. (1988) Probabilistic models of genome shuffling. *Bull. Math. Biol.*, **51**, 117–124.

Sankoff, D. and Rousseau, P. (1975) Locating the vertices of a Steiner tree in an arbitrary metric space. *Math. Program.*, **9**, 240–246.

Schmitt, W. and Waterman, M.S. (1991) Multiple solutions of DNA restriction mapping problems. *Adv. Appl. Math.*, **12**, 412–427.

Schmitt, W.R. and Waterman, M.S. (1994) Linear trees and RNA secondary structure. *Disc. Appl. Math.*, **51**, 317–323.

Schöniger, M. and Waterman, M.S. (1992) A local algorithm for DNA sequence alignment with inversions. *Bull. Math. Biol.*, **54**, 521–536.

Schwartz, D.C., Li, X., Hernandez, L.I., Ramnarain, S.P., Huff, E.J. and Wang, Y.-K. (1993) Ordered restriction maps of saccharomyces cerevisiae chromosomes constructed by optical mapping. *Science*, **262**, 110–114.

Seed, B., Parker, R.C. and Davidson, N. (1982a) Representation of DNA sequences in recombinant DNA libraries prepared by restriction enzyme partial digestion. *Gene*, **19**, 201–209.

Seed, B. (1982b) Theoretical study of the fraction of a long-chain DNA that can be incorporated in a recombinant DNA partial-digest library. *Biopolymers*, **21**, 1793–1810.

Sellers, P. (1974) On the theory and computation of evolutonary distances. *SIAM J. Appl. Math.*, **26**, 787–793.

Sellers, P. (1980) The theory and computation of evolutionary distances: Pattern recognition. *J. Algorithms*, **1**, 359–373.

Skiena, S.S., Smith, W.D. and Lemke, P. (1990) Reconstructing sets from interpoint distances. *Proc. of the Sixth ACM Symp. Computational Geometry*, 332–339.

References

Skiena, S.S. and Sundaram, G. (1994) A partial digest approach to restriction mapping. *Bull. Math. Biol.*, **56**, 275–294.

Smith, T.F. and Waterman, M.S. (1981) The identification of common molecular subsequences. *J. Mol. Biol.*, **147**, 195–197.

Smith, T.F., Waterman, M.S. and Fitch, W.M. (1981) Comparative Biosequence Metrics. *J. Mol. Evol.*, **18**, 38–46.

Staden, R. (1979) A strategy of DNA sequencing employing computer programs. *Nucleic Acids Res.*, **6**, 2601–2610.

Steele, J.M. (1986) An Efron-Stein inequality for nonsymmetric statistics. *Ann. Statist.*, **14**, 753–758.

Stein, P.R. and Waterman, M.S. (1978) On some new sequences generalizing the Catalan and Motzkin numbers. *Disc. Math.*, **26**, 261–272.

Tang, B. and Waterman, M.S. (1990) The expected fraction of clonable genomic DNA. *Bull. Math. Biol.*, **52**, 455–475.

Tavaré, S. (1986) Some probabilities and statistical problems in the analysis of DNA sequences. *Lectures on Mathematics in the Life Sciences*, **17**, 57–86.

Tinoco, I., Jr., Uhlenbeck, O.C. and Levine, M.D. (1971) Estimation of secondary structure in ribonucleic acids. *Nature, Lond.*, **230**, 362–367.

Tinoco, I., Jr., Borer, P.N., Dengler, B., Levine, M.D., Uhlenbeck, O.C., Crothers, D.M. and Gralla, J. (1973) Improved estimation of secondary structure in ribonucleic acids. *Nature New Biol.*, **246**, 40–41.

Trifonov, E.N. and Sussman, J.L. (1980) The pitch of chromatin DNA is reflected in its nucleotide sequence. *Proc. Natl. Acad. Sci. USA*, **77**, 3816–3820.

Ukkonnen, U. (1985) Finding approximate patterns in strings. *J. Alg.*, **6**, 132–137.

Vingron, M. and Argos, P. (1990) Determination of reliable regions in protein sequence alignments. *Protein Engineering*, **3**, 565–569.

Vingron, M. and Waterman, M.S. (1994) Sequence alignment and penalty choice: Review of concepts, case studies and implications. *J. Mol. Biol.*, **235**, 1–12.

von Haeseler, A., Blum, B., Simpson, L., Strum, N. and Waterman, M.S. (1992) Computer methods for locating kinetoplastid cryptogenes. *Nucleic Acids Res.*, **20**, 2717–2724.

Wagner, R.A. and Fischer, M.J. (1974) The string-to-string correction problem. *J. Assoc. Comput. Mach.*, **21**, 168–173.

Waterman, M.S. (1983) Sequence alignment in the neighborhood of the optimum with general applications to dynamic programming. *Proc. Natl. Acad. Sci. USA*, **80**, 3123–3124.

Waterman, M.S. (1984) Efficient sequence alignment algorithms. *J. Theor. Biol.*, **108**, 333–337.

Waterman, M.S. (1978) Secondary structure of single-stranded nucleic acids. *Studies in Foundations & Combinatorics, Advances in Mathematics Supplementary Studies*, **1**, 167–212.

Waterman, M.S., Smith, T.F. and Beyer, W.A. (1976) Some biological sequence metrics. *Adv. in Math.*, **20**, 367–387.

Waterman, M.S., Smith, T.F. and Katcher, H. (1984) Algorithms for restriction map comparisons. *Nucleic Acids Res.*, **12**, 237–242.

Waterman, M.S. and Griggs, J.R. (1986) Interval graphs and maps of DNA. *Bull. Math. Biol.*, **48**, 189–195.

Waterman, M. and Perlwitz, M. (1984) Line geometries for sequence comparisons. *Bull. Math. Biol.*, **46**, 567–577.

Waterman, M. and Eggert, M. (1987) A new algorithm for best subsequence alignments with application to tRNA-rRNA comparisons. *J. Mol. Biol.*, **197**, 723–725.

Waterman, M.S. and Smith, T.F. (1978) On the similarity of dendograms. *J. Theor. Biol.*, **73**, 789–800.

Waterman, M.S. and Smith, T.F. (1986) Rapid dynamic programming algorithms for RNA secondary structure. *Adv. Appl. Math.*, **7**, 455–464.

Waterman, M.S., Smith, T.F., Singh, M. and Beyer, W.A. (1977) Additive evolutionary trees. *J. Theor. Biol.*, **64**, 199–213.

Waterman, M.S., Eggert, M. and Lander, E.S. (1992) Parametric sequence comparisons. *Proc. Natl. Acad. Sci. USA*, **89**, 6090–6093.

Waterman, M.S. and Vingron, M. (1994a) Rapid and accurate estimates of statistical significance for sequence database searches. *Proc. Natl. Acad. Sci. USA*, **91** 4625–4628.

Waterman, M.S. and Vingron, M. (1994b) Sequence comparison significance and Poisson approximation. *Statist. Sci.*, **2**, 367–81

Watson, J.D. and Crick, F.H.C. (1953) Genetical implications of the structure of deoxyribonucleic acid. *Nature*, **171**, 964–967.

Watson, J.D., Hopkins, N., Roberts, J., Stietz, J.A. and Weiner, A. *Molecular Biology of the Gene*, 4th ed. Benjamin-Cummings, Menlo Park, CA 1987.

References

Watterson, G.A., Ewens, W.J., Hall, T.E. and Morgan, A. (1982) The chromosome inversion problem. *J. Theor. Biol.*, **99**, 1–7.

Whittle, P. (1955) Some distribution and moment formulae for the Markov chain. *J. Roy. Statist. Soc. B*, **17**, 235–242.

Wilbur, W.J. and Lipman, D.J. (1983) Rapid similarity searches of nucleic acid and protein data banks. *Proc. Natl. Acad. Sci. USA*, **80**, 726–730.

Wu, S. and Manber, U. (1992) Fast text searching allowing errors. *Commun. ACM*, **35**, 83–90.

Xu, S. (1990) Dynamic programming algorithms for alignment hyperplanes. Master's thesis, Department of Mathematics, University of Southern California.

Zuker, M. and Sankoff, D. (1984) RNA secondary structures and their prediction. *Bull. Math. Biol.*, **46**, 591–621.

Appendix I

Problem Solutions and Hints

Chapter 1

1.2
- Using AAAA..., conclude $g(\text{AAA})=\text{Lys}$.
- Using CCCC..., conclude $g(\text{CCC})=\text{Pro}$.
- From ACACAC..., conclude (1)$g\{\text{ACA,CAC}\}=\{\text{Thr,His}\}$.
- From AACAAC..., conclude (2)$g\{\text{AAC,ACA,CAA}\}=\{\text{Asn,Thr,Gln}\}$.
- From CCACCA..., conclude (3)$g\{\text{CCA,CAC,ACC}\}=\{\text{Pro,His,Thr}\}$.

The intersection of (1) and (2) implies that $g(\text{ACA})=\text{Thr}$ and, therefore, that $g(\text{CAC})=\text{His}$. It then follows that $g\{\text{AAC,CAA}\}=\{\text{Asn,Gln}\}$ and $g\{\text{CCA, ACC}\}=\{\text{Pro,Thr}\}$.

1.3 If n is odd, there are $S_n = 20^{(\frac{n}{2})+1}$ symmetrical protein sequences. If n is even, there are $S_n = 20^{(\frac{n}{2})}$ symmetrical protein sequences. Except for symmetrical sequences, the count is reduced by a factor of 2. The answer is $(20^n - S_n)/2 + S_n$.

1.4 A start codon occurs with probability $p_A p_U p_G$. $\mathbb{P}(X = k+1) = p^k(1-p)$ when $1 - p = p_U p_A p_A + p_U p_A p_G + p_U p_G p_A$.

1.5 (i) The problem is to find how many words can be written with $E_1 E_2 \cdots E_k$ using at least one letter (or exon). Thus, the answer is $2^k - 1$. (ii) For this problem look at the $k+1$ markers "∧" in the word $\wedge E_1 \wedge E_2 \cdots \wedge E_k \wedge$. Sequences can be made by choosing two markers in $\binom{k+1}{2}^{-1}$ ways.

Chapter 2

2.1
$$\begin{pmatrix} & 1 & 4 & 3 & 2 \\ 3 & 1 & 1 & 0 & 0 \\ 1 & 0 & 1 & 0 & 0 \\ 5 & 0 & 1 & 1 & 0 \\ 4 & 0 & 0 & 1 & 0 \\ 2 & 0 & 0 & 1 & 1 \end{pmatrix}$$

2.2 The solution has three components:

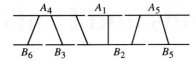 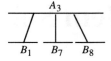

2.3 Hint: Define $X * Y * Z = (\alpha_{ijk})$ by
$$\alpha_{ijk} = \sum_l X_{il} Y_{jl} Z_{kl}.$$

Chapter 3

3.1 (ii) $|A| + |B| - |A \wedge B| = |A \vee B|$.

3.2 (i) If we define a coincident cut site to be a site where A, B, C all cut, the statement is identical.

(ii) $k = -\log(8)/p_A p_B p_C$.

3.4 An example can be produced by using Figure 3.1 with no C cuts until the coincident right-hand end of the DNA.

3.6 Figure 3.3: $2^{3-1} 3!(2!)$.

Figure 3.4: $2^1 1!(2!)(3!)$.

3.8 For example, $(|A|, |B|, |A \wedge B|)$ contains $(7,8,4)$ which is not in $(|A'|, |B'|, |A' \wedge B'|)$.

3.10 $\binom{l+1}{2}$.

Chapter 4

4.1 $\mathbf{c} = (1,1,1,2,2,2)$, $\mathbf{a} = (1,1,3,4)$, $\mathbf{b} = (2,1,4,2)$, $E = \begin{pmatrix} 1 & 0 & 0 & 0 \\ 0 & 1 & 0 & 0 \\ 0 & 0 & 1 & 0 \\ 0 & 0 & 1 & 0 \\ 0 & 0 & 0 & 1 \\ 0 & 0 & 0 & 1 \end{pmatrix}$;

$F = \begin{pmatrix} 1 & 0 & 0 & 0 \\ 1 & 0 & 0 & 0 \\ 0 & 1 & 0 & 0 \\ 0 & 0 & 1 & 0 \\ 0 & 0 & 1 & 0 \\ 0 & 0 & 0 & 1 \end{pmatrix}$. So $\mathbf{c}E = \mathbf{a}$ and $\mathbf{c}F = \mathbf{b}$.

4.2 $R_1 = \{1\}, R_2 = \{1\}, R_3 = \{1,2\}$, and $R_4 = \{2,2\}$. $S_1 = \{1,1\}, S_2 = \{1\}, S_3 = \{2,2\}$, and $S_4 = \{2\}$.

4.3 (i) $P = (p_{ij}) = \begin{matrix} & 1 & 2 & 3 \\ 1 & \\ 2 & \\ 3 & \end{matrix}\begin{pmatrix} 1/2 & 1/2 & 0 \\ 1/2 & 0 & 1/2 \\ 1/2 & 1/2 & 0 \end{pmatrix}$. (ii) $P^3 = \begin{pmatrix} 1/2 & 1/4 & 1/4 \\ 1/2 & 1/4 & 1/4 \\ 1/2 & 3/8 & 1/8 \end{pmatrix}$.

(iii) $\pi P = \pi$ solves to yield $\pi = (1/2, 1/3, 1/6)$.

4.4 Let $\oplus$ and $\ominus$ be addition and subtraction modulo n. (i) $N_{i,j} = \{(i\ominus 1, j), (i\oplus 1, j), (i, j), (i, j\ominus 1), (i, j\oplus 1)\}$. Two points (i,j) and (k,l) can be reached by (assuming $i < k$) $(i,j) \to (i+1, j) \cdots \to (k, j)$ and similarly in the second coordinate. (ii) The neighborhood structure is illustrated for $n = 3$

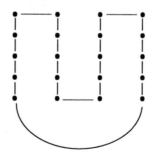

In your solution be sure to handle n even and n odd.

4.5 Let $\oplus$ be addition modulo 2. For $\mathbf{c} = (c_1, c_2, \ldots, c_n) \in C$, generate the neighborhood by changing to the i-th coordinate for each i:

$$N_\mathbf{c} = \{\mathbf{c}\} \bigcup_{i=1}^{n} \{\mathbf{c} + (0, \ldots, 0, 1, \ldots, 0)\}.$$

Clearly, $|N_{\mathbf{u}}| = |N_{\mathbf{w}}|$ for all $\mathbf{u}, \mathbf{w} \in C$, and you can move from $\mathbf{u}$ to $\mathbf{w}$ in $\sum_{i=1}^{n} |u_i - w_i|$ steps.

4.6 (i) Let E and F denote intervals of the map with reversals E^r and F^r, respectively.

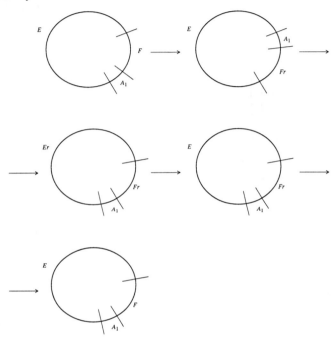

(ii) Repeat (i) to obtain rotation through distances $k_1 a_1$ ($|A_1| = a_1$). Thus, rotations of size $\sum k_i a_i$ for any integers k_i are possible and, therefore, rotations through any multiple of $g_A = \gcd\{a_1, a_2, \ldots, a_n\}$. Clearly, rotations through any multiple of $g_B = \gcd\{b_1, b_2, \ldots, b_m\}$ are also possible. Relative to each other the two circular arrangements can rotate through distance $kg_A + lg_B$ for any integers k and l, therefore through multiples of $g = \gcd\{g_A, g_B\}$.

4.7 (i)

```
input:     x₁x₂ ··· xₙ
output:    x_{i₁} ≤ x_{i₂} ≤ ··· ≤ x_{iₙ}
upper ← n
test ← 0
for i = 1 to upper−1
    if x_{i+1} < x_i
        exchange x_i and x_{i+1}
        test ← 1
    end
    if test = 0, stop
    if test ≠ 0, upper ← l, return to for
end
```

(ii) After the first iteration, the $\max\{x_1 x_2 \cdots x_n\}$ will be in the top position; after the second iteration, the second largest is in the next to top position and so on. The running time for the worst case $(x_1 > x_2 > \cdots > x_n)$ is $n \times n = O(n^2)$.

Chapter 5

5.1 Because the sum is a differentiated geometric series,

$$\sum_{l=0}^{\infty} lp^2 e^{-p(l-1)} = p^2 \sum_{l=0}^{\infty} l(e^{-p})^{l-1} = \frac{p^2}{(1-e^{-p})^2}.$$

Also,

$$p^2 \int_0^{\infty} xe^{-p(x-1)} dx = e^p$$

from the last line of the proof of Theorem 7.1 ($L = 0$, $U = \infty$). The difference is $e^p - p^2(1-e^{-p})^{-2}$.

5.2 To clone the gene, all g bps must be covered by a clone of length l, $L \leq l \leq U$. As in Theorem 7.1,

$$P(\text{gene} \in \text{frag of length } l) = (l - g + 1)p^2(1-p)^{l-1}$$

$$f(p) = p^2 \sum_{l=L}^{U}(l - g + 1)e^{-p(l-1)}$$

$$\approx e^p\{(p(L+1) - (g-1)p)e^{-pL} - (pU + 1 - (g-1)p)e^{-pU}\}.$$

5.3 $P(\text{gene is cloned}) = 1 - P(\text{gene is not cloned})$

$$= 1 - \prod_{i=1}^{K} P(\text{gene is not cloned in digest } i)$$

$$= 1 - \prod_{i=1}^{K} (1 - f(p_i)).$$

5.4 $P(F = k) = p(1-p)^{k-1}, k \geq 1.$

$$\frac{1}{n} \sum_{i=1}^{n} F_i \to \mathbb{E}(F)$$

$$= p^{-1} \times \frac{1}{n} \sum_{i=1}^{n} F_i I\{L \leq F_i \leq U\} \to E(FI\{L \leq F \leq U\})$$

$$= \sum_{k=L}^{U} kp(1-p)^{k-1}.$$

Therefore, the limit of the ratio is $f = \sum_{k=L}^{U} kp^2(1-p)^{k-1}$.

5.7 (i) $P(b \in \text{fragment of length } l) = l(p_1 + p_2 - p_1 p_2)^2 \left((1-p_1)(1-p_2)\right)^{l-1}$
$= lq^2(1-q)^{l-1}$, $q = p_1 + p_2 - p_1 p_2$. $f_3 = \sum_{l=L}^{U} lq^2(1-q)^{l-1}$. (ii) $(1-f_1)f_2$. (iii) $1 - f_1 f_2$.

Chapter 6

6.1 $c = (1-\theta)^{-1}$ with $(1-\theta)^{-1} L^{-1} e^{-1} G$ islands.

6.2 With $\theta = 0$, there are Ne^{-c} islands of average length $L(e^{-c} - 1)/c$, so the fraction covered is $Ne^{-c}L(e^{-c} - 1)c^{-1}/G = 1 - e^{-c}$.

6.3 DNA sequenced $= NL$; DNA mapped $= G(1 - e^{-c})$, so the ratio is $c/(1 - e^{-c})$.

6.5 Let $\sigma = 1 - \theta$:

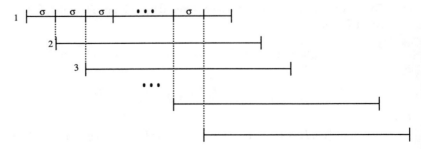

6.7 Fragment i is from clone 1; fragment j is from clone 2. Let W_k be the waiting time to site of type k, $k = 1, 2$, for fragment i.

$$\mathbb{P}(\text{fragment } i \text{ has at least one process \#2 site})$$
$$= \mathbb{P}(W_2 < W_1)$$
$$= \int_0^\infty \lambda_2 e^{-\lambda_2 y} \int_y^\infty \lambda_1 e^{-\lambda_1 x} dx dy$$
$$= \frac{\lambda_2}{\lambda_1 + \lambda_2} = \xi.$$

Similarly, $\mathbb{P}(\text{fragment } i \text{ has } k \text{ \#2 sites}) = \xi^k(1-\xi)$.

$$\mathbb{P}(i \text{ and } j \text{ are matching landmarks})$$
$$= \sum_{k=1}^\infty \mathbb{P}(i \text{ and } j \text{ match} \mid k \text{ \#2 sites in } i \text{ and } j)$$
$$\times \mathbb{P}(k \text{ \#2 sites in } i \text{ and } j)$$
$$= 2\sum_{k=1}^\infty \left(\frac{\beta}{2}\right)^{k+1} (\xi^k(1-\xi))^2 = \frac{(\beta\xi)^2(1-\xi)^2}{2 - \beta\xi^2}.$$

To complete the exercise note that there are an expected $\lambda_1 L$ fragments with \#1 ends per clone. The answer is then $(\lambda_1 L)^2 \frac{(\beta\xi)^2(1-\xi)^2}{2-\beta\xi^2}$.

6.12

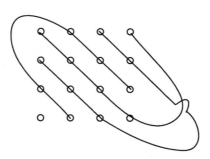

6.13 The first cell in row $k' + 1$.

6.14

(i)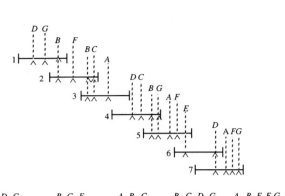

(ii)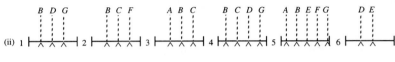

6.15 Theorem 6.5*. With the notation from Theorem 6.5, we have the following:
(i) The probability q_1 that a clone contains no anchors is
$$q_1 = \int_0^\infty e^{-bl} f(l) dl.$$
The expected number of unanchored clones is Nq_1.
(ii) The probability p_1 that a clone is the rightmost clone of an anchored island is
$$p_1 = \int_0^\infty be^{-bu} J(u) \mathcal{F}(u) du.$$
The expected number of anchored islands is then Np_1.
(iii) The expected number of clones in an anchored island is $(1 - q_1)/p_1$.
(iv) The probability p_2 that a clone is a singleton anchored island is
$$p_2 = \int_0^\infty \int_0^1 \int_0^{l-v} b^2 e^{-b(u+v)} \frac{J(u)J(v)}{J(l)} f(l)\, du\, dv\, dl$$
$$+ \int_0^\infty \int_0^l be^{-bl} \frac{J(u)J(v)}{J(l)} f(l)\, du\, dl.$$
The expected number of singleton anchored islands is Np_2.
(v) The expected length of an anchored island is $\lambda \mathbb{E} L$, where
$$\lambda = \left\{ 1 + \int_0^\infty (b^2 u - 2b) e^{-bu} J(u) du \right\} / ap_1.$$

(vi) The expected proportion r_0 of the genome not covered by anchored islands is

$$r_0 = \int_0^\infty \int_0^\infty b^2 e^{-b(u+v)} \frac{J(u)J(v)}{J(u+v)} \, du \, dv.$$

(vii) The probability that an anchored island is followed by an actual ocean of length at least $x(\mathbb{E}L)$ is $e^{-a(x+1)}(1-q_1)/p_1$. In particular, taking $x = 0$, the formula gives the probability that an anchored island is followed by an actual ocean rather than an undetected overlap.

Chapter 7

7.1
f_1=GCTCG,
f_2=CTCG,
f_3=CTC,
f_4=TC,
f_5=T.

For the second example, take the 2^k examples A $\{C,G\}^k$T.

7.2 For $k \geq 2$, the orientations to avoid are all $+$ or all $-$. For each k, the probability is $\frac{e^{-c}c^k}{k!}\left(1 - \frac{1}{2^{k-1}}\right)$. Summing on $k \geq 2$, obtain $1+e^{-c}-2e^{-c/2}$.

7.3
(i)

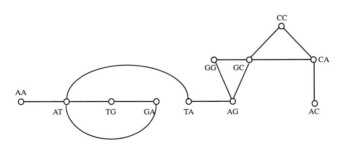

(ii) $(3-1) \times (3-1) \times (3-1) \times (3-1) \times (3-1) = 32$

7.7 (i) $S(\mathbf{a}) = \{b_1 b_2 \cdots b_k : b_1 b_2 \cdots b_k = x_{0 \oplus i} x_{1+i} \cdots x_{k \oplus i}$, where $\oplus$ is addition mod r and $i = 0, 1, \ldots, r-1\}$.

(ii) $\mathbf{b} = b_1, b_2 \cdots b_n$ has spectrum $S(\mathbf{a})$ if $\mathbf{b}$ is repetition of $x_0 x_1 \cdots x_{r-1}$ starting at any of the r positions. Therefore, there are r such sequences $\mathbf{b}$.

Chapter 9

9.1 $\binom{m}{n}$.

9.2 $\binom{n}{2}\binom{m}{n-2}$.

9.3 Look at m white balls, and n black balls. To choose n balls from $n+m$ balls, we have $n-k$ white ($0 \leq k \leq n$) and k black balls, so

$$\sum_{k=0}^{n}\binom{n}{n-k}\binom{m}{k} = \sum_{k=0}^{n}\binom{n}{k}\binom{m}{k} = \binom{n+m}{n}.$$

9.4

	-	A	A	G	T	T	A	G	C	A	G
-	0	-1	-2	-3	-4	-5	-6	-7	-8	-9	-10
C	-1	-1	-2	-3	-4	-5	-6	-7	-6	-7	-8
A	-2	0	0	-1	-2	-3	-4	-5	-6	-5	-6
G	-3	-1	-1	1	0	-1	-2	-3	-4	-5	-4
T	-4	-2	-2	0	2	1	0	-1	-2	-3	-4
A	-5	-3	-1	-1	1	1	2	1	0	-1	-2
T	-6	-4	-2	-2	0	2	1	1	0	-1	-2
C	-7	-5	-3	-3	-1	1	1	0	2	1	0
G	-8	-6	-4	-2	-2	0	0	2	1	1	2
C	-9	-7	-5	-3	-3	-1	-1	1	3	2	1
A	-10	-8	-6	-4	-4	-2	0	0	2	4	3

CAGTATCGCA−
AAGT−TAGCAG.

9.5 In addition to the boundary conditions for $S_{i,j}$ when $i \cdot j = 0$, set $R_{i,j} = 0$ when $i \cdot j = 0$. Then,

$$S_{i,j} = \max\{S_{i-1,j-1} + s(a_i, -), S_{i,j-1} + s(-, b_j),$$
$$S_{i-1,j-1} + s(a,b) + (\eta + \xi)(1 - R_{i-1,j-1}) + \xi R_{i-1,j-1}.$$

If $S_{i,j} = S_{i-1,j-1} + s(a,b) + h(R_{i-1,j-1} + 1) - h(R_{i-1,j-1})$, then $R_{i,j} \leftarrow R_{i-1,j-1} + 1$. Otherwise, $R_{i,j} \leftarrow 0$.

9.6 The new optimal alignments are

```
-CAGTATCGCA-
A-AGT-TAGCAG,

C-AGTATCGCA-
-AAGT-TAGCAG,

CA-GTATCGCA-
-AAGT-TAGCAG,

CAG-TATCGCA-
AAGTTA--GCAG,

CAGT-ATCGCA-
AAGTTA--GCAG,

CAGTAT-CGCA-
AAGT-TA-GCAG,

CAG-TATCGCA-
AAGTTA--GCAG,

CAGT-ATCGCA-
AAGTTA--GCAG,

CAGTATC-GCA-
AAGT-T-AGCAG.
```

9.8 $S_{0,j} = -\delta j$ and $S_{i,0} = -i\delta$, $0 \leq i \leq n$, $0 \leq j \leq m$. Define $P_{i,j} =$ the number of alignments ending at (i,j), so $P_{0,j} = P_{i,0} \equiv 1$. Then $S_{i,j} = S_{i-1,j-1} + s(a_i, b_j)P_{i-1,j-1} + S_{i-1,j} - \delta P_{i-1,j} + S_{i,j-1} - \delta P_{i,j-1}$. $P_{i,j} = P_{i-1,j-1} + P_{i-1,j} + P_{i,j-1}$.

9.7 The distance $d = \mathbb{D}((i,j), \mathcal{D})$ is obtained by drawing a perpendicular line from (i,j) to $\mathcal{D}$. Wlog, $i \geq j$.

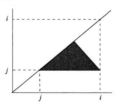

The sides of the indicated triangle have lengths $1, 1, \sqrt{2}$ times d. Therefore $d = |i - j|/\sqrt{2}$. The required distance is then $\sqrt{2}d_A = \max\{|i_k - j_k| : 1 \leq k \leq K\}$.

9.9 See Proposition 9.1.

9.10 $f_k(l) + f_k(m) \geq f_k(l+m)$ is equivalent to

$$f(l+k) - f(k) \geq f(l+k+m) - f(k+m).$$

9.12

	-	A	T	T	G	A	C
-	0	-1	-2	-3	-4	-5	-6
C	0	-1	-2	-3	-4	-5	-4
A	0	1	0	-1	-2	-3	-4
G	0	0	0	-1	0	-1	-2
T	0	-1	1	1	0	-1	-2
A	0	1	0	0	0	1	0
T	0	0	2	1	0	0	0
C	0	-1	1	1	0	-1	1
G	0	-1	0	0	2	1	0
C	0	-1	-1	-1	1	1	2
A	0	1	0	-1	0	2	1

The alignment is

ATCGCA
ATTB−A.

9.14 (i)

	-	A	A	G	T	T	A	G	C	A	G
-	0	0	0	0	0	0	0	0	0	0	0
C	0	0	0	0	0	0	0	0	1	0	0
A	0	1	1	0	0	0	1	0	0	2	1
G	0	0	0	2	1	0	0	2	1	1	3
T	0	0	0	1	3	2	1	1	1	0	2
A	0	1	1	0	2	2	3	2	1	2	1
T	0	0	0	0	1	3	2	2	1	1	1
C	0	0	0	0	0	2	2	1	3	2	1
G	0	0	0	1	0	1	1	3	2	2	3
C	0	0	0	0	0	0	0	2	4	3	2
A	0	1	1	0	0	0	1	1	3	5	4

The alignment is

AGTATCGCA
AGT−TAGCA.

Declump:

Problem Solutions and Hints

	-	A	A	G	T	T	A	G	C	A	G
-	0	0	0	0	0	0	0	0	0	0	0
C	0	0	0	0	0	0	0	0	1	0	0
A	0	1	0	0	0	0	1	0	0	2	1
G	0	0	0	0	0	0	0	2	1	1	3
T	0	0	0	0	0	0	0	1	1	0	2
A	0	1	1	0	0	0	2	1	0	2	1
T	0	0	0	0	1	0	1	1	0	1	1
C	0	0	0	0	0	0	0	0	2	1	0
G	0	0	0	1	0	0	0	0	1	1	2
C	0	0	0	0	0	0	0	0	0	0	1
A	0	1	1	0	0	0	1	0	0	0	0

The alignment is

$$\text{CAG}$$
$$\text{CAG.}$$

9.16 Suppose the repeat is S, so that $S_1 S_2 \cdots S_k$ is the tandem repeat of k approximate copies of S. The first alignment is

$$\begin{array}{c} S_2 \cdots S_k \\ S_1 \cdots S_{k-1}. \end{array}$$

S_1 can be found by the offset of the initial matched letters of the alignment. Then S_2 is located by alignment with S_1, S_3 aligned with S_2, and so on. Therefore, the matches between all the repeat units can be estimated by following the matched letters (from S_1 to S_2 to S_3 ...). These matches are then used to declump the matrix.

9.21

$$\begin{aligned} H_{i,j} &= \max\{0, H_{i,j-1} - \delta_P, H_{i-1,j} - \delta_D, \\ &\quad H_{i-3,j} + s(g(a_{i-2}a_{i-1}a_i), b_j), \\ &\quad H_{i-4,j} + \max\{s(g(a_{i-3}a_{i-1}a_i), b_j), \\ &\quad s(ga_{i-3}a_{i-2}a_{i-1}), b_j\} - \delta_D\}. \end{aligned}$$

Chapter 10

10.1 Each column has exactly one letter, so there are a total of $\sum_{i=1}^{r} n_i$ columns which can be ordered in $\left(\sum_{i=1}^{r} n_i\right)!$ ways.

10.2 $\sum_{i=0}^{d} \binom{k}{i} 3^i$.

10.3 Time: $O\left(2^r \prod_{i=1}^{r} n_i\right)$. Space $O\left(\prod_{i=1}^{r} n_i\right)$.

10.4 Consider $\dfrac{\text{ATGTA}}{\text{CTC}-\text{A}}$ and $\dfrac{\text{ATGTA}}{\text{CT}-\text{CA}}$.

10.5 From the theorem, $D(\mathbf{c}(\lambda_1), \mathbf{b}) = D(\mathbf{c}(\lambda_1), \mathbf{d}(\lambda)) + D(\mathbf{d}(\lambda), \mathbf{b})$ and, therefore, $\lambda_1 D(\mathbf{a}, \mathbf{b}) = D(\mathbf{c}(\lambda_1), \mathbf{d}(\lambda)) + \lambda D(\mathbf{c}(\lambda), \mathbf{b}) = D(\mathbf{c}(\lambda_1), \mathbf{d}(\lambda)) + \lambda \lambda_1 D(\mathbf{a}, \mathbf{b})$. Thus, setting $\lambda = \lambda_2/\lambda_1$, the result follows.

10.6 The i-th deleted letter either begins a new deletion or extends an earlier deletion:
$$g_{\text{pro}}(i) = \max\{g_{\text{pro}}(i-1) - \gamma_i, g_{\text{pro}}(i-1) - \delta_i\}.$$

10.7 Set $\mathbf{B} = \{\mathbf{b}_1, \mathbf{b}_2, \ldots, \mathbf{b}_r\}$. Let M_j be the best value of $SM_1(a_1 \cdots a_j, \mathbf{B})$. Then
$$M_j = \max\left\{M_{j-1}, \max_{\substack{1 \le i \le j, \\ 1 \le k \le r}} M_{i-1} + S(a_i \cdots a_j, \mathbf{b}_k)\right\}$$

The algorithm has running time $O(rn^3)$.

10.8 Define $\mathbf{S}^1 = S_1^1 S_2^1 \cdots S_n^1$, where
$$S_i^1 = \max\{S(a_1 \cdots a_i, \mathbf{b}_{j1}) : j = 1, 2, \ldots, r\}.$$

From $\mathbf{S}^{k-1}$, define $\mathbf{S}^k$ by
$$S_i^k = \max\{S_{l-1}^k + S(a_l \cdots a_i, \mathbf{b}_{jk}) : 1 \le l \le i, \ j = 1, 2, \ldots, r\}.$$
$$SM_2 = \max\{S_l^m : 1 \le l \le r\}.$$

Chapter 11

11.1 If $Y'_n = C_1 + \cdots + C_{i-1} + C'_i + C_{i+1} + \cdots + C_n$, then $|Y_n - Y'_n| \leq 1$, so
$\mathbb{P}(Y_n \geq \alpha n) = \mathbb{P}(Y_n - np \geq (\alpha - p)n) \leq e^{-n(\alpha-p)^2}/2$.

11.4 Set $X = (S_n - \mathbb{E}(S_n))^2$; then $\mathbb{P}(X > x) \leq 2e^{-\frac{x}{2c^2n}}$. Recall that $\mathbb{E}(X) = \int_0^\infty \mathbb{P}(X > x)dx$, so that

$$\mathbb{E}(X) \leq 2\int_0^\infty e^{-\frac{x}{2c^2n}} dx = 4c^2 n.$$

11.8 (i) Set $p = 1 - p_A^2 p_U - 2p_A p_G p_U$; then we have the usual longest head run problem which has solution

$$\mathbb{P}(L_1 \geq t) \approx 1 - e^{-\lambda_1}$$

with $\lambda_1(1-p)np^t$.

(ii) Set $A_i = \{L_i \geq t\}$; then

$$\mathbb{P}(U A_i) \leq \sum_i \mathbb{P}(A_i) = \sum_{i=1}^6 (1 - e^{-\lambda_i}).$$

(iii) Here, the declumping event is more complex. The open reading frame of length at least t with probability $(p_A p_G p_U)p^{t-1}$. To declump, consider that each codon until the first stop codon must **not** be AUG, so there is a geometric random variable determing the declumping event. Set $p_* = p - p_A p_G p_U$. Then

$$\lambda = \left(\frac{p_*}{1 - p_*}\right) np^t.$$

11.9 (a) $\mu_{a,a} = \dfrac{\xi_a^2}{\sum_{b \in \mathcal{A}} \xi_b^2}$. (b) $s(a,a) = \dfrac{\log\left(\frac{\xi_a}{|\mathcal{A}|}\right)}{\sum_{b \in \mathcal{A}} \log\left(\frac{\xi_b}{|\mathcal{A}|}\right)}$.

11.11 Note that $\log_{a^k}(x) = y$ implies $y = (\log_a(x))/k$. We have

$$t = \log_{a^{N-1}}(n^N) = N\log_{a^{N-1}}(n) = \frac{N}{N-1}\log_a(n).$$

For N large, $t(n, N) \approx \log_a(n)$.

Chapter 12

12.1 $\beta_{HHH,HT}(i) = 0$ unless $i = 3$. $\beta_{HHH,HH}(i) = 1$ for $i = 1, 2, 3$.

12.2 Using Equations (12.2) and (12.3),

$$\mathrm{cov}(\mathbb{I}\mathbf{u}(i), \mathbb{I}\mathbf{v}(j)) = \mathbb{E}(\mathbb{I}\mathbf{u}(i))\mathbb{I}\mathbf{v}(j) - \mathbb{E}(\mathbb{I}\mathbf{u}(i))\mathbb{E}(\mathbb{I}\mathbf{v}(j))$$

$$= \begin{cases} \mathbb{E}(\mathbb{I}\mathbf{u}(i))\left\{\beta_{\mathbf{u},\mathbf{v}}(j-i)P_{\mathbf{v}}(j-i+l-k) - \mathbb{E}(\mathbb{I}\mathbf{v}(j))\right\}, \\ \quad \text{if } 0 \leq j - i < k, \\[4pt] \mathbb{E}(\mathbb{I}\mathbf{u}(i))\left\{p_{u_k,v_1}^{(j-i+k-1)}P_{\mathbf{v}}(l-1) - \mathbb{E}(\mathbb{I}\mathbf{v}(j))\right\}, \\ \quad \text{if } k \leq j - i, \\[4pt] \mathbb{E}(\mathbb{I}\mathbf{v}(j))\left\{\beta_{\mathbf{v},\mathbf{u}}(i-j)P_{\mathbf{u}}(i-j+k-l) - \mathbb{E}(\mathbb{I}\mathbf{u}(i))\right\}, \\ \quad \text{if } 0 \leq i - j < l, \\[4pt] \mathbb{E}(\mathbb{I}\mathbf{v}(j))\left\{p_{v_l,u_1}^{(i-j+l-1)}P_{\mathbf{u}}(k-1) - \mathbb{E}(\mathbb{I}\mathbf{u}(i))\right\}, \\ \quad \text{if } i - j \leq l. \end{cases}$$

Now,

$$\mathrm{cov}(N_{\mathbf{u}}, N_{\mathbf{v}}) = \sum_{i=1}^{n-k+1}\sum_{j=1}^{n-l+1} \mathrm{cov}(\mathbb{I}\mathbf{u}(i), \mathbb{I}\mathbf{v}(j))$$

and $k \leq l$. We rearrange the terms:

$$\sum_{j=0}^{n-l}\sum_{i=1}^{n-l-j+1} \mathrm{cov}(\mathbb{I}\mathbf{u}(i), \mathbb{I}\mathbf{v}(i+j)) + \sum_{j=0}^{n-l}\sum_{i=1}^{n-l-j+1} \mathrm{cov}(\mathbb{I}\mathbf{u}(i+j), \mathbb{I}\mathbf{v}(i))$$

$$- \sum_{i=1}^{n-l+1} \mathrm{cov}(\mathbb{I}\mathbf{u}(i), \mathbb{I}\mathbf{v}(i)) + \sum_{i=n-l+2}^{n-k+1}\sum_{j=1}^{n-l+1} \mathrm{cov}(\mathbb{I}\mathbf{u}(i), \mathbb{I}\mathbf{v}(j)).$$

Substituting into this formula gives Proposition 12.1.

12.3 Only two terms change, those involving the transition probabilities:

$$\vdots$$

$$+ \pi_{\mathbf{u}} P_{\mathbf{v}}(l-\nu) \sum_{j=0}^{\infty} \left(p_{u_{k-\nu+1}\cdots u_k, v_1\cdots v_k}^{(j+\nu)} - \pi_{v_1\cdots v_k} \right)$$

$$\vdots$$

Problem Solutions and Hints

$$+ \pi_{\mathbf{v}} P_{\mathbf{u}}(k-\nu) \sum_{j=0}^{\infty} \left(p_{\nu_l-\nu+1\cdots\nu_l,u_1\cdots u_k}^{(j+\nu)} - \pi_{u_1\cdots u_k} \right)$$

$$\vdots$$

12.4 $F(1) = \sum_{i=1}^{\infty} \leq 1$. If $F(1) < 1$, then $\mathbb{P}(\mathbf{w}$ does not occur in $\mathbf{A} = A_1 A_2 \cdots) > 0$. In fact, $\mathbf{w}$ occurs infinitely often with probability 1 by the Borel-Cantelli lemma.

12.9 Let $M_{\mathbf{w}}(n) = \sum_{i=1}^{n} Y_i$, where $Y_i = 1$ if a renewal occurs at i, and 0 otherwise. The Y_i are dependent random variables.

$$\mathbb{E}(M_{\mathbf{w}}(n)) = \sum_{i=1}^{n} u_i$$

and $u_i \to 1/\mu$ by the renewal theorem.

12.10
$$\mathbb{P}(\mathbf{w}_1) = \sum_j u_{i-j}(\mathbf{w}_1)\beta_{\mathbf{w}_1,\mathbf{w}_1}(j)\mathbb{P}_{\mathbf{w}_1}(j) + \cdots$$
$$+ \sum_j u_{i-j}(\mathbf{w}_m)\beta_{\mathbf{w}_m,\mathbf{w}_1}\mathbb{P}_{\mathbf{w}_1}(j)$$

Multiply both sides by s^i, $i \geq |\mathbf{w}_1|$, and sum:

$$\frac{s^{|\mathbf{w}_1|}\mathbb{P}(\mathbf{w}_1)}{1-s} = (U_1(s) - 1)G_{1,1}(s) + \cdots + (U_m(s) - 1)G_{m,1}(s).$$

12.11 Apply the renewal theorem to the first equation in the above solution.

12.12 One approach is via $U(s) = \sum_{i=1}^{m}(U_i(s) - 1) + 1$. Another is by noting $M(n) = M_1(n) + \cdots + M_m(n)$ and $\frac{M(n)}{n} \to \mu$, $\frac{M_i(n)}{n} \to \mu_i$.

12.14 You bet \$1 on H. If T appears you lose. The return for H to make a fair game is $x\mathbb{P}(H) + 0\mathbb{P}(T) = 1$ or $x = 1/p$. Then $-\mathbb{E}(X) + 1/p = 0$.

Chapter 13

13.1

n	1	2	3	4	5	6	7	8
$S(n)$	1	1	2	4	8	17	37	82

13.3 $\sum_{k \geq 0} \binom{a-2}{1}\binom{b-2}{1} = \binom{a+b-4}{a-1}.$

13.4 $L(n)$ is the sum of the number of hairpins with stems joining $[1, i]$ and $[j, n]$ with $i \bullet j$ a basepair and $j - i - 1 \geq m$ (to satisfy the loop constraint). Using Problem 13.3,

$$L(n) = \sum_{i=1}^{n-m-1} \sum_{j=i+m+1}^{n} \binom{n-j+i-1}{n-j}$$

$$= \sum_{i=1}^{n-m-1} \sum_{l=0}^{n-i-m-1} \binom{l+i-1}{i-1}$$

$$= \sum_{i=1}^{n-m-1} \binom{n-m-1}{i} = 2^{n-m-1} - 1.$$

13.5 $S_{m+j} = S_{m+j-1} + S_{m+j-2} + \cdots + S_{j-1} + \sum_{i=0}^{m+j-2} S_j S_{m+j-2-i}$, where $j \geq 1$ and $S_0 = S_1 = \cdots = S_{m-1} = 0$, $S_m = 1$.

13.6 $S_n = S_{n-1} + \sum_{j=0}^{n-2} S_j S_{n-2-j}$ where $S_0 = 1$.

13.7 **Lemma:** Using $y = \varphi(x)$, the equation is
$$F(x, y) = x^2 y^2 - y(1 - x) + 1 = 0.$$

Theorem: $S(n) \sim \sqrt{\frac{3}{4\pi}} n^{-3/2} 3^n$.

13.8 **Lemma:** $F(x, y) = x^2 y^2 - y(1 - x - x^2 - x^3) + x^3 = 0$

Theorem: $S(n) \sim \sqrt{\frac{1+\sqrt{2}}{\pi}} n^{-3/2} (1 + \sqrt{2})^n$.

13.9 $R(n+1) = R(n) + \sum_{k=0}^{n-2} R(k) R(n-k-1) \rho(A_{k+1}, A_{n+1})$. Take expectations and follow the proof of Theorem 13.2.

13.10 If $n + 1$ is not in a basepair, there are $S(n, k+1)$ structures with $k + 1$ basepairs. Otherwise, $n + 1$ is paired with j, and k basepairs must be split among $[1, j-1]$ and $[j+1, n]$ of length $n - j$.

Chapter 14

14.1 If two L_k cannot be ordered by inclusion, show that a contradiction results.

Problem Solutions and Hints

14.3

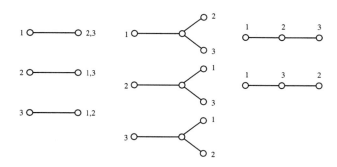

14.4

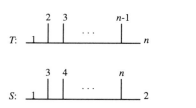

have disjoint set of splits.

14.6
Splits distance $= 6$ and nni distance $= 4$.

14.7
For $\{i, j, k, l\}$, there are $\binom{4}{2} = 6$ pairwise distances. Let $d(i, j)$ be the minimum. Applying the ultrametric inequality $d(i, j) \le d(i, k) = d(j, k)$, and $d(i, j) \le d(i, l) = d(j, l)$. Therefore, $d(i, k) + d(j, l) = d(i, l) + d(j, k)$. Of the distances $d(i, k), d(i, l), d(j, k), d(j, l)$, and $d(k, l)$, suppose $d(k, l) \le d(j, k)$ (for example). Then $d(i, j) + d(k, l) \le d(i, l) + d(j, k)$ and the four point condition follows. If, instead, $d(k, l)$ is strictly larger than one of the other distances, $d(i, l)$ for example, it follows from the ultrametric inequality that $d(i, k) = d(k, l)$. A contradiction follows: $d(k, l) > d(i, l)$ implies $d(k, l) = d(i, k)$. But $d(i, k) = d(j, k), d(k, l) = d(j, k) \ge d(j, l)$ and $d(j, l) = d(i, l)$.

14.9
At each parent, child pair (v, w), by Problem 14.8 it must hold that $O(v) \cap S(w) = \emptyset$.

14.10
When $a \in F(v)$, $\|f_v(a)\| = L(v)$ and $M_v(a) = \{a\}$. When $a \in S(v)$, then $\|f_v(a)\| = L(v) + 1$ and $M_v(a) = F(v) \cup \{a\}$. It is not necessary to know $\|f(v)(a)\|$ or $M_v(a)$ for $a \notin F(v) \cup S(v)$.

14.11
The proof follows that of Theorem 11.4. Let $S_n = \|f(t)\| = S(A_1 \cdots A_n)$. Define $Y = S_n - \mathbb{E}(S_n)$, $\mathcal{F}_i = \sigma(A_1 \cdots A_i)$, and $X_i = \mathbb{E}(Y | \mathcal{F}_i)$. Now

$$\mathbb{E}(S|\mathcal{F}_i) = \sum_{a_{i+1} \cdots a_n} S(A_1 \cdots A_i a_{i+1} \cdots a_n) \mathbb{P}(A_{i+1} = a_{i+1} \cdots A_n = a_n)$$

and

$$\begin{aligned}|X_i - X_{i-1}| &\leq \sum_{a'_i a_{i+1} \cdots a_n} |S(A_1 \cdots A_i a_{i+1} \cdots a_n) \\ &\quad - S(A_1 \cdots A_{i-1} a'_i a_{i+1} \cdots a_n)| \\ &\quad \cdot \mathbb{P}(A_i = a'_i, A_{i+1} = a_{i+1}, \cdots A_n = a_n) \\ &\leq \max |S - S'| \leq 1.\end{aligned}$$

The last inequality follows because changing a letter changes the overall fit by at most 1. Azuma-Hoeffding implies

$$\mathbb{P}(S_n - \mathbb{E}(S_n) \geq \gamma n) \leq e^{-(\gamma^2 n^2)/2n} = e^{-\gamma^2 n}.$$

14.12 $p_{\{a\}}(h) = p_{\{a\}}^2(h-1) + 2p_{\{a\}}(h-1)p_{\{a,b\}}(h-1)$, $p_{\{b\}}(h) = p_{\{b\}}^2(h-1) + 2p_{\{b\}}(h-1)p_{\{a,b\}}(h-1)$; $p_{\{a,b\}}(h) = p_{\{a,b\}}^2(h-1) + 2p_{\{a\}}(h-1)p_{\{b\}}(h-1)$.

Appendix II

Mathematical Notation

Functions

$\lfloor x \rfloor$	the greatest integer $\leq x$
$\lceil x \rceil$	the least integer $\geq x$
$x \wedge y$	the minimum of x and y
$x \vee y$	the maximum of x and y
x^+	$x \wedge 0$
$a_n \sim b_n$	$\lim_{n \to \infty} a_n/b_n = 1$
$f(x) \approx g(x)$	$f(x)$ is approximately equal to $g(x)$
O	$f(x)$ is $O(x^3)$ as $x \to \infty$ if there is a constant c such that $|f(x)| \leq cx^3$
o	$f(x)$ is $o(x^3)$ as $x \to \infty$ if $f(x)/x^3 \to 0$ as $x \to \infty$
A^T	the transpose of the matrix A

Subsets of the Reals

$\mathbb{N}$	the natural numbers: 1, 2, ...
$\mathbb{Z}$	the integers
$\mathbb{R}$	the reals

Set Notation

$\emptyset$	the empty set
$A \cup B$	the union of A and B
$A \cap B$	the intersection of A and B
A^c	the complement of A
$A \sim B$	$A \cap B^c$
$\limsup A_n$	$\cap_{n \geq 1}(\cup_{m \geq n} A_m)$
$\liminf A_n$	$\cup_{n \geq 1}(\cap_{m \geq n} A_m)$
$|A|$ or $\#A$	the number of elements in A
$\mathbb{I}_A, \mathbb{I}(A)$	the indicator function of A

Probability

$\mathbb{P}(A)$	the probability of A
$\mathbb{E}(X)$	the expectation of the random variable X
$\text{Var}(X)$	the variance of X
$\text{cov}(X, Y)$	the covariance of (X, Y)
$\text{cor}(X, Y)$	$\text{cov}(X, Y)/\sqrt{\text{Var}(X)\text{Var}(Y)}$
$X \stackrel{d}{=} Y$	X and Y have the same distribution
$X_n \stackrel{d}{\Rightarrow} Y$	X_n converges in distribution to Y
iid	independent and identically distributed
$\mathcal{B}(n, p)$	binomial with n trials and success probability p
$\mathcal{P}(\lambda)$	Poisson with mean λ
$\mathcal{N}(\mu, \sigma^2)$	normal with mean μ and variance σ^2

Algorithm Index

Algorithm 2.1 Restriction Map 37
Algorithm 7.1 Greedy 141
Algorithm 7.2 Overlap 144
Algorithm 7.3 Accuracy 147
Algorithm 7.4 Assembly 157
Algorithm 8.1 Bubble Sort 169
Algorithm 8.2 Chain 170
Algorithm 8.3 Suffix 171
Algorithm 8.4 Repeats 171
Algorithm 8.5 Fast 174
Algorithm 8.6 Fastgap 175
Algorithm 8.7 $l$fast 175
Algorithm 8.8 Filtration 178
Algorithm 8.9 Double Filtration 179
Algorithm 9.1 Shortest Path 190
Algorithm 9.2 Global Distance 193
Algorithm 9.3 $\hat{g}$ 196
Algorithm 9.4 Repeat 207
Algorithm 9.5 Nonoverlapping Repeat 207
Algorithm 9.6 Wrap 209
Algorithm 9.7 S 209
Algorithm 9.8 S^* 209
Algorithm 9.9 SL 210
Algorithm 9.10 All Inversions 215
Algorithm 9.11 Best Inversions 216
Algorithm 9.12 Map 221
Algorithm 9.13 Map* 221
Algorithm 9.14 Multiple Match 222
Algorithm 10.1 HMM Alignment 248
Algorithm 14.1 PGM 359
Algorithm 14.2 Parsimony Score 363
Algorithm 14.3 Parsimony 364
Algorithm 14.4 Peeling 373

Algorithm 14.5 Model Test 375

Author Index

Aho, A.V., 4, 380
Alberts, B., 3
Aldous, D., 281, 383
Alexander, K., 382
Alizadeh, F., 379
Altschul, S.F., 380, 383
Apostol, T., 4
Argos, P., 381
Armen, C., 379
Arratia, R., 378, 383

Baase, S., 4
Baeza-Yates, R.A., 380
Bafna, V., 381
Baltimore, D., 3
Bandelt, H.-J., 385
Bellman, 185
Bender, E.A., 385
Benham, C.J., 385
Benzer, S., 33, 377, 378
Berg, P., 3
Biggins, J.D., 384
Billingsley, P., 312, 384
Blum A., 379
Borodovsky, M., 384
Bray, D., 3
Breen, S., 384
Brutlag, D., 381
Bull, J.J., 385
Buneman, P., 385
Burge, C., 384

Cannings, C., 384
Carbon, J., 85, 102, 378
Carillo, H., 382

Caskey, C.T., 378
Cavalli-Sforza, L.L., 385
Chang, W.I., 380
Chastain, M., 385
Chen, L.H.Y., 93, 278–282, 287, 289, 291, 292, 300, 321, 383, 384
Chung, K.L., 4
Churchill, G.A., 379, 384
Chvátal, V., 257, 299, 382
Clarke, L., 85, 102, 378
Collins, F.S., 234, 235
Corasick, M., 380
Coulson, A., 378
Crick, F.H.C., 5, 7–9, 327, 334, 377
Crochemore, M., 4

Dančik, 382
Darnell, J., 3
Dayhoff, 234
Dress, A., 385
Drmanac, R., 379
Dumas, J.P., 380
Durrett, R., 4, 383
Dyer, M., 379

Edwards, A., 378
Edwards, A.W.F., 385
Eggert, M., 381
Erdös, P., 264, 270, 272, 383
Evans, G.A., 379

Fasman, K.H., 380
Feller, W., 4, 314, 315, 384
Felsenstein, J., 385
Fischer, M.J., 381

Fitch, W.M., 381, 385
Foulds, L.R., 385

Galas, D.J., 382
Galil, Z., 381
Gallant, J., 137, 379
Gamow, 9
Garey, M.R., 378
Geman, D., 74
Geman, S., 74
Giancarlo, R, 381
Gilman, M., 3
Goldman, N., 385
Goldstein, L., 378, 382–384
Goldstein, M., 381
Golumbic, M.C., 36
Gordon, L., 383
Gotoh, O., 381
Gribskov, M., 381, 382
Griggs, J.R., 378, 380
Guibas, L.J., 384
Gusfield, D., 381, 382, 385

Hall, T.E.; P., 379
Hartigan, J.A., 385
Hillis, D.M., 385
Hirschberg, D.S., 381
Hopcroft, J.E., 4
Hopkins, N., 3
Howe, C.M., 379
Howell, J.A., 385
Huang, X., 379, 381
Hunkapiller, T., 379

Ibragamov, I., 312
Ibragimov, I., 384
Idury, R., 379

Johnson, D.S., 378

Karlin, S., 4, 383, 384
Karp, R.M., 380
Kececioglu, J., 379, 381, 382
Kingman, J., 44, 46, 48, 255, 383
Kleffe, J., 384

Knuth, D.E., 65
Kohara, Y., 112, 114, 118, 378, 384
Kosaraju, S.R., 379
Kotig, A., 53
Kozhukhin, C.G., 384
Krogh, A., 382

Lagrange, 273
Landau, G.M., 381
Lander, E.S., 378
Laquer, H.T., 380
Lawler, E.L., 380
Lehninger, A., 3
Lemke, P., 378
Lewin, B., 3, 377
Lewis, J., 3
Li, S.-Y., 307, 314, 318, 320, 321, 384
Lipman, D.J., 380, 382
Lipshutz, R.J., 379
Lodish, H., 3
Lundstrum, R., 383

Macken, C., 384
Magarshak, Y., 385
Manber, U., 380
Margoliash, E, 385
Martin, D., 378
Martinez, H.M., 380
Meacham, C.A., 385
Mendel, G. J., 5
Metropolis, 70–76
Miller, W., 381
Myers, E.W., 380, 381

Naor, D., 381
Nathans, D., 377
Needleman, S.B., 185, 186, 380, 381
Nei, M., 385
Neyman-Pearson, 274
Ninio, J., 380
Nussinov, R., 384

Odlyzko, A.M., 384
Olson, M.V., 111, 378

Author Index

Paterson, M., 382
Pauling, L., 377
Pearson, W.R., 380
Penner, R.C., 385
Perleberg, C.H., 380
Perlwitz, M., 382
Pevzner, P.A., 378–381, 383–385
Port, E., 378, 379
Prum, B., 384

Rényi, A., 264, 270, 272, 383
Rabin, M.O., 380
Raff, M., 3
Riordan, J.R., 382
Roberts, J., 3
Robinson, D.F., 385
Rosenblatt, J., 378
Ross, S., 4
Rousseau, P., 385
Rudd, K.E., 378
Rudin, W., 4
Rytter, W., 4
Rzhetsky, A., 385

Sankoff, D., 257, 299, 381, 382, 384, 385
Schöniger, M., 381
Schmidt, J.P., 381
Schmitt, W., 378, 385
Schwartz, D.C., 273, 284, 378
Seed, B., 378
Sellers, P., 381
Seymour, P.D., 378
Singer, M., 3
Skiena, S.S., 378
Smith, H.O., 377
Smith, T.F., 381, 384, 385
Smith, W.D., 378
Staden, R., 379
Steele, J.M., 260, 383
Stein, C., 93, 278–282, 287, 289, 291, 292, 300, 321, 379, 384
Stein, P.R., 385

Steitz, J.A., 3
Stryer, L., 3
Sundaram, G., 378
Sussman, J.L., 384

Tang, B., 378
Tavaré, S., 385
Taylor, H.M., 4
Tinoco, I.J., 384, 385
Trifonov, E.N., 384

Ukkonen, U., 380
Ulam, S., 185
Ullman, J.D., 4

Vingron, M., 381–383
von Haeseler, A., 382

Wagner, R.A., 381
Ward, E.S., 379
Waterman, M.S., 378–385
Watson, J.D., 3, 5, 8, 327, 334, 377
Watterson, G.A., 381
Weiner, A., 3
Whittle, P., 384
Wilbur, W.J., 380
Witkowski, J., 3
Wu, S., 380
Wunsch, C.D., 185, 186, 380, 381

Xu, S., 382

Zoller, M., 3
Zubay, G., 3
Zuker, M., 384, 385

Subject Index

accordions, 285
additive tree, 353, 385
algorithm, 65–67
alignment, 144, 145, 184–188, 194, 195, 198, 200, 203–205, 233, 238, 380–383
    linear space, 209
    multiple-sequence, 233
    parametric, 223
    similarity, 198
    statistical, 253
    sum of pairs (SP), 237
    weighted-average, 238
allele, 234
analysis by pattern, 249
analysis by position, 248
ancestral sequence, 345
anchors, 119, 124, 378
anticodon, 14
aperiodic, 72
approximate matching, 177, 288, 290, 301, 322, 380, 383
ATP, 235
Azuma-Hoeffding, 255, 257, 259, 294, 367, 382

backtracking, 363, 367
bacteriophage lambda, 77
balanced graph, 53
ballot theorem, 290
Bayes, 146, 247
Bernoulli process, 85
bijection, 332
binary chip, 155
binary tree, 238, 345

bipartite graph, 32, 36
birth process, 359
BLAST, 251, 380
blunt ends, 29
border block graph, 56
Borel-Cantelli, 264, 297, 383
bulges, 331

cassette exchange, 59
cassette reflection, 59
cassette transformations, 58, 59
Cauchy-Schwartz, 273, 284
center of gravity, 236, 242
central dogma, 7
Central Limit Theorem, 180, 307, 314, 383, 384
Chen-Stein, 93, 278, 383
chromosome, 21, 33, 83, 101, 219, 381
clone, 83, 85, 87, 92, 101, 111–114, 119, 127, 233, 378, 379
cloverleaf, 14, 331, 340
clumps, 205, 280, 281, 288, 300, 306, 321
codon, 9, 12, 16
communicate, 72
complexity, 66
concatenation, 141, 168
concave, 196, 223, 225, 381
consensus
    folding, 341
    word, 250, 261, 341
convex, 196, 228, 257
    polygons, 229
convolution, 291, 315

Subject Index

cosmids, 87
crabgrass, 284
cumulative distribution function (cdf), 255, 301
cycle, 52, 139, 345
cystic fibrosis (CF), 382

declumping, 205, 281, 283, 284, 300, 321
degree, 32, 52, 151
deletion, 143, 183
destabilization function, 338
directed graph, 32
distance, 148, 192, 242, 359
dominated convergence theorem, 310
double digest problem (DDP), 42, 65, 67, 378
duality, 126, 334
dynamic programming, 144, 183, 236, 248, 380, 382

edge, 31, 52
edge color, 52
electrophoresis, 38
end loop, 331
equilibrium or stationary distribution, 72
Euclid's algorithm, 142
eukaryotes, 7, 17
Euler
    constant, 276, 325
    theorem, 151
Euler-Macheroni, 276, 325
Eulerian, 52
Eulerian path, 56
exact matching, 263, 282
exons, 17
expectation maximization (EM), 146
exponential distribution, 117, 275
extremal statistics, 305
extreme value distribution (type 1), 245, 275

fingerprint, 102

four point condition, 355

gamma distribution, 97
gapped chip, 156
gcd, 79, 142
gel, 38, 146
generating function, 314, 318, 320, 334, 383
genetic diseases, 233
genetic mapping, 101, 219, 233
geometric distribution, 86
Gibbs distribution, 71
global alignment, 193, 200, 254
global distance, 193
gradient method, 70
graph, 31–33, 52–54
graph isomorphism, 346

hairpin, 13, 331
Hamiltonian path, 145, 150
hemoglobin, 183, 214, 253, 379
hidden Markov models (HMM), 246
hybridization, 114, 127, 148
hydrophilic, 235
hydrophobic, 261
hyperplane, 228–230

identity, 184
incidence matrix, 34
inclusion-exclusion, 42
indel, 143, 184
    function, 194
    multiple, 195
infinitely often, 264
infinitesimal, 226
initial distribution, 71, 72, 307
insertion, 143, 183
integer programming, 69
interior loop, 331
intersection graph, 32
interval graph, 33, 377
intractable, 66, 67
intron, 17
inversions, 215, 270, 345, 381

irreducible, 268, 307
islands, 91, 102

join, 41
jumping genes, 18
junk DNA, 17

Kingman's Theorem, 255
Kolmogorov-Smirnov, 324
Kullback-Liebler, 261

labeled tree, 346
large deviations, 257, 259, 261
lattice path, 130
law of rare events, 279
layout, 129, 143
Li's method, 318
library, 84
LIFO (last in, first out), 213
likelihood ratio, 274, 375
line geometries, 238
linear tree, 332
linkage mapping, 233
local alignment, 145, 203, 263
loci, 162
lognormal, 39
longest common subsequence, 257

map alignment, 220
Markov chain, 71
Markov renewal processes, 314
Markov's inequality, 258
martingale, 257
match, 184
maximum likelihood, 185, 247, 367
meet, 41
memoryless, 87
metric space, 185
minimum mutation fit, 363
mismatch, 143, 184
molecular evolution, 18, 183
Monte Carlo, 70, 71
morphology, 345
multibranch loop, 331, 339, 340
multiple matching, 222, 381

near-optimal, 213, 381
nearest neighbor interchange metric (nni), 352, 385
neighbor joining (NJ), 360
network, 190
normal distribution, 255
NP-complete, 67

oceans, 104
order exchange, 53
order reflection, 54
order statistics, 293
overlap bit, 309
overlap equivalence, 51
overlap polynomial, 309
overlap size equivalence, 51

pair group method (PGM), 359
palindrome, 29
PAM, 234
parsimony, 361, 385
partition problem, 67
path, 52, 53, 56
peeling algorithm, 373
period, 72, 142
Perron-Frobenius, 268
phase transition, 294
physical map, 101
plasmid, 31, 83
Poisson approximation, 93, 278, 321
Poisson clumping heuristic, 280
Poisson distribution, 275
Poisson process, 89, 103
pools
    clone, 128, 379
    SBH, 154
position-dependent weights, 197
prefix, 138
primary structure, 327
prior, 247
profile, 233, 243, 382
prokaryotes, 17
promoter, 16

RAM, 209

Subject Index

reading frame, 9
recombinant, 83
reflections, 48
relative entropy, 261
renewal theory, 314
repeats, sequence, 206
replication, 8
restriction endonuclease (enzyme), 29
restriction map, 29
restriction site, 29
reversible Markov chain, 371
ribosome, 15
RNA
    16SRNA, 379
    mRNA, 10–12
    tRNA, 12, 327, 379
root, 167, 345

Schur product, 373
secondary structure, 327
selfish DNA, 18
sequence reconstruction problem (SRP), 137, 379
sequence, weighted-average, 239
sequencing by hybridization (SBH), 148, 154, 156, 379
sequencing chip, 148, 379
Sheppard's
    continuity correction, 277
    variance correction, 277
shortest common superstring (SSP), 137, 138, 379
shortest path, 190
shotgun sequencing, 135
site distributions, 384
    intersite, 324
spectrum, 150
splits, 347, 385
stacking energy, 335
stationary distribution, 72
Stirling's formula, 188
stochastic matrix, 268, 369
stopping time, 106

strong law of large numbers (SLLN), 44
subadditive, 44, 196
subadditive ergodic theorem, 44, 255
substitution, 183, 184
suffix tree, 167, 380
synthesis, 7

tandem repeats, 207, 381
total variation distance, 279
trace, 184
traceback, 194
transcription, 7
transition matrix, 72
transition probabilities, 71
translation, 7
transposable elements, 18
transpositions, 345
traveling salesman problem (TSP), 67
triangle inequality, 186

ultrametric condition, 357
unmethylated, 29
unweighted pair group method (UPGMA), 360

vector, 83–85
vertex, 31, 32, 53

Watson-Crick pairs, 327
weighted pair group method (WPGMA), 360
weighted parsimony, 366, 385

yeast artificial chromosome (YAC), 101, 111
Yule process, 359